The European Smart City Movement

Christian Schachtner
Editor

The European Smart City Movement

Case Studies from Around Europe

Springer

Editor
Christian Schachtner
DCSM
RheinMain University of Applied Sciences
Wiesbaden, Hessen, Germany

ISBN 978-3-031-96699-6 ISBN 978-3-031-96700-9 (eBook)
https://doi.org/10.1007/978-3-031-96700-9

This Springer imprint is published by the registered company Springer Nature Switzerland AG
The registered company address is: Gewerbestrasse 11, 6330 Cham, Switzerland

I extend my gratitude to everyone who picks up this book, appreciating your willingness to consider my ideas. Let these concepts serve as a compass, helping you reimagine and transform public environments into smart, interconnected landscapes.

Foreword

Towards Global Strategic Smart City Alliances

In comparison with international developments, the smart city topic gained traction relatively late in Germany. While other countries and regions had extensive programs already in place, Germany took good time to position itself in the overall landscape of smart city initiatives. Now—due to concerted large-scale public funding especially by the federal government—the topic is in full swing. German smart city development has turned into a broad movement and may upscale (if the built-in replication mechanisms work out) into mainstream in the near- and middle-term future.

This position of a late-mover has certain advantages. It allows the reflection of previous experiences made in other locations, the adaptation of most recent findings in scientific research, and the application of state-of-the-art technologies. Importantly, it enables the smart city community to pick up challenges for specific consideration that haven't been resolved yet elsewhere. Doing so—despite being a late-comer—the German smart city community can still claim a pioneering position in certain respects.

For example in Germany great importance is attached to privacy and data protection and strict compliance with the corresponding regulations—more than in many other countries. This is due, among other things, to the demand of civic society organisations for the law-abiding application of the General Data Protection Regulation (GDPR), which has led to a broad discourse on data sovereignty and the digital sovereignty of the individual. This resulted in the growth of trust among the population and acceptance of digitalisation solutions and shows how the initially feared competitive disadvantage can shift to an understanding of a positive location factor.

Similarly, the overall interpretation of how smart cities are understood stands out. Reflecting the evolution of the concept through its three phases over the past decades—from (1) a corporation-driven approach via (2) a government-oriented perspective towards a notion that (3) views civic society as another key driver in smart city developments—the current understanding of smart cities in Germany is certainly a comprehensive and complex one. It acknowledges the intricate multiplicity of stakeholders and dimensions that must be managed to make the new urban paradigm operative in the long run—on all levels of technological development, urban planning and development, as well as citizen engagement and participation.

On such premise, related key challenges can be outlined that the smart city community—not only in Germany—should address.

First, reliable **conceptual and scientific fundaments** are yet to be established. Key terminologies and notions must be clarified and agreed upon, such as: What does "open source" mean in the context of smart cities? Basic principles of design and development must be outlined such as: How to plan and integrate physical and digital infrastructures? Finally, impact measures must be created in order to estimate the overall social benefit, as well as the by-effects of smart city developments.

Secondly, recognised **platforms and marketplaces** for the exchange of smart city knowledge are much needed. The multiplicity of activities on national and international levels calls for public fora where technological solutions, organisational innovations, process knowledge can be effectively transferred and disseminated. In order to reach beyond the level of individual projects or solutions, the design, setup, and operation of such marketplaces emerges as an eminent task. This is inherently connected to open source principles and IP sharing, as well as to the topic of forming strategic alliances between public institutions, civic society organisations, and private corporations. Here, suitable models are yet to be created.

Thirdly, effective mechanisms must be built that foster **science and business spinoffs from smart city projects**. The current initiatives are genuine greenhouses of innovation whose creative impulses should be channelled into venture building pipelines for citytech/govtech solutions. The establishment of a vital smart city startup-ecosystem is one of the best guarantees for improving cities in the long run.

The above-mentioned issues are already being tackled by committed partnerships in Germany. Close cooperations between players from the civic society like the German Smart City Association, actors in research and education like the Digital City Science Department of HafenCity University Hamburg or Technical University of Dresden's Laboratory of Knowledge Architecture, as well as public city administrations in Germany's digital model cities like Dresden or Hamburg pave the way in this direction. The productive exchange with international actors in the smart city field is an essential catalyst for progress on the local level in Germany. Vice versa—especially for the topics highlighted above—Germany may become a resource of fresh knowledge and insights for the advancement of smart cities on the global level.

Prof. Dr.-Ing. Joerg Rainer Noennig
Chair of Digital City Science at
HafenCity University Hamburg

Mirko de Paoli
Chairman of the Board of the German
Smart City Association
(Bundesverband Smart City)

Preface

It is a great pleasure for me to present to you the reference work *The European Smart City Movement—Case Studies from Around Europe.* At a time when digital transformation is fundamentally changing our cities and regions, this book offers a comprehensive insight into the innovative world of smart cities in Europe. Digitization is omnipresent today and determines our everyday lives. From smartphones to big data, artificial intelligence and smart city concepts, we are in the midst of a technological revolution that offers innovative control options for the use of resources and linkable data assets on the way to a digital future. It is a global vision of smart, sustainable urban development—for the benefit of the people, of course!

This book aims to strengthen the capacity of municipalities to act by presenting proven smart city concepts and implementation measures in numerous European cities, districts and municipalities of all sizes by 2030. Our goal is to make the solutions developed in the context of these projects secure, interoperable and transferable, while using the financial and human resources in the municipalities as efficiently as possible. For two decades, digital innovation has been at the heart of the discourse around "smart cities", with the aim of creating more efficient and livable urban environments. Initially, the concept of "smart cities" was largely supply-side and sector-driven, with the private sector playing a leading role in defining both the problem and the solution to digital innovation to create new economic opportunities, improve service delivery, and facilitate civic engagement.

In Europe, there are already numerous examples of cities that are well on their way to becoming smart cities. These include metropolises such as Barcelona and Amsterdam, but medium-sized cities and regions are also already implementing exciting concepts. Various key tasks in public spaces are also to be designed to be smart and liveable. This can be done in various ways, which is why the contributions are divided into chapters. The focus is in particular on pilot projects ranging from intelligent spaces to data management and smart participation. Each case study is thus one of the many municipalities that are embarking on the path to a smart future and have already developed or implemented the first complete urban districts and smart buildings on the basis of integrated urban planning. One of the initial examples of a smart city in Europe is Santander in Spain. Since 2009, the northern Spanish port city has been developing into a prime example of the future of modern cities. With 20,000 sensors that constantly process data on a wide variety of processes, the project aims to make the coexistence of the approximately 175,000 inhabitants more efficient, sustainable and comfortable. Another reference example for the future of smart cities in Europe is the Ellinikon project in Greece. Within 15 years, a picture-perfect smart city will emerge on the Greek coast, where everything can be reached within 15 min on foot or by bike. This unique project shows the possibilities that open up when you have the chance to build a city from scratch.

This reference work offers you an exemplary collection of implementation examples that describe the challenges, solutions, participants, financial expenditure and methodological procedures for practical transfer. The ambitions of the implementation vary and depend on the local requirements of the target group, which should be able to start its own initiatives more easily through an overview of the projects.

I invite you to be inspired by the innovative ideas and solutions presented in this book. May they serve as a catalyst for your own smart city initiatives and help make our cities and regions fit for the challenges of the future.

I wish you inspiring ideas and insightful transfer through reading this publication!

Prof. Dr. Christian Schachtner
Professor for Business Informatics with
Focus on Digitization in the Public
Sector
RheinMain University of Applied
Sciences
Wiesbaden, Germany
christian.schachtner@hs-rm.de

Acknowledgements I would like to thank all my fellow professors, practice partners, experts and project staff for their valuable suggestions over the years of work in the field of digitization of Smart Regions. A special "thank you" goes to my post-doctoral researcher for her support in terms of content and form in the process of writing this publication.

Contents

Editors and Contributors

About the Editor

Prof. Dr. Christian Schachtner is Professor for Business Informatics with focus on Digitization in the Public Sector at the RheinMain University of Applied Sciences in Wiesbaden/Germany. In March 2025 he was elected as Vice President for Education and Sustainability at RheinMain University of Applied Sciences. Till 2023 he was Head of Public Management at IU University of Applied Sciences, Bad Reichenhall/Germany. After completing a diploma in public administration (FH) at HfoeD Bavaria/Germany, he completed master's degree in European administrative management at the HWR Berlin/Germany. Subsequently, he completed his Ph.D. at the Faculty of Economic Sciences of the Catholic University Eichstätt-Ingolstadt/Germany at the Chair of Business Administration, Human Resources and Organization. In addition to 10 years of experience as a Chief Digital Officer and manager in public administration and smart city affairs, he works as an in-house consultant, independent organizational consultant and in adult education. His research focuses on public governance, smart city and regions, digital transformation, learning analytics, competency-based frameworks, upskilling, automated systems, and AI-based systems as well as data literacy. He is a member of the Germany Society of Organizational Matters (Gfo) and Society for Informatics (GI) as well as Research Fellow at Stein-Hardenberg Institute Berlin/Germany and University for Continuing Education Krems.

Contributors

Luca T. Bauer University of Bremen, Bremen, Germany

Lasse Berntzen School of Business, University of South-Eastern Norway, Horten, Norway

Leslie A. Carr University of Southampton, Southampton, Great Britain

Franziska Cecon University of Applied Sciences Upper Austria, Linz, Austria

Mustafa Eruyar Hinda Smart City Technologies, Istanbul, Türkiye

Brian Fabrègue Faculty of Law, University of Zurich, Zurich, Switzerland

Licia Felicioni Czech Technical University in Prague, University Centre for Energy Efficient Buildings, Buštěhrad, Czech Republic

Falk Fernbach LuxMobility, Luxembourg, Luxembourg

Niccolò Fiorini Department of Business and Law, University of Siena, Siena, Italy

Christian P. Geiger Bern University of Applied Sciences, Bern, Switzerland

Fatih Gundogan Streadlab, Istanbul, Türkiye

Silvia Hackl Municipality of the City of Linz, Linz, Austria

Frederik R. Heinz Hochschule Hildesheim, Hildesheim, Germany

Barbora Hejtmánková Czech Technical University in Prague, University Centre for Energy Efficient Buildings, Buštěhrad, Czech Republic

Adam Jabłoński WSB Merito University, Poznań, Poland

Marek Jabłoński WSB Merito University, Poznań, Poland

Fatih Kafali Hinda Smart City Technologies, Istanbul, Türkiye

Elisavet Koutsiana Department of Informatics, King's College London, London, Great Britain

Michal Kuzmič Czech Technical University in Prague, University Centre for Energy Efficient Buildings, Buštěhrad, Czech Republic

Milena Krumova Founder and Director, European Institute for Technologies, Education and Digitalisation, Sofia, Bulgaria

Björn Niehaves University of Bremen, Bremen, Germany

Pavla Pelčíková Prague Institute of Planning and Development, Prague, Czech Republic

Nohemi Ramirez Aranda Faculty of Architecture and the Built Environment, Department of Urbanism, Environmental Technology and Design Section, Delft University of Technology (TU Delft), Delft, Netherlands

Christian Schachtner Vice President Education and Sustainability, RheinMain University of Applied Sciences, Wiesbaden, Germany

Kristýna Schulzová Czech Technical University in Prague, University Centre for Energy Efficient Buildings, Buštěhrad, Czech Republic

Elena Simperl Department of Informatics, King's College London, London, Great Britain

Mahulena Svobodová Prague Institute of Planning and Development, Prague, Czech Republic

Gefion Thuermer Department of Informatics, King's College London, London, Great Britain

Serhan Ünalan Berlin School of Business and Innovation, Berlin, Germany

Tomáš Vácha Czech Technical University in Prague, University Centre for Energy Efficient Buildings, Buštěhrad, Czech Republic

Martin Volf Czech Technical University in Prague, University Centre for Energy Efficient Buildings, Buštěhrad, Czech Republic

Catalin Vrabie Faculty of Public Administration, National University of Political Studies and Public Administration, Bucharest, Romania

Johanna Walker Department of Informatics, King's College London, London, Great Britain

Klára Zámečníková Czech Technical University in Prague, University Centre for Energy Efficient Buildings, Buštěhrad, Czech Republic

Ana Zuljevic Municipality of the City of Linz, Linz, Austria

Introduction

Christian Schachtner

In recent years, there has been increasing competition between metropolises and regions for the settlement of companies from new industries. These companies depend on a good digital infrastructure and skilled workers for future technologies. In addition to traditional digital service providers such as hosting companies, application developers or platform providers, developers of artificial intelligence, blockchain technologies and security and encryption tools are increasingly focusing on use cases in the public sphere (BNP Paribas Real Estate 2024).

The technologies that enable these urban transformations are diverse and include cloud computing, blockchain, AR and VR, in addition to the Internet of Things (IoT), artificial intelligence (AI) and big data. Cloud technologies offer the necessary scalability and flexibility for data processing and analysis in smart cities. Blockchain technology is used for secure and transparent transactions and data storage, while AR and VR technologies create immersive and interactive experiences for citizens. The deployment of 5G networks plays a crucial role in the realization of smart city applications that require real-time communication and processing, such as autonomous driving or intelligent traffic control (Get in IT 2024).

Ideas for intelligent energy solutions, smart waste management, smart transport solutions or intelligent urban development that are realised through sensors or data transmission networks hold great potential for new possibilities in plant design. Making successfully completed pilot projects visible also opens up the possibility of transfer to other regions and metropolises, which offer new opportunities for adaptation under different legal and social conditions (Inovex 2024).

C. Schachtner (✉)
RheinMain University of Applied Sciences, Wiesbaden, Germany
e-mail: christian.schachtner@hs-rm.de

C. Schachtner (ed.), *The European Smart City Movement*,
https://doi.org/10.1007/978-3-031-96700-9_1

1 Objectives of This Book

This reference work presents an exemplary collection of implementation examples that describe the challenges, solutions, participants, financial expenditure and methodological procedures for practical transfer for the readers. The demand for implementation varies and depends on the local requirements of the target group, which should be able to start its own initiatives more easily through an overview of the projects.

The projects presented in this reference work make it clear that the implementation of smart city concepts requires not only technological innovations, but also close cooperation between various actors such as municipal utilities, technology companies, research institutions and the urban population. The examples also show that successful smart city initiatives are often tailored to local needs and challenges, while at the same time having the potential to serve as models for other cities and regions.

By providing detailed information on challenges, solutions, stakeholders, financial expenditures, and methodologies, this reference book provides a valuable resource for decision-makers, urban planners, and other interested parties who want to initiate similar projects in their own cities or regions. The diversity of the projects presented underlines the flexibility and adaptability of smart city concepts to different urban contexts and local priorities.

2 The Structure of the Book

This anthology bundles various articles into overarching chapters. In the first chapter, "Smart Identity and Vision", five articles are presented that deal with smart city initiatives in different European countries. These contributions create synergies by highlighting different aspects and approaches to the development of smart cities and regions.

The first article by Zuljevic, Hackl and Cecon describes the "innovation capital" Linz in Austria as an anchor example for long-term smart city development. The vision of this type of urban design is explained as a central element of urban development, citizen participation and as a hub for innovation. This approach shows how a city can increase its innovative strength through targeted measures and citizen participation.

Schachtner's contribution on Slovakia establishes a link between academic expertise and practical implementation by addressing the role of clusters in smart city development through statements by local experts and addressing national specificities.

Bulgaria's contribution from Krumova focuses on transparency and e-governance, highlighting the importance of digital infrastructures and services for smart cities.

This is in synergy with European model initiatives, which rely on digital platforms for citizen participation, among other things.

Based on insights from an international conference, Schachtner's Hungarian article highlights the importance of climate neutrality and shows how cities can pursue concrete emission reduction targets. This complements the previous contributions of the chapter with the important aspect of environmental sustainability, which is essential for modern smart cities.

Finally, Schachtner's contribution on Albania offers an interesting perspective on the development of smart villages. This expands the focus from urban centers to rural areas and shows how digital transformation can also take place outside of major cities.

The chapter "Smart Strategy-building" fits into this context by dealing with the challenges and approaches of digitization in municipalities. It highlights the need for a holistic approach to digital transformation that goes beyond the mere delivery of online services and IoT services.

Schachtner's contribution on the independently developed "Germany Smart City Framework" emphasizes the conceptual importance of a comprehensive framework for the digital town hall and emphasizes the importance of acceptable "service levels" in public administration. It addresses the problem of waiting times and an administration that is perceived as overwhelmed, which can lead to dissatisfaction among the population.

The Estonian example of Tallinn demonstrates perspectives and solutions for the successful implementation of digital services with a focus on accessibility, interoperability and user-friendliness. This complements the challenges identified in the German context and offers potential transfer effects.

The Swiss example of St. Gallen emphasizes the importance of participatory approaches in smart cities, which underlines the need for citizen-oriented and accessible administration. The compilation of this urban development strategy also shows different perspectives on the development and implementation of smart city project modules.

The chapter "Smart Co-creation and Interaction" deals with the synergies between concepts such as digitalization, sustainability and civic participation in the context of urban development.

The article on Latvia highlights the country's progress in the field of information and communication technologies (ICT). It highlights investments in telecommunications technology, cloud computing and artificial intelligence (AI) as well as the digitization of public services. These aspects are closely related to the topics in the other chapters, in particular with regard to the use of technology to improve urban services.

The contribution on the Irish Smart Cities Forum defines the concept of the smart city and emphasises the importance of innovation, economic development and cooperation between different actors. These aspects are also reflected in the other chapters, with a focus on creating an ecosystem that attracts and nurtures entrepreneurs, investors and businesses.

Fabrègue's chapter on the Swiss Smart City Hub examines the comprehensive networking approach of Swiss cities in the integration of smart city concepts. It emphasizes the importance of sustainability, technological innovation, citizen participation and quality of life. These topics form a bridge to the other contributions and show the evolution from a purely technical understanding to a holistic approach that puts citizens at the centre.

Ünalan's contribution focuses on Germany's considerations to define the role of artificial intelligence (AI) in the implementation of smart city concepts. It explores how AI can be used to analyze data and improve urban processes to address complex challenges such as transport, energy consumption, and environmental management. This aspect is closely related to the technological innovations and sustainability goals discussed in the other chapters.

The chapter "Smart Data Governance and Data Management" offers a comprehensive insight into the development and implementation of data-based concepts in various European countries.

Vrabie's article on Romania examines the digital transformation of services between 2014 and 2023. The research shows significant progress in the digital transformation of services, with most municipalities now offering online functionalities such as property tax payments, public transport information and civil status documentation. However, the study also identifies inconsistencies in service quality and availability, highlighting the need for standardized digitization practices.

The article on Cyprus presents various core projects that show the framework of the smart city movement in the country. Based on project reports and expert assessments, relevant activities in Cyprus are presented. These projects demonstrate the practical implementation of smart city concepts and could serve as a source of inspiration for similar initiatives in other countries.

The chapter on Great Britain deals with the use of open data in rural areas. It is argued that the predominantly urban focus of the smart city paradigm reinforces the existing urban–rural divide and thus perpetuates inequality. The availability and use of rural open data will be assessed using two parameters: data on rural populations and on key sectors of the rural economy (agriculture, fisheries and forestry). In addition, the relationship between key statistics of rural economies and rural open data, as well as their use and users, will be explored.

Bauer et al.'s contribution to Germany's GovTech scene in promoting smart city solutions examines innovative procurement practices and their impact on the digitalization of cities. However, the complexity of smart city technologies often exceeds internal capabilities, leading to dependencies that challenge digital sovereignty. GovTech—the technology-based collaboration between the public and private sectors—provides a solution by enabling governments to procure tailored, innovative solutions.

The chapter "Smart Mobility and Transport" serves as a framework for various country examples. The individual articles form the main part of the anthology and show different aspects and approaches to intelligent mobility solutions in different European contexts.

Fernbach's article on Luxembourg focuses on the national mobility plan 2035 and the initiative for free public transport. He emphasizes the state's holistic approach to tackling future mobility challenges and the involvement of various stakeholders.

The article on North Macedonia is based on conversations with project managers and sheds light on the complex challenges in the development of the country's transport infrastructure, both nationally and internationally.

Smart City Implementation Leaders' Contribution to Smart City Istanbul presents a methodology for creating a smart city roadmap implemented in Istanbul. It describes a top-down approach to analysing the current state and includes the development of a roadmap and architecture.

The article from Finland focuses on the AiRMOUR project, which investigates the public and social acceptance of drones in urban airspace, especially in the context of emergency services.

The chapter "Smart Urban Development" offers a comprehensive insight into innovative approaches to urban development in various European countries. The individual articles shed light on different aspects of intelligent urban planning and management, with numerous synergies and overarching themes emerging.

The article on the Czech Republic, written by the research community of the Technical University of Prague, focuses on sustainable urban planning strategies at the neighbourhood level in Prague. It highlights the need to balance the interests of residents, developers and local authorities to create neighbourhoods that are sustainable, livable and functionally diverse. Particularly noteworthy is the analysis of sustainability assessment tools such as BREEAM, LEED, DGNB and SNTool in the Czech context.

The Dutch contribution presents the innovative open-source Netherlands3D project, a platform for three-dimensional data visualization. This ambitious project based on unity aims to make complex spatial information accessible and understandable for the whole of the Netherlands. The detailed presentation of the goals and vision of Netherlands3D underlines the importance of advanced visualization technologies for modern urban planning and management.

Berntzen's Norwegian presentation is dedicated to the topic of smart energy and sheds light on the new threat landscape created by the increasing technology dependence of urban infrastructures. Critical areas such as transport, energy, waste management and water supply are discussed. The post highlights the growing role of the Internet of Things (IoT) in urban operations, while also highlighting the security risks associated with it. The remarks on the increasingly complex, often cloud-based systems in cities illustrate the need for robust security concepts in smart city development.

The article on South Tyrol presents the SINFONIA project, a pioneering initiative to implement large-scale, integrated and scalable energy solutions in medium-sized European cities. SINFONIA impressively demonstrates how innovative energy concepts can be successfully integrated into existing urban structures.

The Greek example is dedicated to the Ellinikon project, one of the most ambitious urban development initiatives in Europe. A smart city is being built on the site of the former Athens International Airport, which is to serve as a model for future urban developments. The description of this large-scale project illustrates the potential of comprehensive smart city concepts for the redesign of urban spaces.

The French contribution focuses on the "Top 5 Smart Cities in France" project, which identifies and analyzes the most advanced and innovative cities in France in the field of smart city development. This approach shows how important the exchange of best practices and the comparative analysis of smart city initiatives are for the further development of the concept.

The synergies between the individual articles in an anthology can be seen in various aspects. In the present case of the "Smart Culture" anthology, this is made clear by the contributions on Italy and Lithuania.

Fiorini's chapter on Italy examines how historical sites can be transformed into centers of interdisciplinary collaboration and innovation within the framework of the smart city concept, using the example of the Santa Maria della Scala Museum in Siena. It sheds light on the intersections of culture, technology and social inclusion, as well as the potential economic and social impact of such a transformation.

The elaboration on Lithuania, on the other hand, focuses on the challenges and opportunities of the capital Vilnius as an up-and-coming smart city. In particular, it considers the aspects of sustainable and innovative urban development in the context of smart arts and cultural heritage.

References

BNP Paribas Real Estate. (2024). Smart City: The Digital Transformation of Cities. https://www.realestate.bnpparibas.de/blog/trends/smart-city-die-digitale-transformation-der-staedte

Get in IT. (2024). Smart City: The urban network. https://www.get-in-it.de/magazin/arbeitswelt/it-arbeitsmarkt/smart-city-das-urbane-netzwerk

Inovex. (2024). Smart City—Innovative Cities for a Sustainable Future. https://www.inovex.de/de/blog/smart-city/

Innovationshauptplatz Linz—Or How to Become a Leading Innovation City

Ana Zuljevic, Silvia Hackl, and Franziska Cecon

1 Introduction

The paper first displays the chronological evolution from 2016 till nowadays. In details describes the creation of an ecosystem for innovation in the city of Linz. A series of examples underline the social innovations, the economic development and the climate protection that are fruitful outcomes of the mission of the Innovationshauptplatz. As it is living a participatory approach the means are shown in "Have a say, help to shape, take part". Finally the "Lessons Learned" reflect the evolvement and activities of the Innovationshauptplatz so far. The foresight of future initiatives linked to the city's strategy underline its vital role in the innovation ecosystem of Linz.

2 Objectives and Methods

This paper shows a journey of the city of Linz to create a place where smart ideas, participation, digital solutions are developed, fostered, distributed. It should be an inspiration for cities and municipalities on how to develop a strategy of becoming a smart city and on how to make it come true. There are a lot of details given (also on resources) to get an idea what is necessary and what can be achieved if people, organisations and ideas work together on the some vision.

A. Zuljevic
Municipality of the City of Linz, Linz, Austria

S. Hackl
Municipality of the City of Linz, Linz, Austria

F. Cecon (✉)
University of Applied Sciences Upper Austria, Linz, Austria
e-mail: Franziska.Cecon@fh-linz.at

C. Schachtner (ed.), *The European Smart City Movement*,
https://doi.org/10.1007/978-3-031-96700-9_2

The authors, who worked—partly from the very beginnings—at the Innovationshauptplatz tell the "success story Innovationshauptplatz" using mainly official documents such as political decisions, strategy papers etc. and by using their personal experiences. It is less a scientific paper discussing the strategy, the organisation and its processes and its outcomes by defined criteria but it is a development that is approved by the recognition of the various stakeholders and numerous awards alike.

3 Linz to Become Austria's Most Innovative City

The City of Linz. Have you ever heard of it? Perhaps. Perhaps not. From this day forward, you will remember Linz and keep it in your mind's eye. Discover how this city transformed from the image of a traditional, old, grimy industrial steel town to a modern Digital City and a genuine hotspot of digital innovation. Explore how an innovation hub, the Innovationshauptplatz Linz (translated as the 'main square of innovation Linz'), played a pivotal role in this remarkable metamorphosis. Join us on the following pages of this publication to uncover this fascinating journey.

Linz stands as a city where tradition and innovation seamlessly converge. Situated in the heart of Upper Austria and straddling both banks of the Danube River, Linz is distinguished by its rich historical legacy and progressive vision for the future, particularly in digitalisation. By harnessing technology and fostering innovation, Linz has emerged as a vibrant hub for startups and entrepreneurs, fuelling regional growth and creativity. The city's emphasis on research and development has attracted some of the world's brightest minds, contributing to a thriving innovation ecosystem. Linz is also renowned for its friendly and welcoming atmosphere, complemented by a dynamic cultural scene with numerous museums, galleries, and performance spaces. The city exudes a palpable sense of excitement and opportunity, underpinned by a strong spirit of community and collaboration among its residents. Despite its modest size, Linz boasts a rich and captivating history, with origins tracing back to Roman times. Over the centuries, it has played a pivotal role in developing Central Europe, evolving into a significant trading hub and cultural centre. Today, Linz is a modern, dynamic city that continues to advance while honouring its deep-rooted cultural heritage. (Startup Guide 2023).

Before we delve into the details, here are some key facts about the City of Linz: With an area of 96 square kilometres and a population of 212,538 in 2024, Linz is the third-largest city in Austria. (linz.at) The city of Linz is situated in the heart of Europe and is Austria's largest industrial region. With 209,000 jobs boosting employment to unprecedented levels, around 110,000 individuals commute to Linz for work each day (see Zahlstadt 2023). For 2024, the city of Linz has a total budget of € 1.074 billion (Magistrat Linz 2024).

The history of Linz is one of remarkable change. Hardly any other city has undergone such an impressive transformation from steel to a digital culture capital in the last 25 years. From a city known primarily for industry and steel, it has become a hotspot for innovation, all thanks to creating open spaces for collaboration and mutual empowerment. We recognise that all stakeholders in the city need to be part of this process to keep this change on track. In Linz, citizens, artists, scientists, and organisations are invited to venture into the new, fostering a climate of constant learning and prosperous development. This attitude keeps the city fresh and on the cutting edge and ensures that it is responsive to people's needs. In short, Linz is an example of how a city can use its past to build an innovative future.

3.1 Background and Initiative for the Innovationshauptplatz Linz (2016–2019)

The genesis of this future-oriented discourse can be traced back to Linz Mayor Klaus Luger's ambitious declaration that Linz should become Austria's most innovative city. Central to advancing these discussions and formulating strategies was a study led by Prof. Robert Bauer at Johannes Kepler University (Bauer et al. 2019). This study, grounded in rigorous academic research, explored how a medium-sized city could position itself as the most innovative region of the future. It engaged in qualitative, open-ended dialogues with key stakeholders across the Linz metropolitan area. Starting in October 2016, approximately 90 innovation leaders from prominent companies and key figures from the innovation ecosystem were actively involved. Initially, qualitative interviews were conducted, with their findings integrated into the study (Bauer et al. 2019).

The Innovation programme comprises 35 measures aimed at fostering innovation across various domains of urban development. Key elements include initiatives to enhance sustainable mobility and transportation infrastructure, promote environmental sustainability through energy efficiency and renewable energy projects, support education and research collaborations to advance knowledge and innovation, bolster cultural initiatives and the creative economy, and drive digital transformation across public services and industries. These efforts have improved urban infrastructure, reduced environmental impact, strengthened cultural vibrancy, and enhanced digital connectivity and capabilities. Linz's Innovation programme underscores its commitment to positioning the city as a forward-looking, sustainable, and innovation-driven urban environment, fostering positive societal and economic outcomes (Bauer et al. 2019).

3.2 Building Up the Innovationshauptplatz Linz (2019–2022)

One of the major achievements and the first implementation measure of the innovation program discussed in the previous chapter, "Measure 6—the establishment of an Innovation Office"—was the opening of the Innovationshauptplatz Linz ('main square of innovation') in the summer of 2019, endorsed unanimous approval by the City Council (Bauer et al. 2019).

Since its opening in 2019, the Innovationshauptplatz Linz has emerged as the core hub and networking node within Linz's innovation ecosystem. It serves as a one-stop shop for various innovation initiatives, connecting all relevant stakeholders and facilitating the introduction and advancement of innovative projects by citizens, companies, startups, and institutions in an open and accessible manner throughout the city. Furthermore, this innovation hub acts as a central connecting point of contact and coordination, facilitating networking, mediation, and collaboration for innovative projects. It provides direct access to policymakers and all relevant stakeholders. Conveniently situated on Linz's main square, it welcomes citizens, companies, start-ups, and institutions in an open and accessible way. The ease of access is evident in the following quote: "When you come to Linz, you can just drop by, tell our team what you need—be it an accelerator program, access to the innovation ecosystem, or a coworking space—and we'll put you in touch with every player," says Silvia Hackl, Head of Participation and Innovation at the City of Linz.

As a catalyst, the Innovationshauptplatz aims to consolidate and promote the existing innovation potential while inspiring citizens to initiate their projects. It is important to emphasise that the Innovationshauptplatz Linz operates not only offline but is also accessible and open online for the city's residents through the citizen participation *platform innovation.linz.at*.

The Innovationshauptplatz Linz launched in 2019 with a budget of € 300,000, including personnel costs and 2.5 full-time equivalent positions (FTEs). Approximately € 160,000 was explicitly allocated for projects. In February 2022, the Innovationshauptplatz Linz merged with the Department of Economy and EU Affairs, from then on, operating under the name "IWE"—Innovation, Business and EU Affairs, resulting in a total of 9 FTEs. All projects are financed exclusively through funds allocated by the City of Linz. For 2023, a "pure" project budget totalling € 239,300,— has been earmarked. In autumn 2023, there was a further merger, this time with the Climate Office. Since then, the department has operated under the name "Economy, Innovation, Climate Protection and EU" and has a project-only budget of € 242,000 for the years 2024 and 2025, along with personnel costs of € 325,000 for a total of 14 FTEs and one part-time employee.

3.3 *What Are the Main Targets of the Innovationshauptplatz Linz?*

First, it is to advance digital literacy among Linz residents, highlight existing offerings in the city, foster connections between the creative scene and businesses, strengthen Linz's overall ecosystem, and deepen synergies further. Linz aims to be Austria's most innovative city, facilitating networking and support among individuals, groups, and initiatives within Linz's administrative sectors and promoting innovative ideas and projects through partnerships, infrastructure, support services, and public relations, particularly focusing on vulnerable groups, fostering future skills accessibility to all citizens. Second, the active support of existing communities and initiatives involves developing and executing projects and events to foster innovation in Linz, particularly in public urban spaces. This includes inviting engagement from the community through the Innovationshauptplatz initiative of the city of Linz.

Third, the Innovationshauptplatz aims to provide consulting, mediation, and support for creative individuals and groups, facilitating their connection with supportive networks and the city of Linz. Its goals include promoting innovative ideas and projects through partnerships, infrastructure, assistance services, and public outreach. The Hub is committed to developing and implementing its initiatives, projects, and events to embed the innovation mindset within the Linz community, particularly in public urban spaces. Additionally, it plays a crucial role in implementing and advancing the first Linz Innovation Program, approved by the city council in May 2018. The primary goal is to act as a translator between the public, the government administration, and other stakeholders, ensuring effective communication and understanding among all parties involved.

3.4 What Sets Us Apart from Other "Business and Innovation Offices"? What Do We Do Differently?

Efficient and agile operations are paramount, ensuring a direct civic interface, streamlined processes, swift project execution, minimal bureaucratic hurdles, and expedited approvals. Our emphasis on accessibility integrates technology literacy and lifelong learning, fostering synergies within our ecosystem. Collaboration extends beyond city borders, drawing expertise from across Austria, notably informing initiatives like the Founder Scholarship. Participants receive sustained support, including mentorship across diverse fields, reflecting our network's expansion and visibility across sectors—from coding communities and innovative formats like "Linz connects" to collaborations with social institutions and educational partners.

Our cross-departmental approach within the City of Linz leverages expertise across initiatives such as the Climate Experience Room and Digital Linz Program, underscoring our collaborative ethos. Notable projects include Sciencity, Linz hACkT, and initiatives like virtual parent evenings and Founder Scholarship. International engagements at, for example, the H22 City Expo in Helsingborg and the expo in Barcelona highlight our accessible approach resonating globally.

The Linz City Council directly operates Innovationshauptplatz Linz as part of the "City Government Office." This direct placement within the City Government facilitates short decision-making processes and enables quick and agile operations. The office is not conventionally located within the City Administration building but occupies a former storefront on the main square of the City of Linz. As a result, the Innovationshauptplatz Linz offers a highly accessible and inviting environment without any barriers. The office environment departs from conventional municipal settings, featuring modern, bright, and inviting spaces. Its consciously open atmosphere also attracts tourists and passersby. Operating nearly paperless, the office exemplifies a modern approach to municipal administration, fostering an open exchange of ideas and networking opportunities, prioritising accessibility and collaboration between administration and citizens. In collaboration with other Austrian and international cities, Linz's innovative office concept represents a pioneering project that stands out in its openness and innovative approach.

The Innovationshauptplatz Linz plays a central transformative role in shaping the city's innovation landscape. It cultivates a strong sense of community involvement, bridges the gap between policymakers and citizens, and fosters an environment conducive to creative and innovative thinking. By focusing on initiatives that span multiple demographic groups and diverse interests, from an e-participation platform for the citizens of Linz to children learning about government in Create Your City to seniors becoming digitally savvy through Smartphone Days and from open administrative engagement with Enter_Tainer to tackling environmental issues in the Climate Experience Space. Overall, Innovationshauptplatz Linz has been and is instrumental in transforming Linz into a dynamic, inclusive, and forward-thinking city that values the contribution of all its citizens. This can serve as a benchmark for democratised urban development in other cities. Insights into the wide-ranging activities are available on the website *innovation.linz.at*, Instagram *(@innovationshauptplatz)* or LinkedIn *(@Innovationshauptplatz der Stadt Linz)*.

This initiative marked a pivotal step towards fostering a dynamic environment for creativity and forward-thinking solutions within Linz. The Innovation Office serves as a hub for interdisciplinary collaboration, facilitating partnerships between public institutions, private enterprises, and academic entities. Its establishment underscores Linz's commitment to cultivating a culture of innovation, aiming to drive sustainable urban development and enhance the city's competitiveness on a global scale.

In the last years, the Innovationshauptplatz Linz implemented various significant formats and projects, such as the Pop-Up Store, the Founders Scholarship and other grants for young entrepreneurs, which are explained in detail within this article in the part "Exemplary Practices in Social Innovation, Entrepreneurship and Climate Protection".

3.5 External Acknowledgements

During this period, the Innovation Hub was awarded numerous prizes and participated in international expos and festivals, such as:

- Twice winner of the Austrian Administration Award in 2021 for the Electronic Citizen Participation Platform of the City of Linz and 2023 for the project

"Climate-Neutral Industrial City Linz 2040/Concept Development through a City-Wide Process." Read more in the chapter "Austrian Administration Awards 2021 and 2023".

- In 2023, the creative installation "Enter_Tainer: A Container as a Communication Tool," which won the Community Voting at the Creative Bureaucracy Festival in Berlin, provided an accessible window into the city's administration. This project invited citizens to engage directly with city operations through initiatives like the "Stadtcafe Linz" podcast. In 2024, the Innovation Hub was selected with the "Create your City!" format and conducted a workshop during the Creative Bureaucracy Festival.
- The Innovationshauptplatz Linz was a finalist for the European Commission's "iCapital Award (The European Capital of Innovation Award)" in 2021 and reached the semifinals for "The European Rising Innovative City" Award in 2023.
- At the H22 City Expo in Helsingborg, the City of Linz was showcased in its pavilion alongside four other cities. This entire project was led and executed by the Innovation Hub Linz. A special highlight was the visit of the Swedish Crown Princess Victoria to the Linz pavilion. Additionally, the robot dog from our partner, qapture, went viral on TikTok during the event in Sweden.
- An additional achievement was the City of Linz securing second place for the first time in the Austrian Startup Monitor 2023/24.
- Finalists iCapital Award 2024— awarded with the title 2nd European Rising Innovative City

The Innovationshauptplatz Linz is also represented in international networks such as the European Capital of Democracy (ECoD) and the European Campus of City-Universities (EC2U) network. This year, the official forum of the EC2U network will be held in Linz for the first time, hosting approximately 300 international guests, including representatives from city administrations, politicians, professors, researchers, students, and various organisations. The ECoD network is also vital and exciting for city exchanges. Through the focus of this network, best practices can be regularly shared between cities.

3.6 Settle in the Department of Economics, Innovation, Climate Protection and EU (2023-Present)

Significant developments have taken place in recent years. After the elections in autumn 2021, the Department of Economic Affairs and Economic Development first joined the Innovation Hub, followed by the Climate Unit and the new focus on climate protection. In autumn 2023, the department was restructured and expanded from an initial team of two to fifteen members. It was renamed the 'Department for Business, Innovation, Climate Protection and the EU'.

The department's responsibilities were divided into three different focus areas: Innovation and Participation (= Innovation Centre Linz), Economy and EU, and

Climate Protection and Hydrogen. By bundling these competencies in one department, the three focal points can be even better interlinked. New projects were launched, and the strategic focus was adapted and evaluated to better align it with the city's current needs.

4 It's All About Strategy

As explained in the previous chapter, the department "Economy, Innovation, Participation and EU" has been divided into three distinct areas. Nevertheless, these areas function as an integrated unit, with each part operating according to the city of Linz's strategic directives.

The "Innovation and Participation" area aligns with the goals of the city's Innovation Program (Bauer et al. 2019) and the Digital Program Linz. This focus includes the digital transformation of the city and active citizen participation in these projects. Early-stage support and awareness-building are also key components of these strategies. The Digital Program Linz was developed through a participatory process, incorporating around 30 out of 55 citizen-submitted ideas into the program. The Digital Program aims to transform Linz into a smart, efficient, and innovative hub. Key objectives include digitising municipal services to enhance efficiency and transparency, developing smart city initiatives such as intelligent traffic systems and energy solutions, expanding digital infrastructure with improved broadband and public Wi-Fi, supporting innovation and startups, promoting digital skills and education, and ensuring robust data security and privacy. These efforts are designed to position Linz as a modern, forward-thinking city, enhancing the quality of life and fostering technological advancement.

The Digital Program is continuously implemented and adapted. The primary goal of the Innovation Program is to foster research and development by supporting projects and providing funding. Additionally, the program strengthens the startup ecosystem through targeted support and infrastructure. It promotes partnerships between businesses, research institutions, and the city, focusing on integrating smart city technologies to enhance quality of life and efficiency. The program also advances education and training in new technologies to build a skilled workforce and drives sustainability and improved living standards through innovative solutions. In 2024, the Innovation Strategy will be realigned for the coming years, and the existing Innovation Program will be evaluated.

The "Climate Protection and Hydrogen" area focuses on climate awareness and promoting hydrogen initiatives in Linz in cooperation with the companies from Linz. The underlying strategies are detailed in the city's "Climate Change Adaptation Concept" and "Climate Neutrality Concept." The "Future Linz" Climate Change Adaptation Concept outlines the risks of heat, warming, and drought for Linz and provides strategies to address these impacts. It acknowledges that some climate effects are unavoidable and emphasises preparing for and managing these changes. The concept aims to integrate climate adaptation into urban planning and management, promoting synergy across city departments to mitigate risks and capitalise on potential benefits effectively. The Climate Neutrality Concept aims for Linz to achieve climate neutrality by 2040, targeting a reduction of at least 90% in city-wide greenhouse gas emissions compared to 2019 levels. Up to 10% of the remaining unavoidable emissions can be offset through natural carbon sinks (forests, wetlands), technical solutions (carbon capture and utilisation, circular economy), and greenhouse gas compensation projects.

The "Economy and EU" area operates under the city strategy, designed to significantly enhance residents' quality of life through a comprehensive urban development plan. This approach emphasises sustainable development by integrating eco-friendly practices to support environmental sustainability. Economic growth is a key focus, with initiatives aimed at bolstering local businesses, fostering innovation, and generating job opportunities. Central to the strategy are improvements in infrastructure, public transport, and recreational facilities, all geared towards elevating community well-being. Social inclusion is prioritised to ensure equitable access to city resources, while cultural and creative initiatives are supported to enrich urban life. Additionally, the plan embraces digitalisation and innovation, using technology to advance urban services and Smart City projects. Transportation and mobility enhancements are also crucial, focusing on better traffic management and sustainable transport solutions. These efforts are set to transform Linz into a modern, inclusive, and dynamic city for the future. The previously mentioned Innovation Program is also integral for the "Economic and EU" area, with a strong emphasis on internationalisation, English proficiency, expanding contacts and fostering cooperation with companies and engaging SMEs in the city.

5 Build Up an Ecosystem for a Vibrant Urban Future

Linz's main goal is to use innovation to shape a responsible and vibrant urban future. The close and trusting cooperation between the various actors in Linz, including citizens, ensures that these innovations are efficiently integrated into the city's holistic development plans, which encompass the pillars of urban economy, urban growth, urban life, urban networks, and urban future (Linz Urban Strategy 2022). Understanding the roles and interactions of these stakeholders is essential for developing a resilient and vibrant urban future.

5.1 Ecosystem, Network, Stakeholders

In the economic sense, stakeholders or stakeholder groups are all internal and external groups directly or indirectly affected by the company's activities. They have the right to assert their interests vis-à-vis the company. (Thommen n.d.)

Successful corporate management considers all stakeholders' interests when making decisions, as the company's success is closely linked to successful stakeholder management. „A stakeholder in an organisation is (by definition) any group or individual who can affect or is affected by the achievement of the organisation's objectives" (Freeman 1984, p. 46).

The following stakeholders can be identified for a city's network:

- **Citizens**: Residents who use services and infrastructure want to improve their quality of life.
- **Businesses**: Local businesses that rely on infrastructure and support and contribute to economic development.
- **Investors**: Individuals or institutions that invest in urban projects and promote economic growth.
- **Employees and trade unions**: Employees who work in city facilities or local companies and want to improve their working conditions.
- **State institutions**: Higher-level authorities that set policies and laws and provide financial and legal support.
- **Public service providers**: Utility companies (e.g. energy, water) responsible for basic infrastructure.
- **Non-governmental organisations (NGOs):** Organizations that represent specific social, environmental or cultural interests.
- **Educational institutions**: Schools and universities that promote education and research and train future professionals.
- **Media**: Local press, radio and television that report on urban affairs and influence public opinion.
- **Tourists**: Visitors who choose the city as a destination and contribute to the local economy.

These groups play a crucial role in the urban network and contribute to the shaping and development of the city. In addition, supra-regional and international networks are important for exchanging ideas with their stakeholders and learning from each other.

As a central hub in the City of Linz's network, the Innovationshauptplatz fulfils several essential functions, which the various stakeholders perform: The primary task as a central interface is communicating information and content between institutions and the population. This is done by simply passing on knowledge and, above all, by translating it into a form understandable for the target group. This gives people access to valuable information that they might otherwise be denied (Su and Lee 2020). Furthermore, a central interface acts as a moderator or even mediator within a network. By translating complex institutional content into a generally understandable language, different stakeholders can co-operate effectively in a coordinated manner. This is particularly important to avoid conflicts and create synergies that strengthen a network's collective ability to act.

However, it is not only the transfer of knowledge within a network that requires moderation but also the formation of the network itself and the mediation of contacts. A central interface can facilitate access to institutions and relevant players by acting as a door opener and enabling network members to make relevant contacts more quickly and efficiently (Lopes and Faria 2021). It is also essential to maintain an overview of the network to pool resources, utilise them effectively, and avoid creating parallel structures.

5.2 *The Urban Innovation Ecosystem of Linz*

Linz's main goal is to use innovation to shape a responsible and vibrant urban future. Several institutions play a crucial role in this experimenting process. The Innovationshauptplatz Linz, the Ars Electronica Linz, the Johannes Kepler University Linz, and the University of Arts Linz are integral parts of this luminous ecosystem. This culture of bringing together participation, technology, art, science, and digitalisation opens new perspectives for individuals and organisations alike. The close and trusting cooperation between the various actors in Linz, including citizens, ensures that these innovations are efficiently integrated into the city's holistic development plans, encompassing the pillars of urban economy, urban growth, urban life, urban networks, and urban future (Stadt Linz 2022).

To realise its ambitions, Linz must continue to foster an environment conducive to innovation and citizen-centred growth. This includes supporting local industry and universities, promoting participatory governance, and maintaining collaborative platforms that cultivate innovative thinking and action. The urban innovation ecosystem of Linz is based on three foundational pillars that drive the city forward:

- Citizen-Centered Innovation: The city's commitment to participation and co-creation and the need to ensure that all citizens have a voice in the city's growth and development.
- Urban Production: The city's history and legacy as an industrial centre and the need to transition to a sustainable, climate-neutral economy.
- Science and Technology and Arts: The city's unique blend of arts, science and technology and the potential for creative collaboration between citizens and private and public actors.

The Science and Technology and Arts sector is of central importance for Linz and characterises the city as a creative and innovative centre. Its designation as a UNESCO City of Media Arts (Linz Kultur 2024) and its role as the European Capital of Culture 2009 have strengthened the city's cultural profile internationally and promoted innovative projects. The Ars Electronica Festival, which has been running for over 40 years, serves as a global forum for media art and digital culture, bringing together artists, scientists and technologists (Ars Electronica Linz GmbH and Co KG, 2024). These initiatives not only promote artistic and technological education but also strengthen the innovative power and economic development of Linz. The combination of art, science and technology creates a unique ecosystem that supports creative processes and knowledge transfer. This makes Linz an attractive location for talent and companies, improves the quality of life and increases the city's international visibility.

6 Exemplary Practices in Social Innovation, Entrepreneurship and Climate Protection

In recent years, Linz has implemented diverse formats and initiatives under the Innovationshauptplatz banner to foster innovation and enhance citizen participation. For instance, "Create Your City" aims to engage youth in municipal affairs through classroom discussions, debates at the Old Town Hall, and educational materials for teachers. Concurrently, the award-winning "Enter_Tainer" installation provides a direct interface for citizens to interact with city operations, exemplified by initiatives such as the Stadtcafé Linz podcast. Selected examples from social innovation, entrepreneurship and climate protection are presented below to illustrate the focus of the Department of Economics, innovation, climate protection and EU on the Innovationshauptplatz of the city of Linz.

6.1 Social Innovation Initiatives: Linz hACkT and Tour of Innovation

The Innovationshauptplatz Linz is home to the annual citizens' hackathon, Linz hACkT. 2023, in its third year, the hackathon invites citizens to propose and develop innovative solutions to social inclusion, climate change, and digitalisation issues. Over 1000 people voted and supported their favourite project online through the e-participation platform. One of the standout successes from these hackathons is Nochba, an app developed by three students from HTL Leonding. Nochba is designed to foster community cohesion, offering a platform for neighbours to request and provide various forms of support. This project has achieved significant recognition, winning numerous awards locally and internationally (#1 Linz hACkT Event X Innovationshauptplatz—#1 mPreneur Austria Contest X ICT4D.at—#1 Platz Immotopia Innovation Award—#3 Spusu Innovation Award). It represents the creative, community-focused innovation the Innovationshauptplatz Linz seeks to inspire.

The Innovationshauptplatz Linz plays a central transformative role in shaping the city's innovation landscape. It cultivates a strong sense of community involvement, bridges the gap between policymakers and citizens, and fosters an environment conducive to creative and innovative thinking. By focusing on initiatives that span multiple demographic groups and diverse interests, from an e-participation platform for the citizens of Linz to children learning about government in Create Your City to seniors becoming digitally savvy through Smartphone Days and from open administrative engagement with Enter_Tainer to tackling environmental issues in the Climate Experience Space. Overall, Innovationshauptplatz Linz has been and is instrumental in transforming Linz into a dynamic, inclusive, and forward-thinking city that values the contribution of all its citizens. This can serve as a benchmark for democratised urban development in other cities. Insights into the wide-ranging activities are available on Instagram or LinkedIn.

The Coding Community in Linz (see Digital Program Linz 2023, p. 64), anchored by the groundbreaking "Coding Community Roundtables" initiative, is a thriving ecosystem that brings together diverse stakeholders in the IT and programming sector. Comprising influential organisations, major companies, universities, private associations, and more, this community serves as a core driver of innovation and collaboration in this area. Regular meetings under this initiative foster an environment that encourages exchanging ideas, industry trends, and best practices. The collaborative nature of these roundtables nurtures innovative projects, which leverage the collective expertise and resources of the participating organisations. For instance, "Linz Codes!" engaged over 500 children and teenagers in 2021. This initiative emerged from the coding community and received support from prominent partners such as Microsoft, Dynatrace, the City of Linz, and the CoderDojo Linz. Numerous volunteers from the coding community actively participated in this event, contributing their time and expertise.

The Coding Community's influence extends beyond the city's digital sphere, positioning Linz as a prominent hub for digital advancement and new work. The roundtables encourage innovation and play a pivotal role in enlarging the free-of-charge offers to learn programming in Linz. They stimulate partnerships and synergies that contribute significantly to the collective growth of new IT talents in the city and enhance the digital understanding of the young generation. As an enabling partner, Linz, in collaboration with the other actors, is creating opportunities for economic growth, attracting investments, and overcoming the challenge of missing skilled workers. The Coding Community Roundtables, therefore, serve as a powerful catalyst for Linz's digital development and transformation.

The City of Linz has been successfully organising the Tour der Linzer Innovation for the past five years, an initiative to bring cutting-edge technologies closer to the local urban communities. This comprehensive program offers interactive experiences for all age groups, providing an opportunity to engage with emerging technologies such as robotics, programming, climate technology, hydrogen experiments, VR experiences, plastic recycling, digitalisation, and more for free. The primary focus of this project is early education and empowerment of children and adolescents while also addressing the concerns of the general population regarding digitalisation. We are proud to collaborate with 30 esteemed partners, such as Open Commons Linz, Female Coders, and Ars Electronica, who actively support and contribute to the success of the tour. The involvement of the partners helps create an enriching experience for everyone involved, fostering curiosity, exploration, sustainability orientation, and knowledge transfer.

By reaching out to young minds, Tour der Linzer Innovation aims to cultivate their interest and passion for innovation, as well as reduce any apprehension towards digitalisation. Additionally, the tour strategically focuses on locations within "challenged" neighbourhoods and areas where parents may not have access to specialised educational institutions such as programming schools. By doing so, we effectively reach marginalised communities, enabling equal opportunities for all.

The impact of the Linz City Innovation Tour is far-reaching. By actively selecting locations, including "challenged neighbourhoods" and "difficult schools", the project ensures that the initiative benefits those who need it the most. Furthermore, it actively encourages the participation of schools by offering dedicated timeslots for their visits, creating an inclusive and collaborative learning environment. The tour has become integral to the city's identity, leaving a lasting impression on the community. By directly engaging with the population, the project has overcome the challenge of effectively reaching all residents, and these major efforts have been highly recognised and appreciated locally. The grand finale of the Linz City Innovation Tour is the Sciencity—Kids Edition event, held on the tour's final day. This event provides children and adolescents with an accessible platform to experience the wonders of science through exciting experiments using everyday objects. The Linz City Innovation Tour and the Sciencity—Kids Edition event showcase the city's dedication to fostering a future-ready society by equipping children, adolescents, and the general population with the necessary skills and knowledge in terms of digital and technological literacy. Through this community-centred approach, Linz aims to inspire a sense of curiosity, eliminate fear, and empower all individuals to thrive in an increasingly interconnected world.

6.2 *Economic Development: Entrepreneurship Scholarship and PopUp Store*

The Innovationshauptplatz Linz is revolutionising its approach to fostering entrepreneurship and innovation through various start-up initiatives. Two of the most prominent initiatives are the Pop-Up Store and the entrepreneurship scholarship. These initiatives reflect Linz's commitment to nurturing a vibrant ecosystem for start-ups while ensuring that local citizens are engaged and given a voice in the process.

The Pop-Up Store, located in the heart of Linz, is a dynamic platform that encourages creativity and innovation, welcoming initiatives, start-ups, companies, communities, ideas, concepts, and projects originating from or contributing to the city. It is a free-to-use space available for a month at a time and has been home to 40+ tenants across diverse fields since 2020. The selection process to occupy the space is community-driven, with Linz residents being able to vote for their favourite projects through the e-participation platform. This novel approach fosters collaboration and empowers the residents to have an active say in shaping the city's business and industrial landscape.

To complement the Pop-Up Store initiative, Linz has also introduced the Entrepreneurship Scholarship. Designed to bolster the city's start-up ecosystem, this initiative offers significant financial support to early-stage entrepreneurs. https://innovation.linz.at/

In 2022, the City of Linz started the Founders Scholarship after asking Startups what they really need to start their own company. With this Scholarship, founders are eligible for a monthly grant of EUR 1000 for five months, offering them a buffer as they establish their new companies. This grant aims to alleviate the stress of living costs, freeing entrepreneurs to focus on realising their innovative ideas. Beyond the grant, the Entrepreneurship program also offers access to a diverse mentor network of 50 industry professionals, providing new businesses with expert guidance and support for free. Moreover, the initiative includes free press support, preferential access to the Pop-Up Store, and continued assistance from the team of Innovationshauptplatz Linz beyond the initial grant period. This package of resources is designed to provide founders with comprehensive support in their entrepreneurial journey. https://innovation.linz.at/

Linz's Pop-Up Store and Entrepreneurship Scholarship and Rental Subsidies initiatives are testaments to the city's progressive and supportive stance towards local start-ups. By fostering a collaborative environment, engaging local citizens, and providing robust financial and mentorship support, Linz is positioning itself as a nurturing ground for start-ups and a vibrant hub for sustainable economic growth.

In addition, the city of Linz published the Startup Guide Linz in 2023 with the support of aws Austria Wirtschaftsservice, Junge Wirtschaft as a representative of "hub,ert" network for founders in Upper Austria and Dynatrace, a software company. The book presents the most important players in the region on over 200 pages. It gives an overview of the startup ecosystem of Linz and important tips for starting a business in this city. This includes educational institutions, available workspaces, profiles of innovative startups, insights from experienced entrepreneurs and information about the city and its special features. https://www.startupguide.com/ecosystems/linz

The series Startup Guide was founded in 2014 by CEO Sissel Hansen. Her idea arose when she moved from Copenhagen to Berlin to start a company. In this startup process, she noted the difficulty of finding useful information about starting a business. After being rejected by several investors, she tried her luck with bootstrapping and went to market with Startup Guide Berlin, which sold out in less than 48 h. Startup Guide creates books and online content that help entrepreneurs find their way in their chosen region and get in touch with different startup scenes worldwide. The goal is to inspire and empower people to start their businesses. Startup Guide has published books for more than 50 cities and regions worldwide. In 2019, Startup Guide committed to focusing on companies that create positive economic, environmental and social value. (Startup Guide 2023).

6.3 Climate Protection: Climate Fund and Climate Action Days

Linz's Municipal Climate Fund is a pivotal initiative that has been instrumental in promoting climate action and resilience within the community of Linz (involving private, organizational, and corporate initiatives). With an annual budget of 1 million euros, the fund has supported a total of 46 projects since 2020. Today, it highlights the diverse projects and their positive impact on climate mitigation, adaptation, and combined approaches.

The Municipal Climate Fund has empowered stakeholders, such as citizens and organisations, to actively participate in Linz's climate action. The fund has witnessed 16 remarkable climate mitigation projects, 13 innovative climate adaptation projects, and nine visionary projects that address climate neutrality and adaptation. Different entities have undertaken these projects, including 13 projects within the municipal administration, nine projects by associations/organisations, nine projects by institutions, five by private individuals, and two by educational institutions. The remarkable success rate of the Municipal Climate Fund highlights its effectiveness in selecting financially worthy projects.

With a success rate of 65.5% since 2020, the Climate Advisory Board has diligently evaluated and endorsed projects that align with Linz's climate strategy. Nearly 3 million euros have been allocated to these approved projects, ensuring their successful implementation and positive climate outcomes. Notably, 15 projects have already been completed, demonstrating our commitment to delivering tangible results.

The following examples show how the Climate Fund effectively fosters sustainable and citizen-driven projects in Linz. The Klimaoase Linz Lustenau (Climate Oasis Linz Lustenau) project has created a peaceful oasis in the heart of an industrial area. Surrounded by a harbour and factories, the garden oasis provides a calming environment where people from nearby companies can escape the bustle, engage in yoga classes, and enjoy an earth shower. This tranquil space is a testament to Linz's commitment to combining industrial and ecological aspects in its city development.

Two other initiatives focus on promoting active mobility and sustainable energy use. The initiative Linz zu Fuß (Linz by Foot) encourages citizens to walk more by engaging them in improving the walking experience. It includes conducting walking conversations and pedestrian audits, focusing on accommodating people with disabilities. Meanwhile, the Pilotprojekt Kaisergasse (Kaisergasse Project) started by Helwin Prohaska and his team (Verein Energiewende Linz) seeks to refurbish residential complexes environmentally, including the implementation of communal photovoltaic systems. Following successful phases of the project, it has inspired other settlements to adopt similar climate-friendly measures, illustrating the city's efforts to empower its citizens and foster a community-led approach to climate action. For instance, Helwin's current project, Power Playground, is an interactive workshop using photovoltaics, pedal power, and virtual reality to teach Upper

Austrian students about renewable energy and eco-friendly home modifications. Power Playground was first exhibited at the Ars Electronica Festival 2022.

Linz's Municipal Climate Fund and its projects strive to create a lasting impact on climate change and resilience in the city's urban environment. This contributes to a more sustainable future and aligns with the European Union's broader climate objectives. The climate fund is shaping a greener, more resilient city through the collective efforts of all stakeholders involved. The city of Linz, through its Municipal Climate Fund, has demonstrated an unwavering commitment to climate action and resilience.

7 Have a Say, Help to Shape, Take Part

In 2020, the Austrian Federal Ministry for Art, Culture, Public Service and Sport (BMKÖS) published the "Green Book on Participation", which outlines a comprehensive approach to promoting citizen participation in the digital age. It highlights the integration of various participation formats to ensure broad and inclusive engagement. Fundamental principles include transparent communication, easy access, and incorporating results into political decision-making. The integrative concept emphasises combining digital and analogue participation methods to reach a diverse audience by including user-friendly, barrier-free digital platforms, targeted information campaigns, and resource provision to support participation. Feedback loops are also needed to keep citizens informed about the progress and impact of their contributions, enhancing trust and motivation (Rosenbichler and Grünwald 2020).

The city of Linz is setting standards in modern urban development with a digital citizen participation platform. It has enabled the population of Linz to participate actively in the design of their city since 2019 by submitting, discussing, and voting on proposals. Based on an integrative concept, as described in the "Green Book on Participation", the platform uses the potential of information and communication technology to design transparent, inclusive and effective participation processes. As mentioned in the examples above, digital and analogue methods are combined to enable broad and diverse participation and improve the quality of urban decision-making by incorporating personal experiences and suggestions from the population.

7.1 Platform Partizipation.linz.at

The centrepiece of the Innovationshauptplatz has been the platform partizipation.linz.at since the beginning. On the citizen participation platform designed by GoVocal, users can submit suggestions on any topic to improve the city of Linz. All that is needed is a short description, a reference to an example from another city or a well-thought-out concept—the better the idea is described, the more likely it is to find support in the community.

The entry rule for the Innovationshauptplatz platform is as follows: if a proposal receives the required 50 votes within 60 days, the Innovationshauptplatz team consults the relevant colleagues in the municipal office or the Linz corporate group (depending on the proposal's topic) and invites them to a joint discussion with the idea contributors. In this meeting, the proposal is discussed in detail to determine if and how the idea can be implemented, as well as to identify any challenges that might not be easily resolved. Open exchange is a priority, so after the discussion, an official statement is published on the platform as an update to keep the community informed. More than 160 proposals on various topics have been submitted through the platform, and more than 35 such discussions have already taken place. Unfortunately, there is no guarantee that the ideas submitted will be implemented. Still, the Innovationshauptplatz team tries to mediate between the responsible bodies, ensures transparency in the exchange and endeavours to incorporate suggestions from the community into existing concepts.

One example is a proposal for the leisure facility on the northern bank of the Danube: one person who suggested the idea noted that there are no toilets, shower facilities, or drinking water taps nearby at the sports facilities. The proposal was discussed with those responsible in the municipality. During this meeting, the idea provider learnt that a drinking water extraction point directly at the desired location was not possible because, firstly, the required budget (more than 100,000 euros for a lifting system) was not available at the time and, secondly, the sports facilities are located in the flood protection area of the Danube. A sanitary facility in this area would be subject to special water regulations. As an alternative, however, the proposal could be implemented at a nearby facility, which was already planned and ultimately approved by the Linz City Council in late summer 2022. https://partizipation.linz.at/de-DE/initiatives/wasserspender-und-sanitaranlagen-fur-den-sportpark-urfahr

Another example is the Open Stage from 2024: Due to an increase in federal taxes, the fees for street musicians in Linz have risen, so one user suggested creating zones for free performance opportunities in Linz—a city that attracts many visitors every year with the large street art festival "Pflasterspektakel". In cooperation with the Altstadt Linz association, the Department of Economics, Innovation, Climate Protection and EU opened the Open Stage in Linz's old town in May 2024. Interested street artists can use this stage free of charge from Thursday to Saturday after booking in advance and perform in public without having to obtain further permits. https://partizipation.linz.at/de-DE/initiatives/strassenmusik-moglich-machen-durch-streichen-der-50-euro-tagesgebuhr

Since 2019, numerous ideas, applications, and suggestions for participatory projects (such as PopUp-Store and Linz hACkT) have been received. The community submitted more than 150 additional ideas related to special topics of climate change and digitalisation, which are then incorporated into municipal concepts, including the Climate Change Adaption Concept and the Digital Linz program.

7.2 Austrian Administration Awards 2021 and 2023

The department was already honoured twice with the Austrian Public Administration Award. First in 2021 for the participation platform. Vice-Chancellor Werner Kogler presented the award online. "Digitalisation and participation are key areas of strength to the City of Linz. With the e-participation platform Innovationshauptplatz, the people of Linz are always up to date on the latest projects.

Furthermore, this platform offers people an additional opportunity to actively help shape their Linz and contribute ideas for the future. I am delighted that the Federal Ministry has honoured Linz with this prestigious award. It strengthens Linz on its way to becoming Austria's most innovative city, " said Mayor Klaus Luger.

In May 2022 the Innovationshauptplatz participation platform was one of the finalists at the Innovation in Politics Award and was presented at the conference and ceremony in Warshaw. In 2023, the head of the department, Oliver Schrot, accepted the second Austrian Administration Award for developing the city's concept climate-neutral industrial city of Linz 2024. The concept was developed through a city-wide elaboration process with many participation workshops and multi-stakeholder discussions. The Linz Professional Fire Brigade also received an award this year.

7.3 Best Practice: "qapture GmbH"

When a city is willing to let companies use it as a testbed, it fosters innovation. Linz's digital program, its ecosystem, and support services (like rent reduction) helped Startup "qapture GmbH" thrive during COVID-19 and beyond. "qapture GmbH" has become a role model within the Linz innovation ecosystem, supporting key projects such as the Linzer Innovation Tour and the H22 City Expo in Helsingborg, Sweden. They encourage other startups, boasting international partners like IKEA and prestigious projects like the digitisation of the new Austrian Parliament. These accomplishments highlight qapture's position as a best practice example. The company "qapture GmbH" transforms physical environments into digital twins using advanced laser scanning and robotics technologies. Their services cater to construction projects, revitalisation efforts, and production facilities. The team, comprising developers, project managers, and architects, collaborates with research institutes and businesses. "qapture" provides detailed mapping and surveying services, the creation of digital twins, and producing point clouds and CAD plans. For more information, see https://www.qapture.at

> The Innovationshauptplatz was instrumental in launching our idea and service. Founded during COVID-19, we struggled to reach clients, but with support from the Innovationshauptplatz, we tested our services by digitising museums. This initiative, called "Couch Culture," allowed people in lockdown to visit museums virtually, leading to successful projects at Lentos, Mural Harbour, and a zoo. Recognising the need for scalability, we shifted our focus to the technical sector, digitising the University Hospital Schleswig–Holstein facilities. Over

the past four years, we expanded from two founders to a team of 20, significantly broadening our services. Our comprehensive approach to digital twins involves supporting clients from the construction phase to facility management and the cradle-to-cradle process, as seen in projects like the Austrian Parliament. We also founded a second startup, Circular, to analyse building data using AI, aiming to reduce CO_2 emissions in the construction industry. Supported by a 1.5 million Euro FFG (Austrian funding agency for industrial research and development) project, Circular employs six people and has a strong future ahead. The Innovationshauptplatz has supported us throughout the four years since our founding. We've made valuable contacts and given feedback in return. It's great that the hub led to the Founder's Grant and the overall development of the ecosystem infrastructure. I appreciate the Innovationshauptplatz because it helps young startups enter the market, allowing them to test their ideas, gain significant projects, and attract cool clients. Our goal now is to attract investors and scale our second startup globally. We believe that building data will be crucial in reducing the construction industry's environmental impact, which accounts for 40% of global emissions. We aim to use digitalisation to prevent these emissions and contribute to a more sustainable future. Daniel Höller, co-founder of qapture GmbH.

8 Lessons Learned

"The strategies we adopt play a crucial role in shaping the city's know-how, knowledge transfer, and its reputation as an attractive employer and role model. By fostering a spirit of dynamism, involving people, maintaining transparency in communication, and staying close to stakeholders, we can create a thriving urban environment", Ana Zuljevic, Head of the Division Economy and EU in the Department of Economy, Innovation, Climate Protection and EU. In a connected network of actors, it's not necessary to handle everything independently. However, it is imperative to establish frameworks, provide security, and ensure continuity. The Innovationshauptplatz evolution is a testament to this approach, demonstrating that while development and innovation continue, stability and predictability are maintained. Balancing this innovative environment with reliable operations is vital.

Remarkably, the city administration is willing to take on such ambitious initiatives. Integrating these efforts into the bureaucratic structure, which includes business regulations, travel protocols, and working hours, presents unique challenges. The speed of decision-making is a critical factor. Establishing governance frameworks that expedite decisions is essential, especially when you need Municipal Council and City Senate proposals and the local government representatives are not making daily decisions. In this context, it is advantageous that the department is directly subordinate to the mayor, which significantly shortens the decision-making pathways.

Effective internal communication is equally important. Over time, numerous projects have been undertaken in collaboration with other departments. This internal synergy strengthens the organisation's value, even if it wasn't initially recognised. As understanding and appreciation grow, the organisation becomes a flagship entity, symbolising excellence and innovation.

By embracing these strategies and continuously fostering a collaborative and transparent environment, we can ensure that the city not only keeps pace with innovation but also leads by example, creating a sustainable and thriving urban ecosystem.

However, a notable challenge remains: there is no guarantee that proposals submitted through Innovationshauptplatz will be implemented. Despite this, the open exchange with the public, initiatives, and businesses contributes to a better understanding of diverse perspectives and issues. This engagement enhances the perception of the city administration as transparent and modern, committed to involving its citizens in the decision-making process.

8.1 Outcome So Far

Since its beginning in 2019, the Innovationshauptplatz has established itself as an important interface between the citizens, economy, institutions, and the city administration. It has a close-knit network of experts who assist with questions about living and working in the City of Linz.

Some key figures so far:

- The Innovationshauptplatz Linz community submitted more than 600 proposals and ideas via innovation.linz.at to the various projects like Digital program Linz, PopUp Store, Climate Plan or for the City of Linz in general and shared more than 1000 comments and nearly 7500 reactions.
- From 2019 to 2024, the Tour of Innovation Linz impacted more than 6000 children and young people by empowering and low-levelly showing them new technology.
- By the end of 2024, more than 500 pupils had participated in the participation workshop 'Create your City!', where they learned about democracy, debating, and idea development and worked together to shape the city's future.

This open and participatory approach has strengthened the city's image as a forward-thinking and inclusive urban centre.

8.2 The Best is Yet to Come

In the age of digitalisation and globalisation, cities worldwide face the challenge of adapting their economic structures and promoting innovation. The Innovationshauptplatz Linz will continue to take on this task in the coming years and is planning a comprehensive transformation in various key areas. The following initiatives are currently being developed from 2024/25:

8.3 MINT Region Upper Austria Central

The importance of MINT (mathematics, informatics, natural sciences, and technology = STEM) for a region's innovative strength cannot be underestimated. The Innovationshauptplatz Linz is planning to create a STEAM region to promote young talent at an early stage and close the skills gap in these areas. Educational programmes are to be developed, and practical experience will be facilitated through cooperation with schools, universities, and companies. The aim is to build a sustainable pipeline of skilled workers that will contribute to the region's long-term economic stability and competitiveness.

8.4 New Economic and Innovation Programme

Another key project is the development of a new economic and innovation programme for Linz. This programme is intended to improve the framework conditions for companies, accelerate innovation processes, and position the city as an attractive location for national and international investment. The planned measures include promoting start-ups, supporting research and development, and creating a favourable environment for business start-ups. Targeted support and incentives create new technologies and business models to boost the city's economic growth.

8.5 Implementation of the Climate Neutrality Concept

Climate change is one of the most significant challenges of our time, and cities play a crucial role in reducing greenhouse gas emissions. Innovationshauptplatz Linz is committed to implementing a comprehensive climate neutrality concept, published in 2024 under the title 'Climate-neutral industrial city of Linz 2040'. This concept includes measures to increase energy efficiency, expand renewable energies and promote sustainable mobility. The aim is to make Linz climate-neutral by 2040. Implementing these measures requires close cooperation between the city administration, the business community and the population.

8.6 Internationalisation

Linz is focusing on internationalisation to remain attractive as a business location and to strengthen its innovative power. Innovationshauptplatz Linz explicitly promotes the attraction of international specialists and students. By cooperating with international educational institutions, creating an attractive living environment and supporting integration, the aim is to attract talent from abroad. These measures help to increase cultural diversity and promote the exchange of knowledge, which in turn strengthens the city's innovative capacity. The international EC2U and ECoD networks also help in this regard and will, therefore, continue to be maintained.

The Innovationshauptplatz Linz has established itself as a pivotal hub for fostering creativity, collaboration, and cutting-edge technological advancements. Through strategic partnerships, innovative initiatives, and a robust support infrastructure, the Innovationshauptplatz Linz has created a vibrant ecosystem that attracts startups, established companies, and research institutions. This dynamic environment stimulates economic growth and enhances the city's reputation as a centre for innovation and entrepreneurship. Key to its success is the holistic approach taken to integrate various stakeholders, including the public sector, private enterprises, and academic institutions. By facilitating an open and collaborative atmosphere, the Innovationshauptplatz Linz ensures that diverse perspectives and expertise converge to tackle complex challenges and create novel solutions. This synergy is further amplified by state-of-the-art facilities and access to cutting-edge technologies, which provide the necessary tools for experimentation and development.

Moreover, the emphasis on sustainability and social responsibility positions the Innovationshauptplatz Linz as a forward-thinking model for future innovation districts. The initiatives that promote green technologies and sustainable practices contribute to environmental conservation and align with global trends towards more responsible and ethical business operations. Looking ahead, the Innovationshauptplatz Linz is poised to continue its trajectory of growth and influence. With ongoing investments in infrastructure, education, and community engagement, it is well-equipped to adapt to the evolving needs of the innovation landscape. By maintaining its commitment to fostering a culture of innovation, the Innovationshauptplatz Linz will undoubtedly remain at the forefront of technological and entrepreneurial progress, serving as an inspiration and benchmark for other cities and regions worldwide.

References

Ars Electronica Linz GmbH & Co KG (Ed.). (2024) *ARS ELECTRONICA* https://ars.electronica.art/

Bauer, R., Luger, K., Auinger, M., Auinger, F., Langeder, B. & Oberneder, J. (2019). *Innovationsprogramm - Leitlinien für die innovative Zukunft von Linz.* Stadt Linz. https://www.linz.at/images/files/innovationsprogramm.pdf

Büro Stadtregierung Linz Stadtforschung & Linzer Innovationshauptplatz (Ed.). (2024) *Zahlstadt 2023.* https://www.linz.at/zahlen/120_Publikationen/005_Zahlstadt/Zahlstadt_Kennzahlenreport%202023%20DE.pdf

CitizenLab SA/NV. (2024) *Go Vocal.* https://www.govocal.com/

Freeman, R. E. (1984) *Strategic Management. A Stakeholder Approach.* Boston: Pitman.

Stadt Linz (Ed.). (2022). *Linzer Stadtstrategie 2022.* https://www.linz.at/media/stadtentwicklung/Linzer-Stadtstrategie-2022.pdf

Magistrat Linz (Ed.). (2024a) *Innovationshauptplatz Linz.* https://innovation.linz.at

Magistrat Linz (Ed.). (2024b) *Klimaneutrale Industriestadt Linz 2040*. https://www.linz.at/media/umwelt/KN-Konzept.pdf
Linz Kultur (Ed.). (2024) *Unesco City of Media Arts Linz*. https://www.cityofmediaarts.at/
Lopes, A. P., & Faria, J. P. (2021). "Innovation Networks: From Technological Development to Organizational Changes." *Journal of Innovation & Knowledge*, 6(1), 27–34. https://doi.org/10.1016/j.jik.2020.05.002.
Magistrat Linz (Ed.). (2024) *Finanzen Voranschlag 2024*. https://www.linz.at/media/politik_verwaltung/VA2024-2025/VA2024_infografik.pdf
Magistrat Linz (Ed.). (2023) *Programm Digitales Linz*. https://www.linz.at/images/files/Programm-Digitales-Linz.pdf
qapture GmbH (Ed.). (2024) https://www.qapture.at/
Rosenbichler, U. & Grünwald, A. (2020) *Grünbuch: Partizipation im digitalen Zeitalter*. Bundesministerium für Kunst, Kultur, öffentlichen Dienst und Sport (BMKÖS). https://permalink.obvsg.at/AC17209060
Startup Guide World. (Ed.) (2023) https://www.startupguide.com/ecosystems/linz
Startup Guide. (2023). *Startup Guide Linz*. Startup Guide World. ISBN 978–989–54894–9–7
Su, C., & Lee, S. H. (2020). "Knowledge Sharing, Social Capital, and Innovation Capability: A Meta-Analysis." *Journal of Business Research*, 119, 13–27. https://doi.org/10.1016/j.jbusres.2020.06.013.
Thommen. J.-P. (n. d.). Anspruchsgruppen. In Gabler Wirtschaftslexikon. Springer Fachmedien. https://wirtschaftslexikon.gabler.de/definition/anspruchsgruppen-27010/version-250673

References linz.at

Bevölkerung https://www.linz.at/zahlen/040_Bevoelkerung/
Europäische Kulturhauptstadt https://www.linz.at/politik/49574.php
Österreichischer Verwaltungspreis https://www.linz.at/medienservice/2023/202306_120632.php
Stadtgebiet Flächen https://www.linz.at/zahlen/010_Stadtgebiet/015_Flaechen/

References innovation.linz.at

Founders scholarship https://innovation.linz.at/
PopUp Store https://innovation.linz.at/

References partizipation.linz.at

Vorschlag Straßenmusik möglich machen durch Streichen der 50 Euro Tagesgebühr https://partizipation.linz.at/de-DE/initiatives/strassenmusik-moglich-machen-durch-streichen-der-50-euro-tagesgebuhr
Vorschlag Wasserspender und Sanitäranlagen für den Sportpark Urfahr https://partizipation.linz.at/de-DE/initiatives/wasserspender-und-sanitaranlagen-fur-den-sportpark-urfahr

The Slovak Smart City Cluster

Christian Schachtner

1 Introduction

The European Cluster Collaboration Platform (ECCP) is a central initiative of the European Union that aims to promote cooperation and networking between clusters within Europe and beyond. This enhanced cooperation is intended to create synergies to strengthen industrial ecosystems and increase the innovative capacity of the companies and organisations involved in the implementation of the EU industrial strategy (Ketels 2017).

The strategic orientation of the ECCP is based on the overarching objectives of EU industrial policy and innovation strategy. It plays a key role in supporting Europe's industrial transformation by fostering cooperation in areas such as digitalisation, green technologies and the circular economy. A comprehensive online platform serves as a central hub for information exchange, networking and collaboration opportunities. It offers functions such as cluster profiles, partner search and event calendar (European Cluster Collaboration Platform 2024).

In addition, the ECCP aims to support European clusters in their internationalisation efforts. The ECCP is faced with the challenge of continuously adapting to the changing needs of European industry and global economic developments. Future priorities are expected to be in the areas of digital transformation, green technologies and resilient supply chains (European Commission 2022).

A central task of the platform is also to act as an information hub. It provides clusters and their members with access to relevant information, best practices, and success stories. This exchange of knowledge is intended to promote the development and implementation of innovative strategies (Meier zu Köcker and Müller, 2018).

C. Schachtner (✉)
Vice President Education and Sustainability, RheinMain University of Applied Sciences, Wiesbaden, Germany
e-mail: Christian.schachtner@hs-rm.de

C. Schachtner (ed.), *The European Smart City Movement*,
https://doi.org/10.1007/978-3-031-96700-9_3

A special focus is on the promotion of small and medium-sized enterprises (SMEs). The ECCP offers specific support measures to integrate SMEs into the cluster networks and increase their competitiveness. A dedicated mapping tool makes it possible to visualize and analyze clusters and their activities across Europe. This facilitates the identification of potential partners and the strategic planning of cooperations (Sölvell et al. 2019).

Another important goal is to promote regional development by strengthening local and regional cluster initiatives. The ECCP supports regions in identifying and building on their specific strengths in order to increase regional competitiveness (Porter 2018).

This article about the Slovak initiative regarding implementing the European Vision of the project is based on a notice of Prof. Dagmar CAGÁŇOVÁ, Ph.D., Vice-Rector for Internationalisation Newton University, Prague.

2 Success Stories of the Slovak Platform: An Analysis of Cluster Collaboration

The Slovak platform for cluster collaboration has established itself as a significant catalyst for innovation and economic growth in the region. The Slovak Business Agency (SBA), as the driving force behind this platform, has played a central role in promoting clusters and strengthening the competitiveness of Slovak companies. Since its inception in 1993, the SBA has continuously provided support to small and medium-sized enterprises (SMEs), with a particular focus on cluster development. The achievements of the Slovak Platform are also relevant in the context of EU cohesion policy. By making effective use of EU funding and aligning the cluster strategy with smart specialisation objectives, the platform has helped to reduce regional disparities and strengthen the innovation capacity of less developed regions (Balog 2016).

An outstanding example of the success of these efforts is the IT Valley Košice Cluster. This cluster has developed into a lighthouse project that illustrates the transformation of the Košice region into a centre for information technology and innovation. Through the close cooperation of companies, educational institutions and local authorities, the cluster has not only created highly qualified jobs, but also strengthened the international competitiveness of the Slovak IT industry (Hudec and Procházková 2018).

The platform has continued to make a significant contribution to the development of the automotive cluster in Western Slovakia. This cluster, which works closely with major car manufacturers such as Volkswagen, PSA Peugeot Citroën and Kia Motors, has made Slovakia one of the leading car manufacturers in Europe. The successful integration of local suppliers into global value chains has not only increased productivity, but also promoted technology transfer and the innovative capacity of the companies involved (Pavlínek 2018).

Another notable achievement is the development of the Kosice IT Valley Cluster, which focuses on promoting innovation in the field of information and communication technologies. This cluster has successfully built a bridge between academia and industry, thereby accelerating the commercialisation of research results. The close cooperation between universities and companies has resulted in a large number of patents and innovative products that consolidate Slovakia's position as an emerging technology hub in Central Europe (Šebová et al. 2014).

The Slovak platform variant has also played an important role in promoting green technologies and sustainable development. The Slovak Environmental and Energy Cluster (SEEC) has established itself as a pioneer in this field, supporting companies in the development and implementation of environmentally friendly solutions. By promoting energy efficiency and renewable energy, the cluster not only contributes to environmental protection, but also creates new business opportunities in a promising sector (Szabo et al. 2017).

A particularly innovative aspect of the Slovak platform is its role in promoting cross-cluster cooperation. By networking clusters from different industries, synergies are created that can lead to groundbreaking innovations. An example of this is the collaboration between the IT cluster and the automotive cluster, which has led to the development of advanced driver assistance systems and autonomous vehicle technologies (Kováč et al. 2019).

In addition, a significant contribution was made to the internationalization of Slovak companies. By participating in international cluster initiatives and EU-funded projects, Slovak companies have been able to expand their global presence and open up new markets. This has not only led to an increase in exports, but has also encouraged knowledge transfer and the adoption of international best practices (Haviernikova et al. 2017).

Another success of the platform lies in the promotion of start-ups and entrepreneurship. Through dedicated programs and incubators, often operated in collaboration with established clusters, the platform has created a fertile ecosystem for innovative start-ups. This has led to a wave of technology-based start-ups that not only invigorate the local economy, but also have the potential to grow into global players (Jeck 2017).

The continuous development and adaptation of the platform to new challenges and opportunities will be crucial to continue this success in the future and to further strengthen Slovakia's position as an innovative and attractive business location.

3 Cluster Composition and Sectoral Focus Der Slovak Smart City Cluster Plattform

The Slovak Smart City Cluster (SSCC) is an innovative initiative that aims to promote the development of smart cities in Slovakia. As an association of legal entities, the SSCC integrates representatives from business, research, education and public administration to work together on the implementation of the smart city concept. The cluster composition of the SSCC is characterized by its diversity and interdisciplinary orientation. The cluster acts as a bridge between the public and private sectors, between research and practice, and between local and national initiatives. This facilitator role is crucial for promoting public–private partnerships, which are considered one of the main factors for the successful development of smart cities. This composition enables the cluster to develop holistic solutions to the complex challenges of urban development (Angelidou 2015).

The sectoral focus of the SSCC spans several key areas that are central to the development of smart cities. Based on the smart city components defined by the Slovak Ministry of Economy, the cluster focuses on the following main areas: government and local government, buildings, mobility, energy and environment, education, and health (AWEX Export 2022).

This sectoral orientation is in line with the Slovak government's priorities for the 2021–2027 programming period and is in line with the European Smart Cities Directives. By bringing together companies from different sectors and promoting exchanges between industry and research, the cluster creates a fertile environment for the development of new technologies and services in the field of smart cities (Industry Innovation Cluster 2024).

It will be crucial to maintain the balance between technological innovation and social sustainability and to ensure that the solutions developed actually lead to an improvement in the quality of life of citizens.

4 Digital Transformation and Implementation of Policy Objectives

The Slovak Smart City Cluster (SSCC) plays a central role in the promotion and implementation of smart city concepts in Slovakia. As an association of legal entities, the SSCC integrates representatives from business, technology, public administration and academia to work together on the development of sustainable and intelligent urban solutions. The organization recognizes that a smart city is built by people and for people, and that citizen participation is an essential part of successful smart city initiatives. This is in line with the Slovak Act No. 369/1990 on the Municipal Code, which establishes the bases for the participation of citizens in municipal administration (Srebalová and Peráček 2022).

In the context of resource efficiency and CO_2-neutral implementation, the SSCC supports initiatives aimed at sustainable urban development. A key aspect of this is the promotion of projects in the field of intelligent energy distribution and the optimisation of energy consumption in urban areas. The SSCC also attaches great importance to promoting projects that address the circular economy. The organization supports initiatives aimed at reducing waste and using resources more efficiently. This is in line with the principles of sustainable urban development and helps to improve the environmental sustainability of Slovak cities. This is in line with the Slovak Environmental Policy Strategy until 2030, which sets out a framework for environmental sustainability (Némethyová 2011).

Digital transformation is a core aspect of the SSCC's activities. The organization promotes the development and implementation of digital solutions that aim to improve the quality of life in urban areas while increasing the efficiency of urban infrastructure. An example of this is the support for projects in the field of intelligent traffic management, as regulated by the Slovak Act No. 317/2012 on Intelligent Transport Systems in Road Transport (Špirková et al. 2014).

An important contribution of the SSCC to the digital transformation of Slovak cities is the creation of experience sharing platforms and the development of generally accepted standards and practices. These activities not only promote the transfer of knowledge between different actors, but also contribute to the development of a coherent framework for the implementation of smart city solutions (Finka and Belčáková 2002).

5 Conclusion

In summary, the Slovak Smart City cluster structures, the European idea of networked actors, plays a central role in the promotion and implementation of smart city concepts in Slovakia. Through its diverse activities in the areas of resource efficiency, CO_2-neutral implementation, circular economy and digital transformation, the SSCC contributes significantly to the sustainable and smart development of Slovak cities. Durch seine vielfältigen Aktivitäten in den Bereichen Ressourceneffizienz, CO_2-neutrale Umsetzung, Kreislaufwirtschaft und digitale Transformation trägt das SSCC wesentlich zur nachhaltigen und intelligenten Entwicklung slowakischer Städte bei. Diese Bemühungen stehen im Einklang mit der europäischen Idee vernetzter Akteure im Smart-City-Bereich und fördern die Integration innovativer Lösungen in städtischen Umgebungen.

References

Angelidou, M. (2015). Smart cities: A conjuncture of four forces. Cities, 47, 95–106.
Balog, M. (2016). Clustering as a tool for enhancing innovation and competitiveness in Slovakia. Journal of Applied Economic Sciences, 11(3), 362–370.
European Cluster Collaboration Platform. (2024). About ECCP. https://clustercollaboration.eu/about-eccp
European Commission. (2022). Clusters of change: European cluster cooperation projects. Publications Office of the European Union.
AWEX Export. (2022). SMART CITIES. https://www.awex-export.be/files/library/Fiches-Pays/Slovaquie/Slovaquie-fiche-sectorielle-smart-cities-2022(1).pdf

Finka, M., & Belčáková, I. (2002). Sustainable urban development in Slovakia - Approaches and experiences. In H. D. Kammeier & H. Detlef (Eds.), New Markets for Old Cities: An Integrated Approach to Inner City Regeneration (pp. 123–136). SPRING Centre, Asian Institute of Technology.

Haviernikova, K., Lemańska-Majdzik, A., & Mura, L. (2017). Advantages and disadvantages of the participation of SMEs in tourism clusters. Journal of Environmental Management & Tourism, 8(6), 1205–1215.

Hudec, O., & Prochádzková, M. (2018). The evolution of innovation networks in Slovakia: Disintegration and slow recovery. In Knowledge Spillovers in Regional Innovation Systems (pp. 133–161). Springer, Cham.

Industry Innovation Cluster. (2024). Strategy. https://www.industryinnovationcluster.sk/strategy/

Jeck, T. (2017). Small and medium-sized enterprises in the Slovak innovation system. Ekonomický časopis, 65(1), 59–77.

Ketels, C. (2017). Cluster mapping as a tool for development. Institute for Strategy and Competitiveness, Harvard Business School.

Kováč, M., Lešková, A., & Kováč, J. (2019). Innovative approaches in the automotive industry in Slovakia. TEM Journal, 8(2), 426–431.

Meier zu Köcker, G., & Müller, L. (2018). Cluster internationalisation and global mega trends. European Secretariat for Cluster Analysis.

Némethyová, I. (2011). Environmental policy of the Slovak Republic in the context of the EU membership. Acta Universitatis Agriculturae et Silviculturae Mendelianae Brunensis, 59(4), 165–172. https://doi.org/10.11118/actaun201159040165

Pavlínek, P. (2018). Global production networks, foreign direct investment, and supplier linkages in the integrated peripheries of the automotive industry. Economic Geography, 94(2), 141–165.

Porter, M. E. (2018). Clusters and the new economics of competition. In J. Cantwell (Ed.), The eclectic paradigm (pp. 195–213). Palgrave Macmillan.

Šebová, M., Hudec, O., & Džupka, P. (2014). The impact of clusters on regional innovation and competitiveness. In 17th International Colloquium on Regional Sciences. Conference Proceedings (pp. 168–175).

Sölvell, Ö., Lindqvist, G., & Ketels, C. (2019). The cluster initiative greenbook 2.0. Ivory Tower Publishers.

Špirková, D., Ivanička, K., & Finka, M. (2014). Sustainable urban development in Slovakia: Factors and development trends. In M. Schrenk, V. V. Popovich, P. Zeile, & P. Elisei (Eds.), REAL CORP 2014 Proceedings/Tagungsband (pp. 1055–1064). CORP—Competence Center of Urban and Regional Planning.

Srebalová, M., & Peráček, T. (2022). Smart cities and their legal environment in Slovakia. In M. Vochozka, J. Váchal, & J. Horák (Eds.), SHS Web of Conferences (Vol. 129, p. 01027). EDP Sciences. https://doi.org/10.1051/shsconf/202212901027

Szabo, S., Ferencz, V., & Pucihar, A. (2017). Trust, innovation and prosperity. Quality Innovation Prosperity, 21(2), 1–8.

A Smart City and the Advancements of Bulgaria

Milena Krumova

1 Introduction

A smart city is where traditional networks and services are made more efficient using digital solutions to benefit its inhabitants and businesses (European Commission 2024). Several sectoral policies and initiatives in the period 2014–2020 contribute to the definition and implementation of the EU strategy for R&D in the field of smart cities. The most significant of them include: "Horizon 2020", which set common goals and provided funding for innovative projects in the field of urban development research; The European Innovation Partnership "Smart Cities and Municipalities" (and its successor, the Smart Cities Marketplace), which brings together EU industries and cities to align public and private R&D programs and promote collaborative initiatives. The European Energy Technology Strategic Plan promotes cooperation between EU countries, companies and researchers in low-carbon technologies and climate-neutral energy systems (European Court of Auditors 2023). Types of smart city solutions include: energy-efficient buildings; use of renewable energy sources; energy efficient lighting; energy management/smart grids; environmental monitoring; waste and resource management; sustainable mobility; mobility; city governance and citizen participation; information and communication technologies and processes; education and development of human capital; providing support for businesses and economic development; public safety; social inclusion; healthcare.

The digital transformation is of key importance for smart city development. According to research related to the Digital Transformation of Bulgaria for the period 2020–2030, it is a necessary process of technological development to create conditions for innovation and business growth, increase the efficiency of the workforce, a competitive digital economy, and a high citizens' standard (Ministry of Transport

M. Krumova (✉)
Founder and Director, European Institute for Technologies, Education and Digitalisation, Sofia, Bulgaria
e-mail: mkrumova.phd@gmail.com

C. Schachtner (ed.), *The European Smart City Movement*,
https://doi.org/10.1007/978-3-031-96700-9_4

and Communications of the Republic of Bulgaria 2020). Smart City as a concept is among the most actively developed for the last several years and many analyses and research have been done. For example, the International Institute for Management Development (IMD) produces a yearly Smart City Index offering a balanced focus on economic and technological aspects of smart cities on the one hand, and "humane dimensions" of smart cities (quality of life, environment, and inclusiveness) on the other. Smart City Index is an evaluation framework with five criteria used to appraise smart city indices. It is a tool for action, an instrument for better lives for all citizens (Smart City Observatory 2024). The total number of participating countries, including Bulgaria, is 142.

2 Smart City Projects

Smart city development is encouraged by the European Commission. Through WiFi4EU initiative (WiFi4EU countries, 2024), EC promotes free Wi-Fi for citizens and visitors in public spaces such as parks, squares, public buildings, libraries, health centers and museums throughout Europe. In this way, everyone can be connected regardless of where they live. According to the European Commission, 232 out of a total of 265 municipalities in Bulgaria have received a voucher for WiFi4EU (European Union 2024).

The practice shows that there are many barriers and problems to be solved for real advancements and the impact on the citizens' quality of life and cities' efficiency ecosystems. During the last few years, the transport network is becoming increasingly congested due to the excessive use of private cars, and it is observed inefficient use of public transport. Another problem is related to the air quality. As an example of Air pollution in Sofia there are visual map with an air quality index in real time to observe the situation. The temperature in the city increases and air quality deteriorates, as a result of global and local climate changes and people's activity, the carbon footprint on the environment increases, and many other urban problems that lead to the search for solutions for the development and implementation of innovative, energy-saving and low-carbon technologies (Vitanova 2023; World Air Quality Index project 2024). Despite the problems, there are a number of good practices that have been implemented in Bulgaria in recent years. For example, Bulgarian company Oblik Studio created Mission Emission project to raise awareness of the environmental impact of traveling. Mission Emission allows the calculation of the environmental cost of travel. The website helps offset the CO_2 and NO_x footprint by suggesting alternative journeys of citizens to desired destinations (Mission Emission 2024).

Another good examples of smart city development in Bulgaria are the implementation of the Internet of Things (IoT) solutions related to air monitoring, garbage disposal, agriculture, and management of the support networks for water supply and the energy sector. Bulgarian authorities and communities of practice have actively participated in the improvement of the smart city infrastructure, services, and lives of the citizens and startups involved. Some of the very successful Bulgarian startup companies producing smart devices (Smart Homes) are Shelly, MTechnologies and ThingsLog.

- Shelly is a provider of smart home automation solutions. The product portfolio includes smart plugs, bulbs, gas leak detectors, air quality monitoring devices, motion sensors, smoke detectors, and more. It also provides a cloud-based solution to monitor and control smart home devices.
- MTechnologies is a provider of systems design, software, hardware, and firmware development solutions. The offering of the company includes Mcube, a system for lighting control and heating control. The company also offers microprocessor-based embedded systems, software, and devices for multiple industries such as multimedia, cars, networking, telecom, and wireless.
- ThingsLog is a provider of smart monitoring solutions for industries. It provides solutions such as water and liquid metering and monitoring, gas metering and monitoring, power monitoring, remote pressure monitoring, and tank level monitoring. It uses low-power LPWAN data loggers for smart monitoring with IoT platforms. Furthermore, it provides products such as GSM data loggers, NB-IoT data loggers, LoRa data loggers, LoRa gateways, and a platform for consumption monitoring as a service.

Healthcare is another domain of advancement of smart cities. An excellent example of this is Superdoc services. Superdoc (SuperDoc 2024) helps citizens in Bulgaria find good doctors and book an appointment right. The service includes selecting a specialist and a city to view the list of specialists. The citizen can filter by the doctor's gender, work with the NHS, and publish examination prices. Everyone can get to know the opinions of other patients, the qualifications and experience of the specialists, and choose the doctor. Users can view the doctor's calendar and immediately book an appointment convenient for everyone. Another example is the e-Health service. It provides all Bulgarian e-health services in one place: examinations, referrals, prescriptions, vaccines, laboratory test results, personal health record. The eHealth mobile application gives all citizens access to their personal patient file with all electronic health documents registered in the National Health Information System. The activation of a mobile device is authenticated with an electronic signature. The user can enable additional biometric data protection. Every access to the electronic patient record is recorded. (E-Zdrave 2024).

Citizens' involvement in smart city activities can be diverse and impact many communities. In Bulgaria, some industrial companies invest much in social inclusion and encourage citizens for healthier lives through the use of technology. Such a company as a good practice is Devin (Devin company 2024). The company creates a campaign for creating a map of fitness in open spaces. Devin is the producer and distributor of bottled water in Bulgaria. The company encourages citizens to be active in adding finesses to the map. On bottles of water produced for this campaign, there is a QR code. This way, the citizens can use a digital channel throughout the scanning of a QR code and add new fitness. Sofia's open fitness spaces are findable at a map based on geo-data.

Another good example of social inclusion and community development is Plovdiv (Visit Plovdiv 2024) and Sofia (Visit Sofia 2024a, b). City Street Fitness (Street Workout) sites on the territory of the city. With them, everyone can find equipment for rehabilitation and street fitness exercises. The platforms include: a barbell, parallel bar, bicycle equipment, and rehabilitation equipment for movement.

Digital tourist network is another service created by a Yettel company (Yettel 2024) for the use of citizens and the tourist attraction visits in different cities in Bulgaria. This is the digitization of the 100 tourist sites with first distinction for contribution to the development of smart tourism. The innovative project to digitize the 100 sites enables fans of the National Movement "Know Bulgaria—100 National Tourist Sites" to receive digital stamps from the sites, equivalent to those collected on paper. This is possible in Yettel's mobile application, which is open to all users, access is completely free, and all content can also be found in English.

In the digital version of the 100 tourist sites, profiles of all 176 sites, part of the Movement, can be found, which provide important and useful information for visitors such as location, opening hours, etc. In the profiles of some of the objects, curious audio narrations as well as impressive video stories have been added. The videos include specially created illustrations by some of the most talented contemporary Bulgarian artists, as well as a short story with interesting facts about the site. The first video and audio stories created are for some of the most popular and visited sites, such as the Kazanlak Tomb, Ancient Nessebar Museum, Yagodinska Cave Ancient Theater, and new ones are planned to be added gradually.

Another good example that impact the improvement of Smart City transport system related to parking in Bulgaria is the Upass Ltd. company, which develops URBO City app. It is a service that combines the convenience of a parking service app and a tourist map with access to cultural sites, attractions, and the most current events in the city. When parks, the users can pay for to up 4 h and will be notified when their parking time is about to expire (Urbo 2024).

3 Sofia Smart City

Sofia is the capital of Bulgaria and the largest city with population of 1, 286, 965 citizens and area of 492 km^2. In Sofia there are 607,473 dwellings and 101,696 buildings. Sofia City Province has an area of 1344 km^2, while the surrounding and much bigger Sofia Province is 7059 km^2 and has a population of 231,989 people according to 2021 census data (Infostat 2024). Sofia Province has a different number of population by municipalities (Capital BG 2024; National Statistical Institute 2024).

During the last several years, Sofia city Municipality has worked for smart city development and put efforts and investments into administrating this process, starting with the developed strategic documents of high importance (Knowledge Sofia 2024a, b):

- **Strategy for the smart specialization of Sofia**—it aims to increase the degree of intelligent specialization of the capital by establishing Sofia as an Intelligent City that builds an environment with a high quality of life for citizens and good governance that supports the corporate sector by stimulating the creation of better conditions for synergy and partnership in the innovation ecosystem and through effective use of ICT (Sofia Municipality 2024a, b; Stolichen obshtinski savet 2016).
- **Vision for Sofia**—a long-term strategy with shared goals for the development of Sofia, assigned by the Sofia Municipal Council to the Operational Program Softproject; document and an open process of inclusion in the development and analysis of people and organizations such as municipal authorities, NGOs, investors, researchers, experts and citizens. It can serve as a basis for all future city development strategies until 2050;
- **Sofia's digital transformation strategy**—a platform for intelligent growth—a document developed and completed this year by consultants under the EU Digital Cities Challenges initiative, in which Sofia participates together with 40 other European cities.
- **Marketing strategy** with a plan for turning Sofia into a smart city—a document developed by order of the Sofia City of Knowledge Cluster in implementation of a project under the Operational Program "Innovations and Competitiveness".
- **Road map** for integrated planning and implementation of projects for smart cities—Smart City Guidance Package (Borsboom et al. 2019). The handbook helps to plan and implement smart city and low energy projects in an integrated way.
- **Other strategic documents**—Municipal Development Plan of the Capital Municipality (2014–2020) with program period 2020–2027, retrospection of the Integrated Plan for Urban Reconstruction and Development of Sofia (2014–2020), District Development Strategy Sofia-city (2014–2020) and National Spatial Development Concept 2013–2025.

Despite the efforts in smart city development, according to the Smart City Index, Sofia continues to lose positions in the ranking. In 2024, Sofia fell by two places to 113th out of 142 countries in IMDs Smart City Index 2024 (IMD 2024). For comparison, in 2021 it was ranked 103rd, and in 2020—87th.

Sofia has invested in many smart city services in the areas of transport, parking, and digital services. Such example is the service for shared trips with electric scooters, which is available with the idea of freeing up and reducing traffic in the central parts of the city, which allows the use of fewer private cars, which in turn would improve the environment and the urban way in the long term of life.

Sofia Plus allows users to pay by bank card the fee for removing a clamp from a vehicle and also provides other functionalities related to parking and living in the city of Sofia (Sofia Plus 2024). Electronic vignette sticker for local parking—After receiving the electronic vignette sticker, the citizen can park within the "Blue" or "Green" zone, without time restrictions (Sofia Urban Mobility Centre 2024).

In 2022 Sofia Municipality received an award for Smart City of the Year for Southeast Europe. Sofia was chosen as one of four cities alongside Brussels, Dublin, and Tirana to develop the European Union's Digital Rights Management Framework. Sofia also ranks as a city leader in the European Mission for Carbon–Neutral and Smart Cities until 2030, which provides new opportunities for Bulgaria and unlocks additional mechanisms for EU funding (Novinite 2022).

4 Burgas Smart City

Burgas is the fourth-largest city in Bulgaria after Sofia, Plovdiv, and Varna (Burgas Municipality 2024a, b). According to data from the GRAO, as of December 15, 2024, 210,467 people live in the city at their current address. Burgas city total area is 253,644 km^2 (Glavna direktsia 2024). Burgas Province is a province in southeastern Bulgaria on the southern Bulgarian Black Sea Coast, it is the largest province by area, with a territory of 559,151 km.

Burgas has long been an industrial, transport, and cultural center of national importance. During the last several years a lot of investments have been made for improvements in the city, infrastructure, services, education, and digitalisation of the communications and services for the citizens. Burgas is among the most advanced smart cities in Bulgaria. For example, Burgas is famously the first city in Bulgaria to entirely transform and modernise its public transport system. Sustainable mobility has been on the city's agenda since the mid-2000s, when Burgas launched an integrated public transport plan, Back in 2015, the Municipality of Burgas implemented a project to improve urban mobility (European Investment Bank 2024). For this purpose, the entire vehicle fleet with diesel and methane buses and trolleybuses was renewed under the project "Integrated Urban Transport of Burgas" (Atanasova 2021). Part of the smart mobility in the city of Burgas is the smart parking system introduced in 2018 in the central area. For this purpose, sensors are installed under the bitumen pavements, which, through the implemented sensors, provide information in real time about whether there is a car parked in the designated parking space. The information is fed to the installed boards in the city and the created smart offer Smart Burgas, where every resident can get information (Obshtina Burgas 2024a, b).

In 2019, the smart platform Smart Burgas was developed. The platform provides access to data received from all sensors, cameras, and sensors in the city to monitor air conditions, water levels, real-time traffic, road conditions, free parking spaces, bicycles, information on cultural events, sports objects, etc. Through this platform, the residents of the city, as well as the guests, and the local administration have access to information in real time, which allows on the one hand quick interaction and collaboration between all interested groups and the other hand an opportunity to analyze, forecast various situations (Smart Burgas 2024a, b). Solutions for intelligent waste management have already been implemented in many European cities. A proposal in this direction for the city of Burgas is related to the purchase of garbage containers operating with solar power, equipped with sensors that monitor the level of garbage.

The Solar energy drives compressors that compact the waste in the containers so that the maximum volume is collected. The sensors can provide information to the Municipal Company, which provides garbage collection and garbage removal in Burgas.

5 Conclusion

The smart city will develop faster in the coming decades. Bulgarian local communities of practice show many examples of progress. But, of course, there are still many obstacles to the digitalization of all smart urban areas. Lessons learned from the already implemented smart city projects (Sofia and Burgas) can provide knowledge for other city authorities in Bulgaria. A trend in Bulgaria, which may be similar to other EU countries is that the biggest cities are advancing more in Smart City development—Sofia, Burgas, Plovdiv, Varna, etc., in comparison with the province towns and small cities—Vratsa, Sozopol, etc. Among the many priorities for further development should be more active citizen engagement in public decisions related to smart cities, more transparent investments, and digital access for all.

References

Atanasova, A. (2021). Inteligentnite gradove—nova vazmozhnost za planirane i razvitie na gradskata infrastruktura. Université de Sofia. https://shorturl.at/4ZTMH

Borsboom, J., Kallaos, J., Alpagut, B., Bruinenberg, J., Fokaides, P., De Luca, G., Beneicke, J., & Balducci, A. (2019). Smart City Guidance Package. Norwegian University of Science and Technology. https://www.researchgate.net/publication/343615678_Smart_City_Guidance_Package

Burgas Municipality. (2024). Page d'accueil. https://www.burgas.bg/en

Capital BG. (2024). [Image]. https://img.capital.bg/shimg/zx1278_4401365.jpg

Devin Company. (2024). About the company. https://devin-bg.com/en/about-the-company/

European Commission. (2024). What are smart cities? https://commission.europa.eu/eu-regional-and-urban-development/topics/cities-and-urban-development/city-initiatives/smart-cities_bg

European Court of Auditors. (2023). Inteligentni gradove. https://www.eca.europa.eu/ECAPublications/SR-2023-24/SR-2023-24_BG.pdf

European Investment Bank. (2024). Burgas: Planning a Black Sea smart city. https://www.eib.org/en/essays/burgas

European Union. (2024). WiFi4EU countries: Bulgaria. https://digital-strategy.ec.europa.eu/en/news/wifi4eu-countries-bulgaria

E-Zdrave. (2024). Page d'accueil. https://his.bg/ezdrave/

Glavna direktsia "Grazhdanska registratsia i administrativno obsluzhvane". (2024). Page d'accueil. https://www.grao.bg/

IMD. (2024). Smart City Index 2024. https://issuu.com/docs/e7a60c053affbf9e98fcba93afe857af?fr=xKAE9_zU1NQ

Infostat. (2024). Page d'accueil. https://infostat.nsi.bg/infostat/pages/reports/query.jsf?x_2=1962

Knowledge Sofia. (2024). Kak da planirame Sofiya kato inteligenten grad [Comment planifier Sofia comme une ville intelligente]. https://knowledgesofia.eu/bg/blog/230-kak-da-planirame-sofiya-kato-inteligenten-grad

Ministry of Transport and Communications of the Republic of Bulgaria. (2020). Digital transformation of Bulgaria for the period 2020–2030. https://www.mtc.government.bg/sites/default/files/digitaltransformationofbulgariafortheperiod2020-2030f.pdf

Mission Emission. (2024). Page d'accueil. https://missionemission.co/

National Statistical Institute. (2024). Page d'accueil. https://www.nsi.bg/bg

Novinite. (2022). Sofia—with an Award for Smart City of the Year for Southeast Europe. https://www.novinite.com/articles/215426/

Obshtina Burgas. (2024). Page d'accueil. https://www.burgas.bg/bg/posts/view/36394/

Smart Burgas. (2024). Page d'accueil. https://smartburgas.eu/bg

Smart City Observatory. (2024). Smart City Index 2024. https://www.imd.org/smart-city-observatory/home/

Sofia Municipality. (2024). Strategia za inteligentna spetsializatsia na Sofia [Stratégie de spécialisation intelligente de Sofia]. https://www.sofia.bg/w/strategia-za-inteligentna-specializacia-na-sofia

Sofia Plus. (2024). Parking in Sofia. App Store. https://apps.apple.com/bg/app/sofia-plus/id1544084414

Sofia Urban Mobility Centre. (2024). Parkiray barzo i udobno sas Sofia Plus [Garez rapidement et facilement avec Sofia Plus]. https://www.sofiatraffic.bg/bg/parking

Stolichen obshtinski savet. (2016). Strategia za inteligentna spetsializatsia na Sofia [Stratégie de spécialisation intelligente de Sofia]. https://www.sofia.bg/documents/20121/148954/ISIS_Sofia.pdf/f51fcd5a-2973-4679-89fe-62b3dccb6662

SuperDoc. (2024). Page d'accueil. https://superdoc.bg/

Urbo. (2024). Page d'accueil. https://urbo.bg/

Visit Sofia. (2024). Kade da se bavlyavam [Où s'amuser]. https://visitsofia.bg/bg/cityinfrastructure/къде-да-се-забавлявам/спортни-съоръжения

Vitanova, L. (2023). Umen grad—ima li "inteligentni" gradove v Bulgaria? [Ville intelligente - y a-t-il des villes "intelligentes" en Bulgarie ?]. infoz.bg. https://www.infoz.bg/world/10441-umen-grad-bulgaria

WiFi4EU countries. (2024). Bulgaria, European Union. https://digital-strategy.ec.europa.eu/en/news/wifi4eu-countries-bulgaria

World Air Quality Index project. (2024). Zamarsiavane na vazduha v Sofia [Pollution de l'air à Sofia]. https://aqicn.org/map/sofia/bg/

Yettel. (2024). 100 obekta [100 objets]. https://www.yettel.bg/uslugi/za-teb/100obekta

Visit Plovdiv. (2024). Fitnes na otkrito [Fitness en plein air]. https://www.visitplovdiv.com/bg/node/332

Climate-Neutral Vision: Three Hungarian Cities on the Road to Sustainability

Christian Schachtner

1 Introduction

Based on a participation in the CeeeGov Day 2024 in Budapest with the special topic of ICT for smart agriculture, the attached contribution on this discussed questions of the climate-neutral vision for cities will be presented with prime examples.

The European Union has set itself the ambitious goal of becoming climate-neutral by 2050. In this context, cities play a crucial role, as they are responsible for a significant share of greenhouse gas emissions (UN-Habitat 2020). Particularly noteworthy is the commitment of three Hungarian cities—Miskolc, Pécs and Tatabánya—that have joined the EU programme "100 Climate-Neutral and Smart Cities by 2030" (European Commission 2022).

These are examined in more detail below on the basis of their strategic orientation:

1. Miskolc: Industrial city in transition

Miskolc, the fourth largest city in Hungary, faces special challenges due to its industrial past. However, the city has recognized that the transition to a climate-neutral future is not only an ecological necessity, but also an economic opportunity (Nagy et al. 2021). One focus of the efforts is on the transformation of the energy sector. Miskolc plans to increase the share of renewable energy in total energy consumption to 40% by 2030 (Miskolc Város Önkormányzata 2023).

Another important aspect is the promotion of sustainable mobility. The city is investing in the expansion of public transport and the creation of cycle paths. The goal is to reduce private motorized transport by 30% by 2030 (Miskolc Közlekedési Vállalat 2022). These measures not only contribute to the reduction of CO_2 emissions,

C. Schachtner (✉)
Vice President Education and Sustainability, RheinMain University of Applied Sciences, Wiesbaden, Germany
e-mail: Christian.schachtner@hs-rm.de

C. Schachtner (ed.), *The European Smart City Movement*,
https://doi.org/10.1007/978-3-031-96700-9_5

but also improve the quality of life of citizens by reducing air pollution and traffic noise.

2. Pécs: Pioneer in renewable energies

Pécs, Hungary's fifth-largest city, has already made a name for itself as a pioneer in renewable energy. The city has been operating a biomass power plant since 2013, which covers a large part of the city's heating needs (Pécsi Távfűtő Kft. 2021). As part of its climate-neutral vision, Pécs plans to further expand this approach and switch fully to renewable energy by 2030 (Pécs Megyei Jogú Város Önkormányzata 2022).

An innovative project is the development of a "smart grid" system that will enable the efficient distribution and use of renewable energies. This system will not only optimize energy consumption, but also facilitate the integration of electric vehicles into the electricity grid (Pécsi Tudományegyetem 2023). In addition, Pécs is relying on extensive greening measures to improve air quality and reduce the urban heat island effect.

3. Tatabánya: Erbe als Chance Industrial Clubs

Tatabánya, a medium-sized city with a strong industrial heritage, is facing the challenge of diversifying its economy while becoming carbon neutral. The city has developed an ambitious plan to transition to a circular economy (Tatabánya Megyei Jogú Város Önkormányzata 2022). One focus is on promoting cleantech companies and creating "green" jobs.

A particularly innovative project is the conversion of a former coal mine into an energy storage facility. By using the difference in altitude, a pumped storage power plant is to be created that can store excess renewable energy and feed it back into the grid when needed (Magyar Tudományos Akadémia 2023). This project is an example of how industrial heritage can be used for the energy transition.

The experience of these Hungarian cities can serve as a model for other European municipalities. Of particular note is the importance of citizen participation and public–private partnerships for the success of climate neutrality projects (Európai Városok és Régiók Tanácsa 2024).

2 Climate-Neutral Governance and Key Indicators of Success in Hungary

As part of the European ambition to achieve climate neutrality, Hungarian cities have also committed themselves to this ambitious goal. Three pioneers—Budapest, Miskolc and Pécs—exemplify how municipalities in Hungary are taking the path to climate neutrality (Mayors of Europe 2023). This development is in line with Hungary's national efforts to reduce its greenhouse gas emissions and shape a sustainable future.

The Hungarian government has set itself the goal of achieving climate neutrality by 2050, which was legally enshrined in the Climate Change Act adopted in 2020 (Ministry of Innovation and Technology 2020). To achieve this goal, several strategies and measures have been developed to be implemented at both national and municipal level.

A central aspect of Hungary's climate strategy is the decarbonisation of the energy sector. According to the draft update of the Hungarian Climate Plan (NECP), 90 percent of electricity generation in the country is to be carbon-free by 2030 (Clean Energy Wire 2023). This is to be achieved through a combination of renewable energies and nuclear power. Increasing the share of renewable energy in total energy consumption is an important indicator of progress towards climate neutrality. In 2021, this share was already 14.1 percent (Ministry of Energy 2022).

In order to measure and monitor progress in achieving the climate targets, Hungary has developed a comprehensive climate protection monitoring system. This system includes both outcome-oriented and process-oriented indicators. Result-oriented indicators measure the direct impact of climate protection measures, such as energy saved or greenhouse gas emissions avoided. Process-oriented indicators, on the other hand, record progress in the implementation of measures, such as the number of climate protection projects implemented (Agency for Municipal Climate Protection 2024).

Specific indicators have been developed for municipalities to enable standardised assessment and comparability. Key indicators for municipalities include:

1. The share of local renewables in total electricity consumption
2. Household heat consumption per square metre of living space
3. The electricity consumption of households per inhabitant
4. The installed capacity of renewable energy plants per inhabitant
5. The modal split in transport (Agency for Municipal Climate Protection 2024)

Additional indicators have been established for the municipal administration itself, including:

1. Greenhouse gas emissions of the administration per inhabitant
2. Final energy consumption of the administration per inhabitant
3. Heat consumption of administrative buildings per square metre
4. Share of e-vehicles in the total vehicle fleet
5. Share of sustainable procurement in total procurement (Agency for Municipal Climate Action 2024)

These indicators allow municipalities to measure their progress and take targeted actions to improve their climate footprint. The Hungarian government has recognised that the public sector must act as a role model in achieving climate targets. Therefore, an ambitious roadmap has been set for federal administrations to achieve net-zero greenhouse gas emissions within seven years (Detecon 2023).

This represents an enormous challenge and requires consistent action as well as significant investments. To achieve these goals, Hungarian authorities are relying on various measures. These include the energy-efficient renovation of buildings, the modernization of heating and cooling systems, the installation of photovoltaic systems and the conversion of the vehicle fleet to electric vehicles. Some authorities have already achieved considerable success. For example, a state authority was able to permanently reduce electricity consumption at a main location by more than 10 percent through targeted campaigns to raise awareness among its employees (Detecon 2023).

Despite these positive developments, there are also challenges and criticism. Experts doubt the true conviction of the Hungarian government under Prime Minister Viktor Orbán with regard to climate protection. Its policies are described as opportunistic, and the mixed progress in the energy transition points to a lack of real commitment (Clean Energy Wire 2023). It is argued that the new climate-friendly rhetoric does not always go hand in hand with concrete measures and investments.

In order to strengthen the credibility of climate protection efforts and achieve the goals set, it is crucial that Hungary intensifies its efforts and implements them consistently. This requires not only the setting of ambitious targets, but also the consistent implementation of measures and their regular review against the defined indicators.

3 Successful Example Projects for Climate Neutrality in Hungarian Cities

Hungary's cities play a crucial role in the fight against climate change, as 70% of the population lives in urban areas. This concentration leads to increased energy consumption and greenhouse gas emissions in urban areas. To meet these challenges, numerous Hungarian cities have developed ambitious climate strategies and joined the Covenant of Mayors.

On 21 March 2021, the capital Budapest adopted a climate strategy and an action plan for sustainable energy and climate. The goal is to reduce emissions by 40% by 2030 compared to 2015. The focus is on the energy-efficient renovation of residential buildings and the reduction of emissions in the transport sector. Budapest has also declared a climate emergency, which underlines the urgency of action. A particularly symbolic project is the conversion of Hungary's last coal-fired power plant, the Matra power plant. A 200 MW solar park is being planned at its site, which exemplifies how a "clean energy transition" can replace jobs in the coal sector with new opportunities. Financing these ambitious projects is a significant challenge. For Budapest alone, the financing needs of climate action are estimated at around HUF 2500 billion (approx. EUR 7 billion). Recent measures by the national government aimed at cutting funding to cities are jeopardizing these plans (Ámon and Fülöp 2021).

At the national level, Hungary has set ambitious targets. The National Energy and Climate Plan aims to reduce greenhouse gas emissions by at least 40% by 2030, using 1990 values as a benchmark. In addition, Hungary has announced that it will phase out coal as early as 2025 instead of 2030. This is to be achieved by maintaining existing nuclear power capacities and the massive expansion of solar energy In addition to technical solutions, Hungary is also relying on unconventional measures. Prime Minister Viktor Orbán announced that ten trees would be planted for every newborn in Hungary. Through this initiative, the country's forest area is expected to grow by 27% by 2030. Although this measure can be critically questioned, it shows the growing awareness of climate action at the highest political level (Euractiv 2021).

One of the biggest challenges is in the building sector. In order to achieve climate neutrality by 2050, 100,000 to 130,000 properties would have to be modernized annually. This corresponds to a renewal rate of about 3%, which is significantly higher than the previous rate of 1%. Hungary's long-term building renovation strategy aims for 90% of buildings to have almost no energy demand by 2050 (German Energy Solutions 2023).

4 Conclusion

The efforts of Miskolc, Pécs and Tatabánya show that the path to climate neutrality, while challenging, is feasible. Each city has developed its own approach, tailored to the specific local conditions and strengths. What all three cities have in common is a holistic approach that combines the energy transition, sustainable mobility, urban planning and economic transformation. Success will depend on the extent to which it is possible to consistently track the defined indicators, communicate progress transparently and make necessary adjustments to achieve climate neutrality by 2050. The experiences and projects of Hungarian cities can serve as valuable examples for other European municipalities facing similar challenges. It remains to be seen how these initiatives will develop in the coming years and what contribution they will make to achieving national and European climate targets.

References

Agency for Municipal Climate Protection. (2024). Indicators as part of municipal climate protection monitoring. https://www.klimaschutz.de/de/service/meldungen/indikatoren-als-teil-des-kommunalen-klimaschutz-monitorings

Ámon, A., & Fülöp, O. (2021). Budapests Weg in eine grüne Zukunft. Friedrich-Ebert-Stiftung. https://library.fes.de/pdf-files/bueros/budapest/17880.pdf

Clean Energy Wire. (2023). Orbán's opportunistic record belies Hungary's new pro-climate rhetoric. https://www.cleanenergywire.org/news/orbans-opportunistic-record-belies-hungarys-new-pro-climate-rethoric

Detecon. (2023). Klimaneutralität bis 2030 in der Verwaltung: Konkrete Maßnahmen ableiten und Wirksamkeit überprüfen. https://www.detecon.com/de/journal/klimaneutralitaet-bis-2030-der-verwaltung-konkrete-massnahmen-ableiten-und-wirksamkeit-ueberpruefen

Euractiv. (2021, März 4). Ungarn steigt schon 2025 aus der Kohle aus. https://www.euractiv.de/section/energie-und-umwelt/news/ungarn-steigt-schon-2025-aus-der-kohle-aus/

Európai Városok és Régiók Tanácsa. (2024). Klímasemleges városok: A polgári részvétel jelentősége. EVRT Kiadó.
European Commission. (2022). 100 Climate-Neutral and Smart Cities by 2030. https://ec.europa.eu/info/eu-regional-and-urban-development/topics/cities-and-urban-development/city-initiatives/smart-cities_en
German Energy Solutions. (2023). Zielmarktanalyse Ungarn 2023. https://www.german-energy-solutions.de/GES/Redaktion/DE/Publikationen/Marktanalysen/2023/zma-ungarn.pdf
Magyar Tudományos Akadémia. (2023). Innovatív energiatárolási megoldások városi környezetben. MTA Kiadó.
Mayors of Europe. (2023). Meet 3 Hungarian cities aiming for climate neutrality. https://mayorsofeurope.eu/news/meet-3-hungarian-cities-aiming-for-climate-neutrality/
Ministry of Energy. (2022). Hungary—Environmental, Social and Governance (ESG) Report. https://akk.hu/download?path=5da68682-3856-411d-a5aa-95e4024f7bbc.pdf
Ministry of Innovation and Technology. (2020). Eight National Communication and Fifth Biennial Report of Hungary. https://unfccc.int/sites/default/files/resource/Hungary_NC8-BR5_fin_upload.pdf
Miskolc Közlekedési Vállalat. (2022). Fenntartható mobilitási terv 2030-ig. MKV Kiadó.
Miskolc Város Önkormányzata. (2023). Miskolc klímasemlegességi stratégiája. Miskolc Kiadó.
Nagy, B., Szabó, Z., & Kiss, K. (2021). Az ipari örökség újrahasznosítása a klímasemlegesség jegyében. Területfejlesztés és Innováció, 15(2), 23–35.
Pécs Megyei Jogú Város Önkormányzata. (2022). Pécs 2030: Út a klímasemlegességhez. Pécs Kiadó.
Pécsi Távfűtő Kft. (2021). A pécsi biomassza erőmű: 8 év tapasztalatai. PÉTÁV Kiadó.
Pécsi Tudományegyetem. (2023). Smart Grid rendszerek fejlesztése városi környezetben. PTE Kiadó.
Tatabánya Megyei Jogú Város Önkormányzata. (2022). Tatabánya körforgásos gazdasági stratégiája. Tatabánya Kiadó.
UN-Habitat. (2020). World Cities Report 2020: The Value of Sustainable Urbanization. United Nations Human Settlements Programme.

State of Digitization in Albania: Smart Villages as a Driver

Christian Schachtner

1 Introduction

The digital infrastructure in Albania has improved thanks to international support and investment. According to reports, access to internet services was relatively high in urban areas, while rural areas are often still underserved (World Bank 2020).

Another important aspect of digitization in Albania is the introduction of e-government services. The danger remains considerable, as Albania is one of the pioneers in digitization and innovation in the region. Currently, the e-Albania portal offers 1231 electronic services, which represents 95% of all public services in the country. The Albanian government is seeking to use artificial intelligence for proactive government services, which increases the need to remain vigilant against all potential threats to the country (Agjencia Kombëtare e Shoqërisë së Informacionit 2023).

2 Governmental Services

These services are intended to make it easier for citizens to access government services and improve the efficiency of the administration. The Albanian government has launched initiatives to expand internet access across the country, which is crucial for promoting e-learning and online services (European Commission 2020).

Studies show that the use of e-government services in Albania is increasing, but there is still scepticism among the population regarding data protection and data security (UNDP 2022).

C. Schachtner (✉)
RheinMain University of Applied Sciences, Wiesbaden, Germany
e-mail: christian.schachtner@hs-rm.de

C. Schachtner (ed.), *The European Smart City Movement*,
https://doi.org/10.1007/978-3-031-96700-9_6

Albania's cybersecurity landscape is significantly shaped by its geopolitical relations and regional dynamics. The country's alignment with Western institutions and its strategic stance on various international issues make it a target for cyberattacks by both state and non-state actors. As Albania navigates this complex geopolitical situation, the need for a robust, proactive cybersecurity strategy is becoming increasingly apparent. This requires continuous adaptation and international cooperation to counter emerging threats. In 2022, Albania was rocked by a significant cyber incident, mostly attributed to state-sponsored actors from Iran. This attack was part of ongoing geopolitical tensions, largely due to Albania's hosting of Iranian opposition groups, and led to severe disruptions to the country's digital government services. This incident not only highlighted the vulnerabilities in Albania's cybersecurity infrastructure, but also attracted international attention. This led to widespread condemnation and support from Albania's allies, including the United States and NATO members (Miller 2022).

3 Digital Society

Despite this progress, Albania continues to face challenges. These include insufficient digital skills among the population and a lack of investment in digital education (Government Albania 2022). These factors not only hinder the use of digital services, but also the country's innovative strength.

The future of digitalization in Albania depends heavily on further investments in digital infrastructure and education. To realize the full potential of digital transformation, comprehensive strategies are needed that take into account both technical and social aspects (World Economic Forum 2023).

The Western Balkans Investment Framework (WBIF) has proven to be a significant catalyst for the development and integration of the Western Balkans. A particularly noteworthy project in this context is the technical support for the digital infrastructure in Albania, known under the project abbreviation WB23-ALB-DII-02. This project aims to bridge the digital divide in Albania and support the country on its digital transformation journey.

The Smart Cities/Smart Villages program in Albania is an innovative project to ensure this future viability, aimed at improving the quality of life and promoting sustainable development in selected communities. The overall goal of the project is to support communities that use "smart" solutions to provide basic infrastructure services, thereby improving the quality of life and maintaining a clean and sustainable environment (WBIF 2023).

4 Pilot Regions Durres, Vlora and Korca

A specific project objective is the implementation of energy-efficient and environmentally friendly investment programs in the selected pilot municipalities of Durres, Vlora and Korca (WBIF 2023).

These programs are based on integrated strategies and the efficient use of information and communication technologies (ICT). This is in line with the European Union's objective to promote and support digital transformation in the Western Balkans (European Commission 2018).

The selection of these three municipalities in central and southern Albania as pilot projects makes it possible to test the effectiveness of the measures in different geographical and socio-economic contexts (Ministria e Infrastruktures̈ dhe Energjisë 2020).

An implicit, but no less important, goal of the project is to serve as a model for similar initiatives in other parts of Albania and beyond. The successful implementation in the pilot municipalities aims to gain experience and develop best practices that can be transferred to other cities and municipalities (European Commission 2020).

5 Optimization of Infrastructure

Another key objective is to improve basic infrastructure in the areas of transport, energy, public lighting and the environment (WBIF 2023).

The use of intelligent technologies aims to make these infrastructure areas more efficient, sustainable and user-friendly. This is in line with the concept of the "smart city", which aims to make urban spaces more livable and resource-efficient through the use of technology (Albino et al. 2015).

A key aspect of the project is the improvement of broadband connections in rural and remote areas of Albania. This is crucial as access to fast internet not only opens up economic opportunities, but also improves access to education and health services (European Commission 2020).

The technical support includes not only the physical infrastructure, but also the development of skills and competencies in the field of digital technology.

6 Local Impact

The preparation of a conceptual design, a feasibility study and an environmental and social impact assessment is another important project objective (WBIF 2023).

These studies are crucial to assess the technical, economic and environmental feasibility of the project and to identify potential impacts on the local population and environment. The conduct of such studies is in line with project management best practices and the requirements of international financial institutions (World Bank 2020).

In addition, the project aims to strengthen local expertise in smart city technologies and management. Through cooperation with international experts and the practical implementation of innovative solutions, local experts and institutions are to be empowered to independently plan and implement similar projects in the future (UNDP 2022).

A particularly impressive example of the success of the project is the digitization of the healthcare system in a pilot region. Here, a telemedicine service has been introduced, allowing patients in remote areas to consult specialists without having to travel long distances (Albanian Ministry of Health and Social Affairs 2024). This has not only improved medical care, but has also led to relief for overcrowded hospitals in cities.

After all, an important goal of the project is to increase citizen participation and satisfaction. Smart city solutions are not only intended to improve the efficiency of urban services, but also to facilitate and intensify the interaction between citizens and city administration (Cardullo and Kitchin 2019).

This can lead to a stronger identification of citizens with their city and to a more active participation in urban life. The modernisation of public services and the introduction of e-government solutions will not only increase the efficiency of the administration, but also increase transparency and fight corruption (Transparency International 2024).

7 Conclusion

In summary, the project objectives of the Smart Cities/Smart Villages program in Albania are diverse and ambitious. They range from the concrete improvement of urban infrastructure to the promotion of sustainability and energy efficiency to the strengthening of local capacities and citizen participation. Despite the successes to date, the project still faces considerable challenges. These include Albania's geographical conditions, which make it difficult to expand the infrastructure in certain regions, as well as the lack of qualified specialists in the IT sector.

To overcome these challenges, the project relies on close cooperation with local educational institutions and international experts. The successful implementation of these goals has the potential not only to improve the quality of life in the selected pilot communities, but also to serve as a model for future smart city initiatives in the region.

References

Agjencia Kombëtare e Shoqërisë së Informacionit, (2023). TechAccelerator në Bruksel, Karçanaj: Shqipëria plotësisht digjitale në 2030 përmes Inteligjencës Artificiale. https://akshi.gov.al/techaccelerator-ne-bruksel-karcanaj-shqiperia-plotesisht-digjitale-ne-2030-permes-inteligjences-artificiale/.

Albanian Ministry of Health and Social Affairs. (2024). Evaluation report on telemedicine services in rural areas. Tirana: AMGS.

Albino, V., Berardi, U., & Dangelico, R. M. (2015). Smart cities: Definitions, dimensions, performance, and initiatives. Journal of Urban Technology, 22(1), 3–21.

Cardullo, P., & Kitchin, R. (2019). Being a 'citizen' in the smart city: Up and down the scaffold of smart citizen participation in Dublin, Ireland. GeoJournal, 84(1), 1–13.

European Commission. (2018). A credible enlargement perspective for and enhanced EU engagement with the Western Balkans. COM(2018) 65 final.

European Commission. (2020). Western Balkans: An Economic and Investment Plan to support the economic recovery and convergence. https://ec.europa.eu/commission/presscorner/detail/en/ip_20_1811.

Government Albania. (2022). National Employment and Skills Strategy 2023–2030. https://arkiva.financa.gov.al/wp-content/uploads/2023/10/National-Employment-and-Skills-Strategy-2030_EN.pdf.

Miller, M. (2022). Albania weighed invoking NATO's Article 5 over Iranian cyberattack. https://www.politico.com/news/2022/10/05/why-albania-chose-not-to-pull-the-nato-trigger-after-cyberattack-00060347.

Ministria e Infrastrukturës dhe Energjisë. (2020). Strategjia Kombëtare e Energjisë 2018–2030. https://www.infrastruktura.gov.al/strategji/.

Transparency International. (2024). Korruptionswahrnehmungsindex 2024. Berlin: Transparency International.

UNDP. (2022). Digital Strategy 2022–2025. https://www.undp.org/digital/Digital-Strategy.

WBIF. (2023). Smart Cities/Smart Villages in Albania. https://www.wbif.eu/technicalassistancegrants/WB23-ALB-DII-02.

World Bank. (2020). Albania E-Commerce Diagnostic: Leveraging the Digital Trade Opportunity. https://documents.worldbank.org/pt/publication/documents-reports/documentdetail/781641608742366947/albania-e-commerce-diagnostic-leveraging-the-digital-trade-opportunity.

World Economic Forum. (2023). Albania: Stategic Intelligence topics. https://intelligence.weforum.org/topics/a1G0X000006NwUfUAK.

Germany: A Framework for the Digital Municipalities in Administrative Enforcement

Christian Schachtner

1 Introduction

The ability of public administrations in Germany to offer their customers acceptable "service levels" is elementary—this can be seen in the "unrest" that arises in the face of administrative support (waiting times, sluggish digitization and much more) that sometimes seems overwhelmed today. The impression of being overwhelmed is already problematic in itself—combined with a "non-approachable" administration that does not "show a face" and does not invite "digital participation", this becomes a dangerous image factor.

2 Designing a Framework for Digitalized Municipalities

Based on the demands of the German Association of Municipalities on a strategy for an exemplary 'digital town hall' (German Gemeindetag 2024), the following points are taken up:

- Processes without media discontinuity: Internal processes should be completely digitized and made paperless. This requires a redesign of administrative processes and the implementation of appropriate IT systems.
- Citizen-friendly online services: The services must be easy to use and barrier-free. End-to-end digitization enables citizens to handle administrative processes completely online.
- Modern workplace equipment: Administrative staff need up-to-date hardware, including mobile devices, to work flexibly and efficiently.

C. Schachtner (✉)
RheinMain University of Applied Sciences, Wiesbaden, Germany
e-mail: christian.schachtner@hs-rm.de

C. Schachtner (ed.), *The European Smart City Movement*,
https://doi.org/10.1007/978-3-031-96700-9_7

- Data exchange and networking: An improved flow of information between different authorities and administrative levels is essential for efficient services.

The realization of a digital town hall is associated with various challenges: The development and integration of new IT systems requires considerable investment and expertise. This poses financial and personnel challenges for many municipalities. Administrative structures and workflows must be adapted to the digital transformation, which often meets with resistance. This organizational change requires comprehensive change management and can be time-consuming. Ensuring information security and the protection of personal data are key requirements. Strict legal requirements must be complied with and technical safety measures must be implemented. Both citizens and administrative staff must be convinced of the advantages of digitization and supported in using it. This requires targeted communication and training measures to create acceptance and competence in dealing with the new digital offerings.

To overcome the challenges mentioned, it is recommended to turn to best-practice examples. Thus, the strategic concept of the city of Gelsenkirchen can be used as a possible orientation strategy (City of Gelsenkirchen 2021):

1. Strategic planning: A clear digitalization strategy with defined goals and milestones is essential.
2. Cooperation and standardisation: Cooperation between municipalities as well as with the federal and state governments can create synergies and promote the development of uniform standards.
3. Agile development: Iterative approaches enable a step-by-step implementation and continuous improvement of digital services.
4. Training and change management: Comprehensive training opportunities and open communication promote acceptance among employees.
5. Citizen participation: The involvement of users in the development process ensures that the offers meet actual needs.

The digitization of public administration poses major challenges for municipalities. About a third of German municipalities already offer digital administrative services, often under the term "digital town hall". However, these services are often limited to a collection of specialist procedures in accordance with the Online Access Act (OZG), which neglects the complex interactions between citizens, administration and other stakeholders (Ivicos.eu. 2024). Achieving a comprehensive digital transformation of local government requires a holistic approach that goes beyond the mere digitization of processes and takes into account the human component of administration. A promising approach to overcoming these challenges is the development of a comprehensive framework for the Digital City Hall (City of Hamburg 2024). This framework should not only cover technical aspects, but also support organizational and cultural changes. Here's a SaaS platform framework that integrates innovation brainstorming, process optimization, and administrative enforcement (Smartsheet 2024).

3 Approach: SaaS Platform Framework for Innovation Brainstorming, Process Optimization and Administrative Enforcement

A cloud-based infrastructure solution forms the backbone of the Digital Town Hall. It enables the scalability, elasticity and accessibility that are essential for modern administration. The cloud architecture should be designed in such a way that it can flexibly adapt to changing requirements and allow teams to access the resources they need on their own (Rewion 2024).

For change and design performance management, it is important that the infrastructure supports agile methods such as Scrum or Kanban. These promote regular meetings and a transparent way of working, which can break down hierarchical structures and improve collaboration. Implementing SMART (Specific, Measurable, Attainable, Relevant, Time-bound) goals can help measure the success of technology deployments and drive continuous improvement (Flixcheck 2024).

Effective SaaS workflow design is critical for incremental process evolution in Teams. It should offer features to automate tasks, notifications, and alerts, as well as automated actions and integrations. These features allow tasks to be automatically created, assigned, and updated based on defined triggers such as status changes or due dates. For successful collaboration, the use of collaborative tools such as ClickUp, Google Workspace and Slack is essential. These tools promote effective communication and information sharing between team members. Regular brainstorming sessions can help generate new ideas and approaches and actively engage employees in the workflow optimization process (ClickUp 2024).

A central aspect of the Digital Town Hall is the secure and legally compliant storage of data and identities. In its draft law 435/23, the Bundesrat has emphasised the need to meet certain minimum requirements when processing health and personal data using cloud-based information systems (German Bundestag 2023). These requirements are based on the Cloud Computing Compliance Criteria Catalogue (C5) of the German Federal Office for Information Security.

A C5-compliant data and ID storage system must not only meet technical security standards, but also ensure the data protection and privacy of citizens. This is particularly important in the context of enforcement matters where sensitive personal data is processed. The implementation of such a storage system requires careful planning and continuous monitoring to ensure compliance with regulatory requirements.

4 Approach: User-Oriented Design of Digital Government Services According to Overarching Governance

When designing digital government services, it is crucial that they not only function technically, but are also intuitive to use. The user-centered design principle puts the future users with their needs and requirements at the center of the development process. Methods such as the Customer Journey Framework and Design Thinking help to develop digital performance processes sustainably from the user's perspective (Institute for the Public Sector e.V. 2019).

The quality of the services offered must be in line with the expectations of the users. To this end, it is advisable to deploy resources in the administrations in order to determine the exact needs of the users in advance. Integrating citizens into the development process can also provide valuable insights and increase adoption of the finished application. An effective co-design model involves not only citizens but also administrative staff in the development process. This requires a cultural change within the administration, in which decision-makers must be convinced of modern technologies in order to create useful applications for the population (publicplan GmbH 2024).

Quality assurance plays a central role here. Group lines can act as an interface between the technical developers and the end users. They should be equipped with appropriate tools to continuously monitor the quality of the services developed and to ensure that they meet the needs of users. An example of such a quality assurance tool is the "Digitallotse BVS" course, which is offered by the Bavarian School of Administration in cooperation with municipal umbrella associations and the Ministry of Digital Affairs (Bavarian State Ministry for Digital Affairs 2017). Such training courses prepare administrative staff for the challenges of digitization and enable them to take an active role in the co-design process.

To ensure that the digital services developed meet the needs of citizens, representative audience testing procedures are essential. These were to evaluate not only the technical functionality, but also the user performance and user-friendliness of the applications. One possible approach is to conduct user experience (UX) tests, in which representatives of different user groups test the applications in realistic scenarios (publicplan GmbH 2024). Metrics such as time to complete tasks, error rate and subjective user satisfaction can be collected.

In addition, A/B testing can be used to compare different versions of an application and identify the most effective solution. The results of these tests should be used in an iterative process for the continuous improvement of digital services. It is important to note that the digitization of the public sector does not only involve external government processes. Internal processes and the way administrative staff work must also be taken into account (ALE 2024). A holistic approach that takes into account the needs of both citizens and administrative staff is crucial for the success of a digital town hall.

In conclusion, developing a digital town hall framework is a complex task that requires close collaboration between citizens, administrative staff, and technical experts. Through the consistent application of user-oriented design principles, the integration of co-design roles and the use of representative test procedures, a digital town hall can be created that meets the needs of all parties involved and offers real added value for society.

5 Approach: Authentication and Qualified Advice in the Digital Town Hall

In the context of the Digital Town Hall, they are essential for ensuring the identity of users and ensuring a trustworthy exchange between citizens and administration. One of the main challenges in implementing a digital town hall is the secure identity verification of users. Various authentication options are available for this purpose, with the BundID playing a central role. This platform allows users to register once and then use various online services of the public administration. The advantages of this system are the high level of security through central administration, easy integration into nationwide e-government projects and user-friendliness through single sign-on (SSO) (Bamero 2024).

Another important component is the electronic proof of identity (eID). According to Section 2 (3) of the Federal E-Government Act (EGovG), all federal authorities are obliged to offer electronic proof of identity in administrative proceedings that require identification. The eID service, which is provided by Bundesdruckerei, can be easily integrated into existing websites and guarantees reliable and highly secure data exchange for authorities and citizens alike (ID card portal 2024).

In addition to secure authentication, the provision of qualified consulting services is an essential part of the Digital Town Hall. The challenge is to transfer the human component of administration to the digital space while maintaining the quality of advice. A promising approach for this is the implementation of video chat consultations with government employees after successful authentication via eID. This concept combines the advantages of personal advice with the flexibility and efficiency of digital solutions. After successful authentication, citizens can communicate with administrative staff in real time in a virtual citizens' office, exchange documents and discuss complex concerns.

The development of a framework for the Digital Town Hall, especially in the area of authentication and qualified consulting, requires a holistic approach. The combination of secure identity verification through eID systems and innovative consulting solutions such as video chats has the potential to significantly improve the proximity to citizens and efficiency of the administration. For example, it is important for the state capital of Mainz to take technical, organizational and legal aspects into account in order to ensure successful implementation (State capital Mainz 2024).

Developing a framework for a digital town hall is a complex undertaking that encompasses various aspects of administrative digitalization. A central component of this framework is the Service Level Agreement (SLA), which regulates the quality and scope of digital government services. In the following, the concept of the SLA in the context of the Digital Town Hall will be examined in more detail.

6 Approach: Service Level Agreement in the Digital Town Hall

A service level agreement within the framework of the Digital Town Hall serves as an agreement between the administration and the citizens on the expected quality and scope of the digital services offered. It creates transparency and sets standards for the provision of services. The quality and scope of digital government services are crucial for the acceptance and use of the Digital Town Hall. An SLA determines which services are offered in what form and which performance standards must be met. This can include, for example, the availability of online services, the response times to inquiries or the completeness of the services offered. In a network of several municipalities or at the state level, a uniform SLA can help to ensure that citizens can expect comparable services of the same quality, regardless of where they live. This promotes equal treatment and facilitates inter-municipal cooperation (Let's Get Digital 2023).

An essential aspect of the SLA in the Digital Town Hall is the determination and monitoring of processing times. Through continuous monitoring, the actual processing times can be recorded and compared with the goals set in the SLA. This enables the prompt identification of bottlenecks and the initiation of improvement measures. The monitoring can include various key figures, such as (Net orange 2024):

- Average processing time per application type
- Percentage of applications processed on time
- Waiting times for telephone or online inquiries

The publication of these key figures creates additional transparency and strengthens citizens' trust in the administration.

To promote compliance with the standards set out in the SLA, mechanisms can be agreed in case of non-compliance. An innovative approach here is so-called "Service Center Benefits" in the field of non-profit services. This concept provides that in the event of non-compliance with the agreed performance standards, the administration undertakes to provide additional resources for charitable purposes.

The implementation of a Service Level Agreement as part of a framework for the Digital Town Hall is an important step towards improving the quality and reliability of digital administrative services. Clear agreements, continuous monitoring and innovative approaches such as Service Center Benefits can strengthen citizens' trust in digital administration and increase the efficiency of government processes.

7 Approach: Participation and Interaction Channels

The involvement of citizens via various digital channels is a core aspect of the Digital Town Hall. The aim is to offer a variety of communication channels in order to reach as many citizens as possible and make it easier for them to interact with the administration. This can include the following elements:

1. Multi-channel communication: The provision of different digital communication channels such as email, chat, video conferencing and social media allows citizens to choose the path that works best for them (Klemme et al. 2018).
2. Virtual citizens' consultation hours: Regular online consultation hours with administrative staff or political representatives can increase citizen proximity and enable direct exchange.
3. Mobile applications: The development of smartphone apps for urban services and information can significantly improve the accessibility and use of urban services.

Barrier-free access: It is important that all digital services are designed to be barrier-free in order to ensure inclusive participation (Aichholzer and Kubicek 2016).

A central component of the Digital Town Hall should be a citizen participation platform that enables residents to actively participate in shaping their city. This platform can include various features. Citizens can submit, discuss and evaluate proposals for improving public space. This promotes the creative participation of citizens in urban development. The platform should offer opportunities to vote on submitted proposals in order to get an opinion of the citizenry. The progress and implementation of citizens' proposals should be presented in a comprehensible way in order to strengthen trust in the administration. Specific discussion spaces on various urban issues can foster exchanges between citizens and experts (Türken and Eyuboglu 2021).

An example of such a platform is "mitgestalten.trier.de", where citizens can participate in various projects and topics (City of Trier 2023). The federal government's "Digital Administration Dashboard" also shows how such information platforms can be designed to increase transparency and traceability (Federal Ministry of the Interior and Community 2022).

8 Approach: Collaboration and Co-creation

In order to tap into the potential of charitable services for the population, online workshop series are a good idea, and citizen science approaches can be used to carry out detailed analyses of needs, problems and solution scenarios. This participatory research method makes it possible to actively involve citizens in the process of data collection and analysis. For the development of a framework for the digital town hall, the following surveys could be carried out:

1. Need analysis: Identification of the most important needs of citizens in terms of digital government services
2. Pain analysis: Recording the main problems and obstacles in the use of existing digital services
3. Providing scenario analysis: Development and evaluation of different approaches to solving identified problems.

The results of these analyses can be directly incorporated into the design of the digital town hall and ensure that the solutions developed cover the actual needs of the users. These can build on existing platforms and initiatives that already successfully offer digital education in the field of civic engagement. An example of such collaboration is the Co-VAL project, which explores transformative innovation and shared value creation in the public sector (European Commission 2024).

Developing a framework for the digital town hall requires a multidimensional approach that focuses on collaboration and co-creation. Through the active involvement of all stakeholders, the use of online workshop series and the implementation of citizen science projects, a user-centered and effective digital town hall can be created. This approach makes it possible to fully exploit the potential of digitization and at the same time to take into account the needs of all stakeholders by establishing innovation networks between the public and private sectors. The creation of real-world laboratories for testing new concepts and the implementation of feedback mechanisms for continuous improvement are thus possible.

9 Approach: Rating and Idea Generation and Aggregation

An effective rating and suggestion system allows citizens to directly share their experiences with government services. Based on the life situation survey of the Federal Statistical Office, which takes into account various satisfaction factors such as freedom from discrimination, waiting times and comprehensibility of the forms, a detailed evaluation system can be developed (Statista 2024).

This system should be easy to use and allow for both quantitative ratings (e.g. on a scale of 1–5) and qualitative feedback in the form of comments. The suggestion system was intended to enable citizens to submit concrete ideas for improvement. These can relate to specific services, processes or general aspects of administrative work. In order to increase the quality of the proposals, guiding questions or categories could be specified that are oriented towards the main problems identified.

An upvoting system allows the citizenry to prioritize the submitted proposals. Citizens can vote for ideas that they consider particularly valuable or urgent. This helps the administration to assess the relevance and acceptance of proposals. To avoid manipulation, the system should provide for user verification, for example through a one-time registration with an ID card. Implementation foresight reporting complements the upvoting system by providing an assessment of the feasibility and potential impact of the proposals. For this purpose, experts from the administration or external consultants could be called in to evaluate the proposals in terms of technical, legal and financial feasibility. These assessments are presented transparently in the digital citizen network in order to set realistic expectations and promote discourse.

The priority report summarises the results of the evaluation and proposal system and prepares them for political decision-making. It contains:

1. A ranking of the most urgent services to improve, based on citizen reviews.
2. An overview of the most frequently mentioned problems and suggestions for improvement.
3. An analysis of the characteristics of expectations, which shows which aspects are particularly important for citizens (e.g. speed, transparency, user-friendliness).
4. An assessment of the feasibility and potential impact of the top proposals.

This report serves as a basis for the work of political committees, such as the Committee on Foreign Affairs (German Bundesrat 2024). It enables decision-makers to make well-founded and citizen-oriented decisions and to set priorities for the further development of the digital town hall.

The continuous involvement of citizens in the improvement process can not only increase satisfaction with official services, but also strengthen trust in the administration and democracy as a whole. This is particularly important given that satisfaction with government services varies widely in different life situations (German Federal Chancellery 2024).

10 Approach: Tools for Knowledge Management and New Work

A central element here are knowledge management platforms that ensure structured storage and easy findability of information. A promising solution in this context is Bloomfire, a curated intranet platform for knowledge sharing. It supports the distillation of insights from different file formats and offers integrations with popular communication tools such as Slack and Microsoft Teams (Computerweekly 2024). This enables seamless integration into existing workflows and promotes cross-departmental knowledge sharing.

Guru, an AI-powered knowledge orchestration platform, is particularly suitable for the specific needs of public authorities. It enables the curation of knowledge into answers and identifies knowledge gaps that are not covered by existing content (Computerweekly 2024). This is especially valuable for administrations that often have to work with complex regulations and constantly changing regulations.

An internal wiki system is an effective solution for documenting administrative processes and best practices and making them accessible to all employees. It enables collaborative content creation and maintenance, which is especially beneficial in the context of public administration, where knowledge is often spread across different departments and hierarchical levels. However, the implementation of a wiki system in the administration also poses challenges. As Subs and Emis (2025) emphasize, the success of a wiki in administration depends largely on the initiative of the employees. This can be a hurdle in hierarchical structures that are characteristic of many administrations. To overcome these, it is important to foster an organizational culture that values and rewards initiative and knowledge sharing. An example of the successful implementation of a wiki system in the administration is the use of MediaWiki for the creation and updating of administrative regulations. This system allows not only to read and edit content, but also to discuss individual regulations, resulting in improved internal networking and increased efficiency (Subs and Emis 2025).

The development of an internal project structure and visualization engine with AI support represents an innovative approach to capture and analyze the implementation performance of local projects. This engine acts as a metadatabase that organizes indicators into clusters, providing a comprehensive view of the performance of different management areas. AI systems can play a decisive role in this context. They can analyze data from past projects, identify patterns, and make recommendations for the optimal project structure and timeline (Factro 2024). This enables data-driven decision-making and can significantly increase the efficiency of administrative work. Integrating AI into project management tools can also help with task management and resource planning. By analyzing the qualifications and availability of team members, tasks can be assigned efficiently and individual strengths can be optimally used (Factro 2024). This is particularly valuable in administrations, where a large number of projects often run in parallel and resources have to be used optimally.

11 Big Picture of the Framework

Based on the aforementioned aspects, the author suggests that those responsible for digitization of town halls take a look at the following self-developed framework for the establishment of virtual citizen service centers as a living component system:

- SaaS platform framework for innovation brainstreaming, process optimization and administrative enforcement
  - Cloud-based infrastructure solution for change and design performance management
  - SaaS workflow design for incremental process development in Teams
  - C5-compliant data and ID storage over enforcement matters
- Service design of applications
  - User-oriented design of digital government services according to overarching governance
  - Co-design roles with quality assurance instruments of the group leaders
  - Representative target group test procedures on usage performance
- Authentication and qualified advice
  - Secure identity verification of users and expert online advice

  - Example: Video chat consultation with government employees after authentication via eID

- Service Level Agreement
  - Agreement on the quality and scope of digital government services in the network
  - Guaranteed processing time through monitoring in prison
  - Service Center Benefits in the event of non-compliance in the area of non-profit services
- Participation and interaction channels
  - Involvement of citizens via various digital communication channels
  - Citizen participation platform for suggestions for improvements in public space
  - Open information dashboards on investment projects and measures
- Collaboration and co-creation
  - Joint development of solutions by citizens, companies, associations and administration
  - Online workshop series for offering the potential of community service to the population
  - Citizen science surveys on need, pain and providing scenario analysis
- Rating and idea generation and aggregation
  - Evaluation and suggestion system for government services
  - Upvoting system for suggestions for improvement incl. implementation foresight reporting in digital citizen networks
  - Priority report with expression of expectations for policy preparation in committees
- Tools for knowledge management and New Work
  - Digital tools for knowledge management and modern working methods in public authorities
  - Internal wiki system for administrative staff to document processes and best practices
  - Internal project structure and visualization engine with AI support as a metadatabase of local implementation performance in indicator clusters

12 Conclusion

Developing a framework for the digital town hall is a complex but necessary task for modern local governments. Success depends to a large extent on a holistic approach that takes into account technical, organizational and human factors in equal measure. A clear strategy, the cooperation of all stakeholders and the courage to come up with innovative solutions are crucial to driving administrative digitization and creating citizen-friendly, efficient services. Future research should focus on evaluating existing solutions and developing best practices for different municipal sizes to further optimize the transformation process.

The SaaS platform framework presented offers a promising approach to integrate innovation brainstorming, process optimization and administrative enforcement. By combining a cloud-based infrastructure solution, flexible workflow design, and C5-compliant data storage, municipalities can overcome the challenges of digital transformation and create citizen-centric, efficient, and innovative administration. To ensure the success of such a framework, it is important that municipalities take a holistic approach that goes beyond pure technology implementation. The involvement of all stakeholders, continuous training for employees and a culture of continuous improvement are just as important as the technical solutions themselves. This is the only way to exploit the full potential of the Digital Town Hall and achieve a sustainable transformation of public administration.

The challenge is to develop a balanced SLA that sets ambitious goals on the one hand, but also remains realistic and achievable on the other. In addition, the SLA must be reviewed regularly and adapted to changing requirements and technological possibilities in order to maintain its effectiveness in the long term. The development of a framework for a digital town hall with a focus on participation and interaction channels offers enormous opportunities for improved citizen participation and transparency in the administration. By integrating different digital communication channels, implementing a citizen participation platform and providing open information dashboards, municipalities can reshape the relationship with their citizens and create a more effective, citizen-centric administration.

References

Aichholzer, G., & Kubicek, H. (2016). Evaluating Digital Citizen Participation: Tools and Initiatives. In G. Aichholzer, H. Kubicek, & L. Torres (Eds.), Evaluating e-Participation: Frameworks, Practice, Evidence (pp. 15–38). Springer International Publishing.

ALE. (2024). State-of-the-art government operations thanks to a digital workplace. https://www.al-enterprise.com/-/media/assets/internet/documents/digital-workplace-enables-next-gen-public-service-experiences-whitepaper-de.pdf

Bamero. (2024). OZG: Digital Signatures and Authentication. https://www.bamero.de/newswissen/wissen/ozg-digitale-signaturen-und-authentifizierung/

Bavarian State Ministry for Digital Affairs. (2017). Digital Town Hall. https://www.stmd.bayern.de/themen/digitale-verwaltung/digitales-rathaus/

BCG. (2023). Three key demands for the digitization of German government services. https://www.bcg.com/publications/2023/germany-digital-government-citizen-survey

City of Gelsenkirchen. (2021). Digital Municipalities in North Rhine-Westphalia. https://www.gelsenkirchen.de/de/stadtprofil/stadtthemen/die_vernetzte_stadt/_doc/202103_digitalbrosch_re_neuauflage_2021.pdf.

City of Hamburg. Office for IT and Digitalization. (2024). Data Governance Framework—The Data of Our City. https://digital.hamburg.de/digitale-stadt/data-governance-framework-die-daten-unserer-stadt-644522

City of Trier. (2023). Platform for citizen participation. https://www.trier.de/rathaus-buerger-in/buergerbeteiligung/plattform-fuer-buergerbeteiligung/

ClickUp. (2024). Improving team workflows and processes. https://clickup.com/de/blog/163292/wie-man-die-arbeitsablaeufe-im-team-verbessert

Computerweekly. (2024). 8 software tools for knowledge sharing and collaboration. https://www.computerweekly.com/de/tipp/8-Software-Tools-fuer-Wissensmanagement-und-Collaboration

European Commission. (2024). A guide to shared value creation in public administration. Retrieved from https://cordis.europa.eu/article/id/422183-a-guide-for-enabling-co-creation-in-public-administrations/de

Factro. (2024). AI in project management—a comprehensive overview. https://www.factro.de/blog/ki-im-projektmanagement/

Federal Ministry of the Interior and Community. (2022). "OZG Dashboard" becomes the "Digital Administration Dashboard". https://www.digitale-verwaltung.de

Flixcheck. (2024). What is a workflow? Definition and tips. https://www.flixcheck.de/workflow/

German Bundesrat. (2024). Committee on Foreign Affairs. https://www.bundesrat.de/DE/bundesrat/ausschuesse/aa/aa-node.html

German Bundestag. (2023). Bundesrat 435/23 Bill. https://dserver.bundestag.de/brd/2023/0435-23.pdf

German Federal Chancellery. (2024). Citizen satisfaction with government services. https://www.bundesregierung.de/resource/blob/975272/1638358/34ddc9a3d4207ba40a9109606b5e782d/de-zulassung-download-bericht-data.pdf

German Gemeindetag. (2024). Clear strategy for 'digital town hall' demanded. https://www.diegemeinde.de/gemeindetag-fordert-klare-strategie-fuer-digitales-rathaus

ID card portal. (2024). eID service. https://www.personalausweisportal.de/Webs/PA/DE/wirtschaft/technik/eID-service/eid-service-node.html

Institute for the Public Sector e.V. (2019). Digital transformation: Focus on the user. https://publicgovernance.de/media/Digitaler_Wandel_der_Nutzer_im_Fokus.pdf

Ivicos.eu. (2024). Digital Town Hall: A Critical Inventory of Digitization in Municipalities. https://www.ivicos.eu/blog-artikel/digitales-rathaus-eine-kritische-bestandsaufnahme-der-digitalisierung-in-kommunen

Klemme, M., Wiegandt, C.-C., & Wiesemann, L. (2018). Participation online – developing the city together. Information on Spatial Development, 6, 6–11.

Let's Get Digital. (2023). Service Level Agreement. http://www.letsgetdigital.com/terms/sla-de-20230815-3.3.pdf

Net orange. (2024). Service Level Agreement: On the safe side. https://www.netzorange.de/it-ratgeber/service-level-agreement-auf-der-sicheren-seite/

publicplan GmbH. (2024). User-orientation in digital administration. https://www.publicplan.de/blog/nutzerorientierung-verwaltung

Rewion. (2024). Cloud Know-how. https://www.rewion.com/fachwissen/cloud-know-how/

Smartsheet. (2024). Comprehensive Guide to Business Process Management. https://de.smartsheet.com/all-about-business-process-management-expert-insights

State capital Mainz. (2024). Electronic Communications. https://www.mainz.de/verwaltung-und-politik/buergerservice-online/elektronische-kommunikation.php

Statista. (2024). Satisfaction with government services 2023. https://de.statista.com/statistik/daten/studie/1500180/umfrage/zufriedenheit-mit-behoerdlichen-dienstleistungen-nach-lebenslagen/

Subs, E., & Emis, P. (2025). Web 2.0 for the internal networking of the administration: A wiki. EMIS. https://subs.emis.de/LNI/Proceedings/Proceedings162/123.pdf

Türken, S., & Eyuboglu, E. (2021). Digital participation in urban planning: A conceptual framework. Journal of Urban Management, 10(2), 178–188.

Estonia—Citizen Participation in the Capital City

Christian Schachtner

1 Introduction

Tallinn, the capital of Estonia, has become a pioneer in smart city development in recent years. A special focus is on the aspects of accessibility, interoperability and user-friendliness, which contribute significantly to the success of the city's digital transformation.

The accessibility of digital services is a key concern of Tallinn Municipality. The implementation of the "e-Estonia" concept has created a digital infrastructure that allows citizens to carry out almost all administrative services online (Kalvet 2012). This ranges from filing a tax return to participating in elections. Of particular note is the digital ID card, which acts as a key to these services while ensuring the security of personal data (Estonian Information System Authority 2021).

2 Data-Based Infrastructure

The interoperability of different systems and platforms is another pillar of Tallinn's Smart City strategy. The development and implementation of the X-Road platform has created a secure and efficient way to share data between different authorities and service providers (Anthes 2015).

This not only allows for seamless integration of different services, but also encourages the development of innovative solutions by third-party providers who can access this data. A prime example of the successful implementation of accessibility, interoperability and user-friendliness is Tallinn's public transport system. Since 2013, the use of public transport has been free for residents of the city, which is made

C. Schachtner (✉)
RheinMain University of Applied Sciences, Wiesbaden, Germany
e-mail: christian.schachtner@hs-rm.de

C. Schachtner (ed.), *The European Smart City Movement*,
https://doi.org/10.1007/978-3-031-96700-9_8

possible by an electronic ticketing system linked to the digital ID card (Cats et al. 2017). This system not only integrates different modes of transport, but also enables efficient traffic planning by analysing usage data. The user-friendliness of digital services is a top priority in Tallinn. The city has invested significant resources in the development of intuitive user interfaces and accessible applications. An example of this is the "mTallinn" app, which acts as a central access point for city services and is continuously improved based on user feedback (City of Tallinn 2023).

Another important aspect is the digital education of citizens. Tallinn has implemented extensive digital literacy programmes to ensure that all citizens can benefit from digital offerings (Kitsing 2018). This includes training for older people, integration courses for migrants and the integration of digital skills into school lessons. Tallinn's successes in terms of accessibility, interoperability and ease of use have gained international recognition. The city has received several awards for its innovative solutions and serves as a model for other cities worldwide (European Commission 2020).

Despite the impressive progress, Tallinn continues to face challenges. Rapid technological development requires continuous adaptation and improvement of existing systems. In addition, the city must ensure that the focus on digital solutions does not lead to the neglect of traditional services and that it also reaches those citizens who may have difficulties with the use of digital technologies. Tallinn, often referred to as the "digital capital of Europe", has implemented a comprehensive system of e-government that has revolutionized the interaction between citizens and authorities (Estonian Ministry of Economic Affairs and Communications 2023).

3 E-Government Services

At the heart of this system is the digital identity, which is available to every Estonian citizen. This allows residents to access a variety of administrative services online, from filing taxes to setting up a business (e-Estonia 2024). A particularly innovative aspect of e-governance in Tallinn is the "once-only" principle. This concept states that citizens only need to submit their data to authorities once, eliminating redundant data entry and significantly increasing the efficiency of administrative processes (Krimmer et al. 2022).

The implementation of this principle has not only saved time for citizens, but also reduced the susceptibility to errors in data processing. However, citizen participation in Tallinn goes far beyond the mere use of digital services. The city has launched various platforms and initiatives that allow residents to actively participate in urban development and decision-making. An example of this is the platform "AvaLinn" (Open City), which allows citizens to submit and discuss proposals for urban planning projects (Tallinn City Government 2024).

In addition, Tallinn has introduced a participatory budgeting system, where citizens can vote on the use of part of the city's budget. This system, accessible through a user-friendly online platform, has significantly increased citizens' engagement in municipal affairs and led to a more transparent use of public funds (Krenjova and Raudla 2023).

Another notable aspect of e-governance in Tallinn is the focus on data security and transparency. The city uses blockchain technology to ensure the integrity and immutability of government data. This has not only increased citizens' trust in digital systems, but has also positioned Tallinn as a role model for other cities worldwide (Ølnes and Jansen 2022).

Tallinn's successes in e-governance and citizen participation are impressive, but the city is not resting on its laurels. Continuous innovation and improvement are part of the city's digital strategy. Currently, Tallinn is working on integrating artificial intelligence into its e-governance systems to offer personalized services and further increase efficiency (Tallinn Strategy Centre 2025).

Despite the many advantages, digital decision-makers also face challenges. One of them is the digital divide, which can arise when not all citizens have equal access to or use digital technologies. To address this problem, Tallinn has launched various digital education and support initiatives, including free training and public internet access points throughout the city (Tallinn Education Department 2024).

Another critical aspect is data protection. While comprehensive digitization offers many benefits, it also requires robust security measures to protect sensitive citizen data. Tallinn has therefore implemented strict data protection protocols and is continuously working to adapt them to new challenges (Estonian Data Protection Inspectorate 2024).

4 The Estonian Smart City Movement

Estonia, especially its capital Tallinn, has also become a pioneer in smart city development in recent years. Tallinn has set itself the ambitious goal of becoming climate neutral by 2050 (Tallinn City Government 2023).

To achieve this goal, the city relies on a variety of innovative technologies and concepts. One focus is on promoting green mobility. For example, a comprehensive network of charging stations for electric vehicles has been set up, and public transport is gradually being converted to zero-emission propulsion (Estonian Ministry of Economic Affairs and Communications 2024).

Another important aspect of Tallinn's smart city strategy is the digitization of administrative processes. Estonia is considered a global pioneer in e-government, and Tallinn is using this expertise to make citizen services more efficient and accessible (e-Estonia Briefing Centre 2025).

By implementing digital identities and secure online platforms, citizens can handle a variety of administrative matters from the comfort of their own homes. Promoting innovation and start-ups also plays a central role in Estonia's Smart City vision. Tallinn has become a major hub for technology companies, especially in the field of artificial intelligence and the Internet of Things (IoT) (Startup Estonia 2024).

This dynamic start-up scene makes a significant contribution to the development of new solutions for urban challenges. A particularly interesting project is the implementation of smart street lighting systems in Tallinn. These automatically adapt to environmental conditions, significantly reducing energy consumption (Tallinn Smart City Office 2024).

Such initiatives show how technology can be used to both improve the quality of life of citizens and minimize environmental impacts. The integration of environmental data into urban planning is another focus of Estonia's Smart City Strategy. Through a network of sensors, data on air quality, noise pollution and other environmental factors are continuously collected and analysed (Estonian Environment Agency 2025).

This information enables urban planners to make informed decisions and implement targeted actions to improve the quality of life. Another flagship project is the development of "Smart Districts" in Tallinn. Various smart city technologies are being integrated and tested in these neighbourhoods, from smart energy management systems to innovative waste management concepts (Tallinn City Planning Department 2024).

These pilot projects serve as test beds for solutions that can later be implemented throughout the city. Estonia's successes in the field of smart cities have also received international recognition. Tallinn has received several awards for its innovative approaches and is considered a role model for other European cities (European Commission 2024).

Sharing best practices and collaborating with other smart cities around the world are important aspects of Estonia's strategy. Despite the impressive progress, Estonia still faces challenges in implementing its smart city vision. Ensuring cybersecurity in an increasingly connected urban environment is an ongoing task (Estonian Information System Authority, 2025). The involvement of all population groups in the digital transformation also remains an important concern.

5 Conclusion

In conclusion, Tallinn is at the forefront of smart city development through its holistic approach to accessibility, interoperability, and ease of use. The city impressively shows how digital technologies can be used to improve the lives of citizens and at the same time increase the efficiency of city administration. By combining technological innovation, environmentally conscious action and citizen-oriented administration, a model for smart cities is being created that is attracting worldwide attention. Tallinn is taking a pioneering role in the areas of e-governance and citizen participation, demonstrating how digital technologies can be used to make administration more efficient while strengthening the democratic participation of citizens. The experiences and lessons learned from the Estonian projects will undoubtedly make a valuable contribution to the global discussion on the future of the global community of cities. Tallinn's success is based on a clear vision, political will and the active involvement of citizens in the transformation process. Other cities can learn from this example and adapt their own smart city strategies accordingly.

Bibliography

Anthes, G. (2015). Estonia: A model for e-government. Communications of the ACM, 58(6), 18–20.

Cats, O., Susilo, Y. O., & Reimal, T. (2017). The prospects of fare-free public transport: Evidence from Tallinn. Transportation, 44(5), 1083–1104.

City of Tallinn. (2023). mTallinn: Your city in your pocket. https://www.tallinn.ee/en/mtallinn

e-Estonia Briefing Centre. (2025). Digital society of Estonia. https://e-estonia.com/

e-Estonia. (2024). Digital Society. https://e-estonia.com/

Estonian Data Protection Inspectorate. (2024). Annual Report on Data Protection in Estonia. Tallinn: Government of Estonia.

Estonian Environment Agency. (2025). Environmental monitoring in urban areas. https://www.keskkonnaagentuur.ee/

Estonian Information System Authority. (2021). e-Estonia: Digital society. https://e-estonia.com/

Estonian Information System Authority. (2025). Cybersecurity in smart cities. https://www.ria.ee/en

Estonian Ministry of Economic Affairs and Communications. (2023). Digital Agenda 2030. Tallinn: Government of Estonia.

Estonian Ministry of Economic Affairs and Communications. (2024). Green mobility initiatives in Estonia. https://www.mkm.ee/en

European Commission. (2020). European Capital of Innovation (iCapital) Awards. https://ec.europa.eu/info/research-and-innovation/funding/funding-opportunities/prizes/icapital_en

European Commission. (2024). Smart Cities and Communities. https://ec.europa.eu/info/eu-regional-and-urban-development/topics/cities-and-urban-development/city-initiatives/smart-cities_en

Kalvet, T. (2012). Innovation: a factor explaining e-government success in Estonia. Electronic Government, an International Journal, 9(2), 142–157.

Kitsing, M. (2018). Internet voting in Estonia: The digital divide and other challenges. In The Palgrave Handbook of Political Elites (pp. 617–635). Palgrave Macmillan, London.

Krenjova, J., & Raudla, R. (2023). Participatory Budgeting in Estonian Municipalities: The Impact on Financial Decision-Making. Local Government Studies, 49(2), 234–251.

Krimmer, R., Kadak, T., & Rasmussen, M. (2022). The Once-Only Principle: An Assessment of Implementation in European Digital Government. Government Information Quarterly, 39(1), 101606.

Ølnes, S., & Jansen, A. (2022). Blockchain Technology as Infrastructure in Public Sector: An Analytical Framework. In Proceedings of the 55th Hawaii International Conference on System Sciences.

Startup Estonia. (2024). Estonian startup ecosystem overview. https://startupestonia.ee/

Tallinn City Government. (2023). Tallinn 2050—Climate neutral city strategy. https://www.tallinn.ee/en

Tallinn City Government. (2024). AvaLinn: Participatory Urban Planning Platform. https://www.tallinn.ee/avalinn

Tallinn City Planning Department. (2024). Smart Districts initiative. https://www.tallinn.ee/en/ehitus/

Tallinn Education Department. (2024). Digital Literacy Program: Annual Report. Tallinn: Tallinn City Government.

Tallinn Smart City Office. (2024). Smart lighting systems in Tallinn. https://www.tallinn.ee/en/smartcity/

Tallinn Strategy Centre. (2025). Tallinn 2035: Smart City Development Plan. Tallinn: Tallinn City Government.

Smart Participation in a Smart City—Implementation and Experiences Using the Example of the City of St. Gallen

Christian P. Geiger

1 Introduction

Smart cities are often characterized by technology, but also by the promotion and implementation of participatory approaches. Participation or involvement can take place in the area of political decisions, but also in the development of ideas or in the context of participatory project involvement. The Smart City definition of the City of St. Gallen states: "In the Smart City of St. Gallen, technologies and data are linked in order to improve the quality of life of residents and the quality of the location for companies while using fewer resources. Residents are an essential part of the developments and can contribute, test and co-determine solutions for everyday life. They are therefore important partners for future developments" (Stadt St. Gallen 2020b).

Participation in the smart city is therefore both a method and a goal for the smart implementation of projects. Another reason for the necessary strengthening of participation is often the broader support for decisions: In the city of St. Gallen, with a proportion of 33.6% foreigners and 13.7% children under the age of 14 at local level (in 2023), as well as numerous assisted persons, refugees or people with protection status S, there are only around 44,000 voters in the context of political votes (Fachstelle für Statistik Kanton St. Gallen 2024). People out of a total resident population of currently around 83,000 people (Fachstelle für Statistik Kanton St. Gallen 2023) in the year 2024. This corresponds to around 53% of the resident population in 2024.

Against this backdrop, the City Council was asked by the City Parliament on 20 March 2018 with a "Motion" to become more involved in the topic of participation at municipal level and to put the promotion of participation on the political agenda. The conversion of the motion into an item and the simultaneous "declaration of relevance" by the City Council ensured that a fundamental review of the topic could

C. P. Geiger (✉)
Bern University of Applied Sciences, Bern, Switzerland
e-mail: christian.geiger@bfh.ch

C. Schachtner (ed.), *The European Smart City Movement*,
https://doi.org/10.1007/978-3-031-96700-9_9

take place (Stadt St. Gallen 2020a). The background to this was also the questioning of the existing opportunities for participation in the "old" participation regulations at the time. The previous opportunities for participation via migrant associations had hardly been used by the actual target group up to that point. The regulations of 19.09.2006 were the basis for the submission of only 9 petitions, the majority of which found their way into the political apparatus of the city administration as so-called "migrant petitions" via an associated political commission. It became clear that the formal legal instrument of participation cannot be equated with the call for effective participation instruments that are perceived and used as real opportunities for political participation. This also refers to the lack of a culture of participation in the administrative apparatus.

1.1 City of St. Gallen

The city of St. Gallen is the cantonal capital of the canton of St. Gallen with a population of around 83,000. With its population, it is one of the ten largest cities in Switzerland and the center of Eastern Switzerland. St. Gallen is internationally known above all for its embroidery boom in the nineteenth century and its monastery church (UNESCO World Heritage Site). The city administration employs around 2300 people, divided into 5 directorates and a total of 36 departments and 3 staff units of the City Council. In the executive, the 5 elected members of the City Council are also the 5 directors of the respective directorates. The 5 City Council members form the City Council, which functions as a collective body. The "primus inter pares" is the mayor, who is also directly elected. The Town Clerk, who is elected by the legislature, assists the Council in an advisory capacity at all times. The legislative body is the city parliament, which consists of 63 people. The City Parliament and City Council are elected for a four-year term of office. During this period, the electorate is required to vote on substantive matters if they exceed a financial volume of CHF 15 million (Stadt St. Gallen 2005).

1.2 Examples for Participation in City of St. Gallen

Participation has a long tradition in St. Gallen and includes numerous digital and analog opportunities for the population to get involved. It was therefore a matter of course that various analog workshops were held with residents of a neighborhood as part of the Smart City projects. The focus here was primarily on energy and mobility issues, driven by the St. Gallen municipal utilities. After the Smart City topic was implemented more strongly by the Chief Digital Officer, who has been responsible since 2017, participation took place with an internal city working group (Smart City Steering Committee) as part of the strategy development for the Smart City. Two workshops with external participants were open to the public. Participants from the population, political representatives and business representatives were invited. A digital participation platform was also introduced in 2019 very close to the Smart City projects and in line with the new participation regulations. This was primarily used for construction projects, road renovations, the collection of ideas for location promotion and consultations on construction projects.

In the area of participation, St. Gallen has distinguished itself in recent years not only through its more digital smart city activities, but above all by holding numerous analog workshops. For example, participative projects have found their way into everyday life as part of the reformulation of the energy concept or the site planning of school buildings (Stadt St. Gallen 2023a) (Stadt St. Gallen 2022a).

The private initiative of the so-called "open-air parliament" was also publicly visible and particularly noteworthy (Reinhold 2024). Volunteers set up a stand at certain events at various locations in the city and noted down the wishes and concerns of residents and brought them to politicians. This brought politics closer to the people.

The implementation of the first urban hackathons in St .Gallen in 2021, 2022 and 2023 deserves a special mention and is one of the highest levels of participation (if you take the "ladder of participation" as a basis) (Stadt St. Gallen 2021a, 2022b, 2023b). In 2021, for example, the administration formulated urban challenges for the first time, which were to be solved by volunteers with the help of the open government data provided.

2 Objectives of Participation in the "Smart City St. Gallen"

A bouquet of goals—especially in the area of administrative and urban modernization perceptible to the population - should be implemented in St. Gallen with the Smart City theme. The City Council's Vision 2030 states the following in the "Smart City" topic area: "Expanded and digitalized infrastructures enable society and the economy to create new forms of service provision, interaction and participation." The following goal was formulated for participation: "St. Gallen enables effective participation". The 2017–2020 legislative objectives also describe participation with the following ambition: "A participation platform for the involvement of residents and companies has been established." (Stadt St. Gallen 2017). For the 2021–2024 legislative period, it goes on to say "The municipal tasks will be analyzed in relation to the participation regulations, possible areas for participation will be identified and the necessary concepts, measures and digital solutions will be developed. The concerns of special stakeholder groups will be taken into account." (Stadt St. Gallen 2021b).

The smart city strategy published in 2017 was based on Boyd Cohen's Smart City Wheel with its definition of six topic areas: Government, Mobility, Environment, Economy, Living and People (Cohen 2014). The smart city approach should therefore be a more far-reaching approach than the mere "digitalization of administration" or the mere networking of individual directorates or functional silos in the administration.

Not only the city council, but also the city parliament called for greater participation from various stakeholder groups. One of the reasons for this was the dissatisfaction that a large proportion of the resident population was not included in the decision-making process as part of the planned formal or mandatory votes or elections. In addition, a possible right to vote for foreigners would not have found a majority in a cantonal debate by the cantonal parliament or the cantonal government. In 2021, such a right to vote was clearly rejected in the cantonal parliament. One option was therefore participation in the form of informal processes, which are laid down in participation regulations at local level. In addition to the large group of migrants, it is also possible to involve people with disabilities as well as children and young people. In addition to the involvement of other sections of the population, the parliamentarians' demands also focused on the way in which participation takes place - via new, non-threshold channels.

Two key pillars of participation were (I) the introduction of new participation regulations and (II) the introduction of a digital participation platform. These two topics will be presented on the following pages.

3 Implementation of Citizens Participation in St. Gallen

In order to implement the objectives of involving the population more closely in the process of finding ideas and opinions, but also in political decision-making, the motion submitted to the City Council was converted into a postulate and declared "substantial" by the City Council. As a result, a fundamental review of the topic was carried out.

In organizational terms, the Social Affairs and Security Directorate and the Social Affairs Office were in charge of the review. The University of Applied Sciences East (FH Ost) was commissioned as an external expert to review the issue of participation and to draw up a basic report on participation, including conducting interviews with target groups and formulating recommendations for action. The FH Ost also provided support in the formulation of new participation regulations, in which the legal possibilities of participation at the level of the City of St. Gallen were laid down.

Although the topic of participation was originally initiated by a postulate in parliament, it was very quickly implemented by the executive. The political relevance and explosive nature of the topic is also evident from its inclusion in the respective legislative plans and the associated legislative goals. In addition, the topic was continued in the respective annual objectives of the City Council.

The process was started by convening a large steering group with representatives from all relevant departments. The types of participation and the administration's self-image of participation were discussed in this group. The results formed the basis for the interviews conducted by FH Ost with external stakeholders, who were defined as target groups in the area of participation (migrants, etc.). This content was used to formulate the basic report. Parallel to the general participation project, a separate project on child participation was launched. Building on this basis, the aim was now to follow up with activities in the area of participation:

A separate small working group was set up to draw up a modernized version of the participation regulations under the leadership of the Social Affairs department. Moderation and technical support was also provided by the FH Ost. In the working group, the technical discussions and elaborations were also conducted by the Communications department, the Chief Digital Officer and the Organizational Development department. The working group developed into the core group for participation for the city administration. In various discussions, the small working group was guided by the City Council with regard to the basic thrusts. Existing questions from parliament on content-related issues or the current status of the debate in relation to the participation regulations were also answered accordingly.

Parallel to the activities relating to the participation regulations, initial thoughts were given to the implementation of digital participation within the City of St. Gallen. In order to approach a possible technical solution, various brainstorming sessions were conducted by the Communications department and the Chief Digital Officer and mirrored with the core group. The aim of the brainstorming sessions and further discussions was to define the necessary functionalities of the electronic participation platform of the City of St. Gallen. Very quickly, the vague ideas for a participation platform turned into a corresponding specification sheet: the required functionalities were (1) implementation as a cloud service in the form of a SaaS solution, (2) implementation of simple surveys, (3) brainstorming, (4) map-based participation, (5) differentiation between formal and in-formal participation, etc. The platform should be implemented as a minimum viable product. Other necessary functionalities were to be developed in a demand-oriented and agile manner. The requirements for the technical implementation were subsequently defined by the participation core group and also by the city council in conjunction with a budget approval. The activities were also reported on as part of the reporting on the annual objectives, but also in response to parliamentary questions.

In the form of an invitation process, four different providers presented their products, which appeared to be qualified for implementation. After evaluating all the solutions presented, the collaboration with a start-up in the field of participation solutions took place. The collaboration with the company took place in an agile setting. From the administration's perspective, the collaboration was chosen at an ideal time: The young company had a functioning product with the basic functionalities as a SaaS solution on the market. The further development of the product was carried out in coordination with the (so far few) customers. The costs for enhancements were divided among the customers according to the "he who orders, pays" principle. If all customers benefited from extensions, the start-up contributed its own funds. In this way, the platform underwent significant further development with continuous improvements that benefited all customers. This setting is disadvantageous if highly individualized solutions are to be implemented for just one client. The consequences are high costs and possibly even the impossibility of certain implementations.

With the initiation and agile development by the CDO, the topic of digital participation was assigned to the Mayor's staff unit (Finance Directorate), while the actual topic of participation remained in the Social Affairs department (Social Affairs and Security Directorate). This circumstance led to major personnel challenges not so much during the planning and initiation phase, but rather during the operational phase, as digital participation was to be implemented in a budget-neutral manner without additional staff positions.

In addition to the purely technical implementation of the participation platform, the associated processes also had to be implemented. As a basis for the functioning of participation within the city administration, the systematic handling of the topic of participation at city level was to be defined. This includes establishing the necessary operational procedural steps within the administration. For certain projects, e.g. in the construction sector, the necessary internal processes must be adapted so that sufficient time and resources are provided for the implementation of participation projects at the project planning stage.

Communication to publicize the platform to the general public was also a key objective. The idea was also to make it possible to find all of the City of St. Gallen's investments in one place on the Internet.

4 Results of the Project

Subsequently, the results achieved will be presented - based on legal, technical, organizational, cultural and communicative aspects.

4.1 Legal: Results of the Participation Regulations

As a result, after a long debate, a new set of participation regulations (Stadt St. Gallen 2020c) was adopted by the city parliament on September 22, 2020, which came into force on January 1, 2021. In the associated debate in the city parliament, however, there was no majority in favor of the draft of the participation regulations prepared by the working group. The version adopted was a counter-proposal prepared by the parliamentary Social Affairs and Security Committee (KSSI). This took up the demands for greater participation more explicitly than the version that the actual working group and the City Council had advocated. However, there are no operational differences in the implementation of participation between the two variants.

In the city of St. Gallen, the new participation regulations provide numerous different opportunities for participation: These include: formal opportunities for participation, such as referendums, initiatives and petitions, as well as informal opportunities for participation for people without political voting rights, such as citizens' initiatives. Citizens' initiatives have particularly low quantitative hurdles for bringing issues into the political system: Only 15 signatories resident in St. Gallen are required and the minimum age is 13. The initiative is submitted to a political committee or parliament via the City Chancellery or passed on to the City Council for processing (Stadt St. Gallen 2020c).

4.2 *Technology: Introduction of the Digital Participation Platform*

At the technical level, St. Gallen mainly had offerings in the area of social media and a supplementary "city notifier" until 2019. With its help, the population can report infrastructure defects to the city (Stadt St. Gallen 2024b). In addition to the existing tools, a digital participation platform was developed in 2019 to gather the opinions and votes of various stakeholder groups and take them into account when implementing policy. In addition to the normal voting population, this primarily included the non-voting population in the city.

In order to define the relevant stakeholder groups in a city in more detail, a stakeholder map was defined in collaboration with other cities as part of another Smart City project. The roles of individual persons and institutions in the context of possible participation projects were also defined within the stakeholder framework. Stakeholders can be, for example, politicians, residents, association representatives or even tourists and administrative staff. Special attention should also be paid to hyperlocal groups in participation, such as certain interest groups in the neighborhood, neighborhood associations, which have an opinion-forming effect on the population and also on politics. Depending on the stakeholder, the degree of communication in which the actors are to be informed and involved also differs.

4.3 *Organisation and Processes: Adaptation in the Administration*

Numerous different (internal administrative) bodies are involved in the implementation of participations in St. Gallen: The City Council as the organizer of a participation, the specialist departments that provide the topic or subject of the participation, the Communications Unit and the Social Affairs Office, which take on the actors and the overall coordination of a participation, and the CDO, where the digital participation processes are implemented on the platform.

Due to the large number of stakeholders, there were still numerous challenges in the organizational context, especially at the beginning of participation within the administration, which were continuously improved over the following months and years. These included training employees in the basics of participation. The integration of time buffers for the preparation, implementation, evaluation and feedback of participation topics into the urban processes was also of great importance, especially in highly synchronized topic areas (e.g. in the construction sector).

4.4 Culture: Administrative Modernization

While there is a feeling that the top management in particular has demanded and promoted participation, the topic has mostly met with a mixed response and acceptance in the area of administration. Numerous discussions were held with the middle administrative level, which had to make processes and resources available for the implementation of participation in day-to-day operations and equally had to find a way of dealing with the results of the participations.

It became clear that participation must be perceived as a learning process by all sides. Communication, processes and self-image were not yet sufficiently in place in politics, administration and the population for this new form of cooperation.

The dissemination of participation in the administration remains a challenge. Even if administrative staff are required to act in a participatory manner by the respective administrative service, the decision as to where or when to act in a participatory manner often remains the responsibility of middle management. They are not always convinced of the need for more participation. Depending on the political will at top and middle management levels, there are therefore clear differences in the promotion of possible participative topics and projects.

Based on the "House of Participation" (Stadt St. Gallen 2021c), participation was frequently implemented in St. Gallen in the area of projects (e.g. road construction projects, building construction projects), occasionally also in the area of political participation (e.g. consultations) or in the form of informal participation (e.g. brainstorming).

4.5 Communication: Customer Centricity

Looking at the topic of participation as a whole, communication must be particularly emphasized as a fundamental issue for the actual success of participation. All essential information has been bundled on one website (Stadt St. Gallen 2024a). Participants are given good support so that they can achieve their goal of participating: There is a breakdown according to the intention of the participants. A distinction is made between "contributing an idea", "having a say in projects", "giving feedback", "concerns of children", "concerns of neighborhoods or associations" (Stadt St. Gallen 2024a). Based on the respective goal, the appropriate participation channel, the right "instrument" for participation is proposed.

In terms of communication, the city's very serious demeanor was temporarily abandoned in favor of a more relaxed approach. Drawn information sheets in print and online were used to achieve branding within the city in the area of participation.

Based on a "customer journey in the area of participation", it became clear that the initial identification of current relevant participations in St. Gallen in particular represents a pain point for the population. This was attempted to be solved by numerous communicative approaches. These included the use of traditional media coverage, writing directly to property owners in the event of construction projects within a defined perimeter, providing information via newsletters and websites, as well as providing physical information on planned projects in the districts.

If the participants are informed about a planned participation project, they can find out more about the topic or project and give their feedback on the digital participation platform page. This is possible for both in-formal and formal topics. A special feature of the implemented digital participation platform is that joint statements (e.g. from clubs or associations) can also be written. In addition, the roles of the individual participants can be managed in an integrated CRM.

Once a participation process had been completed, the relevant feedback was evaluated by the specialist department of the City of St. Gallen. The results are presented in a general participation report, together with feedback from the specialist department on how the suggestions received should be dealt with. The person submitting the feedback receives information that the corresponding participation report has been published on the website. A personalized response is also possible.

4.6 Side Effects

In addition to the planned developments in the field of digital participation described above, there were numerous unplanned events and developments that occurred during the launch phase of participation:

The so-called "open-air parliament" is particularly worth mentioning here. As part of an initiative by volunteers who want to strengthen democracy, the volunteers collected issues that were particularly important to the population in public and in well-frequented places and then passed these on to politicians and the city council.

Another topic that has an interface with participation was the handling of the topic of "child participation" as part of a municipal project. Where possible, synergies were used (e.g. as part of the participation regulations), but in some cases the requirements were too different to use the same digital solutions (e.g. participation app and participation solution).

One project that was also implemented during the participation period was the so-called Smarthalle. As part of this unique "#smarthalle" project, a vacant building was converted into a pop-up store. Over a period of three months, digital transformation activities were brought closer to the city's residents in the form of projects, workshops and exhibitions (Stadt St. Gallen 2019).

It was not so much a series of projects as a change in the "attitude" of the administration that became apparent with the implementation of its own "hackathons". As part of the hackathons, the administration now uses the "wisdom of the crowds" in a much more targeted way to find solutions to existing challenges at an earlier stage and in a more targeted manner.

5 Recommendations

Based on the project described, there are numerous recommendations for the successful implementation of the topic of participation within a city or municipality:

5.1 *Legal: Establish Legal Certainty*

Establishing the legal basis is essential for the legal certainty of participation and the acceptance of the results of participation by politicians, the administration and the population.

A fundamental distinction must be made between formal and informal participation. These factors are also relevant for the technology decision of the digital participation platform.

Against this backdrop, the implementation of the digital participation platform in conjunction with the formulation of new municipal participation regulations must be viewed very positively. The regulations not only provide the justification for participation in St. Gallen, but also drive the desire for more participation in the political process, as well as in specific projects.

5.2 *Technology: SaaS Solution Versus Individual Software*

In terms of technology, the market for solutions in the field of digital participation platforms is quite large both nationally and internationally. If you compare Switzerland with other countries, the market has grown considerably, especially in recent years. Once the content has been defined, the question of the desired tools and functionalities must be answered when deciding on the technology for the eParticipation solution.

It is therefore worth looking at the existing solutions from a technical and functional perspective. The solution from a Swiss start-up was used in St. Gallen (Konova 2024). In the meantime, other solution providers have also entered the Swiss market, some of which use open source software. These include, for example, the solution from decidim (Decidim 2024), which is used in the cities of Zurich (Stadt Zürich 2024) and Lucerne (Stadt Luzern 2023). The "Consul" solution is also widely used in Germany (Bundesministerium für Wohnen, Stadtentwicklung und Bauwesen 2024). Here, it is necessary to weigh up whether to give preference to a ready-made "all-in-one solution" as "software as a service" or a freely configurable solution, which, however, requires the involvement of other service providers. The choice of solution in St. Gallen also came at a time when the city of St. Gallen was increasingly using SaaS solutions instead of on-premise solutions based on its own cloud strategy.

5.3 *Organisation and Processes: Expectation Management*

In order to increase the number of responses to participations, the timing of participation and the openness of the entire participation process are critical points. The objectives of participation must also be defined more clearly by the clients (politicians). Abstract topics and future issues are difficult for the population to grasp. Topics that affect people themselves (e.g. residents in a street) increase participation. It is always a challenge to meet the expectations of all stakeholders and not always listen to the same people at physical participation events. Ideally, participants can be addressed in such a way that all relevant stakeholder groups feel represented. This requires appropriate stakeholder management.

Depending on the legal situation, the topic of participation is highly complex. There are numerous different legal provisions on participation and a distinction between formal and informal participation. In addition, no two processes are the same: the content differs each time (regardless of whether it is political or project participation). Each time there are different stakeholder groups. And in addition to the actual result of participation, participation as a process is essential - it is very often also about the feeling of being involved, the feeling of participation.

This also means that the basic process steps for participation must be adhered to: Information of those involved, participation of those involved, decision-making by those involved or by politicians. Participation following a decision that leaves no room for maneuver/discretion (placebo participation) is not recommended.

5.4 *Culture: Realization of Open Government*

With the constant will of the political leadership and the recurring implementation of various participation procedures, the acceptance of the implementation of participation projects in the administration has increased. Above all, the early demand for and planning of participation in the classic administrative processes significantly facilitates the acceptance of participation within the administration.

It became clear that the implementation of participation was gradually no longer perceived as "fundamental criticism of the administration" or even of a particular person, but rather as a need on the part of participation participants to share their thoughts and views with the city administration.

However, there is still a need for action in the acceptance and anticipation of participatory proposed solutions. Participants may be offended by the non-acceptance of proposals.

5.5 *Communication: Obligation of the Administration*

Elementary for the success of participation is the question of the same understanding of participation and the fundamental knowledge of basic concepts, such as the ladder of participation, in which the different qualities—in the sense of intensities—of participation are described.

The issue of communication is critical to the success of participation: The basis of participation is information, i.e. one-way communication (often from the state to the population). With participation, a feedback channel from the population to the administration and politicians is established, ideally also established. Communication must be linked to clear expectation management: the stakeholder group involved must know how open the participation process is and how much influence on decisions is possible in order to avoid "placebo" participation.

Information about the participation itself has also proven to be critical. Stakeholders must be brought "on board" and kept informed about the entire process of a project. The appropriate feedback to the population and the groups involved must also be taken into account. Communication in relation to participation (and the participation formats themselves) must be kept digital and analog, in the form of "blended participation". Newsletters, e.g. for the individual neighborhoods, can draw the attention of people with an affinity for participation to participation. Information boards physically set up in the neighborhoods can also provide information on participation activities.

6 Conclusion and Discussion

Looking at the topic of participation in the Smart City St. Gallen as a whole, a lot of time has passed between the first participation regulations of the City of St. Gallen and the reformulation of the participation regulations and the introduction of the digital participation platform. Even before the new participation regulations in St. Gallen, more participation would theoretically have been possible under the old regulations than was actually practiced. With the new regulations, however, the political demand and also the "management attention" for more participation has increased significantly. Participation is now equally demanded by the legislative and executive branches. The administration can now not only involve the population, it is "urged" to do so. The administration, including the city council, now has a greater obligation to apply participatory procedures.

Nevertheless, there are still questions today: In which topics should participation take place? Who should be involved? At what point does participation take place (in relation to the phase in the policy cycle)? What are the expectations of those involved in participation? In addition to these questions to be answered by politicians, the concrete application and implementation of participation within the administration remains a challenge. Fitting the necessary time and financial freedom into the existing

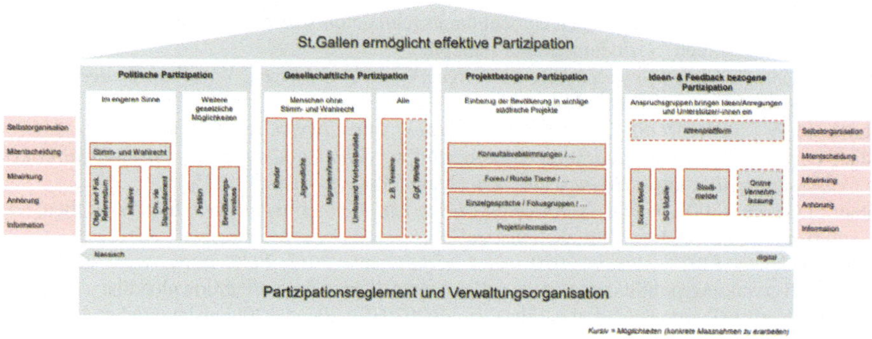

Fig. 1 House of participation, city of St. Gallen

"old" administrative process is not a matter of course. In addition to the legal and technical foundations, it is above all the cultural changes that have not only reached the administration, but also politicians and the population. This is often independent of political color.

It is striking that participation on the digital platform is almost exclusively in the form of consultations on construction issues. An expansion to other subject areas or other formats, e.g. in the direction of "ideathons", would be desirable. However, with the continuation of hackathons or highly topic-specific participation forums (e.g. energy concept, focus groups), the City of St. Gallen uses two good complementary formats for the participation of the population.

With regard to the technical component of participation, further technical developments in the digital participation solutions used are certainly conceivable in the future: These include, for example, the implementation and use of electronic identification solutions (E-ID). This would make it possible to implement more targeted participation (e.g. controlled for a specific geographical perimeter). Already planned for a long time, but not yet implemented due to the restrictions of IT service providers, is the direct input of projects by city residents, as implemented in Zurich or Lucerne.

References

Bundesministerium für Wohnen, Stadtentwicklung und Bauwesen. (2024, März 21). *Digitale Beteiligung einfacher machen | Smart City Dialog*. https://www.smart-city-dialog.de/blogs/digitale-beteiligung-einfacher-machen

Cohen, B. (2014, November 20). *The Smartest Cities In The World 2015: Methodology*. Fast Company. https://www.fastcompany.com/3038818/the-smartest-cities-in-the-world-2015-methodology

Decidim. (2024). *Decidim*. https://decidim.org/

Fachstelle für Statistik Kanton St.Gallen. (2023). *Wohnbevölkerung in St.Gallen*. https://stada2.sg.ch:8443/gpsg/GPSG?type=EXPORT&indikatoren=1&raumtyp=1&export=XLS

Fachstelle für Statistik Kanton St.Gallen. (2024, Mai 30). *Stimmbeteiligung nach Alter, Geschlecht und Quartiergruppe, Stadt St.Gallen, Abstimmung und Wahl vom 14.*

April 2024. https://www.sg.ch/content/dam/sgch/kanton-stgallen/statistik/b17/B17_AB_03_Beteiligung-aktuell_Soziodemografie-Stadt.pdf

Konova. (2024). *Die effiziente Lösung für zeitgemässe Mitwirkungen, Vernehmlassungen und E-Partizipation*. https://e-mitwirkung.ch/

Reinhold, A. (2024). *MOBILE. das Freiluftparlament*. https://www.freiluftparlament.ch/

Stadt Luzern. (2023, Juni 1). *Damit «Luzern glänzt»: Schäden können neu auf dialogluzern.ch gemeldet werden*. https://www.stadtluzern.ch/aktuelles/newslist/1871335

Stadt St.Gallen. (2005, Januar 1). *Anhang 1: Übersicht Finanzkompetenzen*. https://st.gallen.tlex.ch/frontend/annex_document_dictionaries/710

Stadt St.Gallen. (2017). *Vision 2030 und Legislaturziele 2017–2020*. https://www.stadt.sg.ch/home/verwaltung-politik/demokratie-politik/stadtrat/_jcr_content/Par/downloadlist_copy/DownloadListPar/download_0.ocFile/Vision%202030%20und%20Legislaturziele%202017-2020.pdf

Stadt St.Gallen. (2019). *#smarthalle—Die Smarte Stadt im Schaufenster*. https://m.stadt.sg.ch/appl/download.php?objid=1402375503&id=1305070225&nr=1&filename=StadtSG%20Smarthalle%20FlyerA3%20def%20Druck.pdf&usid=572

Stadt St.Gallen. (2020a). *Neuordnung der Partizipation von Bevölkerungsgruppen in der Stadt St.Gallen sowie Erlass eines neuen Partizipationsreglements. Vorlage Stadtparlament*. https://ftp-sg.oca.ch/stadtparlament/cae94e58d77c4bceb211d526b25d592e-332.pdf

Stadt St.Gallen. (2020b). *Smart St.Gallen—Der Mensch als Mittelpunkt der „Smarten Stadt". Strategie*. https://www.stadt.sg.ch/home/wirtschaft-wohnen/smarte-stadt/_jcr_content/Par/stsg_accordion_list/AccordionListPar/stsg_accordion_copy/AccordionPar/stsg_downloadlist_co/DownloadListPar/stsg_download.ocFile/200114%20Smarte%20Stadt%20Strategie%20St.Gallen.pdf

Stadt St.Gallen. (2020c, September 22). *SRS 141.1—Partizipationsreglement—Rechtssammlung der Stadt St.Gallen—Erlass-Sammlung*. https://st.gallen.tlex.ch/app/de/texts_of_law/141.1

Stadt St.Gallen. (2021a). *Open Data Hack St.Gallen 2021*. Open Data Hack St.Gallen. https://opendata-hack-stgallen.devpost.com/

Stadt St.Gallen. (2021b). *Vision 2030 und Legislaturziele 2021–2024*. https://www.stadt.sg.ch/news/stsg_medienmitteilungen/2021/05/stadtrat-praesentiert-legislaturziele-2021-2024/_jcr_content/Par/stsg_downloadlist/DownloadListPar/stsg_download_2083682462.ocFile/StGallen%20Vision-Legistlaturziele%202021-24.pdf

Stadt St.Gallen. (2021c, Januar 5). *Haus der Partizipation St.Gallen*. https://www.stadt.sg.ch/news/stsg_medienmitteilungen/2021/01/neues-partizipationsreglement-ist-in-kraft/_jcr_content/Par/stsg_downloadlist/DownloadListPar/stsg_download.ocFile/210125%20Partizipation%20Haus%20Visualisierung.pdf

Stadt St.Gallen. (2022a). *Mitwirkungsbericht Primarschule Rotmonten.pdf*. https://www2.stadt.sg.ch/od/01%20Open%20Data%20Files/Stadt%20St.Gallen/DIF/OE/Partizipation/Berichte/Mitwirkungsbericht%20Primarschule%20Rotmonten.pdf

Stadt St.Gallen. (2022b). *Open Data Hack St.Gallen 2022*. Open Data Hack St.Gallen 2022. https://opendatahack-stgallen.devpost.com/

Stadt St.Gallen. (2023a). *Gemeinsam wirkt—St.Gallen wird klimaneutral*. stadt.sg.ch. https://www.stadt.sg.ch/home/raum-umwelt/energie/klimaneutrale-stadt.html

Stadt St.Gallen. (2023b, Dezember 1). *Open Data Hack St.Gallen 2023*. Opendata.ch. https://opendata.ch/de/events/open-data-hack-st-gallen/

Stadt St.Gallen. (2024a). *Mitwirkung und Partizipation*. stadt.sg.ch. https://www.stadt.sg.ch/home/verwaltung-politik/demokratie-politik/partizipation.html

Stadt St.Gallen. (2024b). *Stadtmelder | stadt.sg.ch*. https://www.stadt.sg.ch/home/verwaltung-politik/demokratie-politik/partizipation/stadtmelder.html

Stadt Zürich. (2024). *Auf welcher Plattform befinde ich mich? - «Mitwirken an Zürichs Zukunft»*. https://mitwirken.stadt-zuerich.ch/pages/decidim?format=html&locale=de

Latvia –Smart Kommunikationtechnologie Für Smart Cities

Christian Schachtner

1 Introduction

In recent years, Latvia has become a major centre for information and communication technologies (ICT) in Northern Europe. In this respect, a conversation with (the Latvian-type institution) was a hook for writing this country study. In the field of telecommunications technology, cloud computing, and artificial intelligence (AI), the country is showing remarkable progress and innovation (Latvian Investment and Development Agency, 2025).

The article provides an overview of the relevant investment fields of the Latvian government and international initiatives by players with a focus on the digitization of public services. The following descriptions of the state of Latvia are described in order to provide an initial overview of successes, challenges and future goals.

2 The Status Quo of Digital Maturity

The use of telecommunications technology is a key factor in the development and operation of basic digital applications. Latvia's telecommunications infrastructure is one of the most modern in Europe. With nationwide 5G coverage, Latvia offers ideal conditions for the development and implementation of advanced digital solutions (European Commission 2023). The high internet speed and availability have contributed to Latvia becoming an attractive location for international technology companies (OECD 2024).

Cloud computing is an essential future technology for networked working. In the field of cloud computing, Latvia has made significant investments in recent years. The

C. Schachtner (✉)
RheinMain University of Applied Sciences, Wiesbaden, Germany
e-mail: Christian.schachtner@hs-rm.de

C. Schachtner (ed.), *The European Smart City Movement*,
https://doi.org/10.1007/978-3-031-96700-9_10

government has implemented a “cloud-first” strategy that promotes the use of cloud technologies in the public sector while encouraging private companies to develop and use cloud-based solutions (Ministry of Environmental Protection and Regional Development of Latvia 2023).

This initiative has led to an increase in efficiency and innovation capacity in various sectors of the economy. Another technology that is indispensable today is the development of artificial intelligence models. Latvia has positioned itself as a pioneer in the development and application of AI technologies. The country has adopted a national AI strategy that aims to promote research and development in the field of AI and support its use in various fields such as healthcare, education, and public administration (Cabinet of Ministers of the Republic of Latvia 2022).

Companies and research institutions are working on innovative AI solutions ranging from speech processing to predictive analytics. One area of application for these services is the data-based development of public space. This nationwide or regional hub development promotes the Latvian Smart City Initiative. The combination of advanced telecommunications technology, cloud computing, and AI has paved the way for innovative smart city solutions in Latvia. The capital Riga has initiated several pilot projects aimed at improving the quality of life of citizens and making urban infrastructure more efficient (Riga City Council 2024).

These projects include intelligent traffic management systems, energy-efficient building control and digital citizen services. Despite the impressive vision, Latvia faces some challenges. These include the skills shortage in the ICT sector and the need to bridge the digital divide in society (World Bank 2023). However, the Latvian government has taken measures to address these challenges, including investments in digital education and support programs for ICT startups. The future prospects for the ICT sector in Latvia are promising. The country is positioning itself as an ideal location for technology companies looking to benefit from its advanced digital infrastructure and innovation-friendly environment. With continuous investment in R&D and the promotion of industry-academia partnerships, Latvia strives to further consolidate its position as a leading ICT hub in the region (Baltic Institute of Advanced Technology 2024).

Robotics, big data and the Internet of Things (IoT) are playing an increasingly important role in the development of smart cities and information and communication technology (ICT) in Latvia. Latvia has made significant progress in the field of robotics in recent years. Especially in Industry 4.0, robots are increasingly used to optimize production processes and increase efficiency (Ozolina-Ozola 2020). Automation through robotics enables Latvian companies to operate more competitively in the global marketplace while improving working conditions for employees.

Latvia has also made significant developments in the field of big data. The ability to collect, analyze, and leverage large amounts of data is critical for modern cities and businesses. Latvian companies and research institutions are working on innovative data analytics solutions that are used in various sectors such as healthcare, transport, and public administration (Berzins et al. 2022).

The Internet of Things (IoT) is another key area where Latvia is making progress. The networking of everyday objects and industrial systems via the Internet opens up new possibilities for increasing efficiency and innovative services. In Riga, the capital of Latvia, IoT solutions are already being used to optimize urban infrastructure, for example through smart lighting systems and connected traffic control (Klavins and Zariņš 2021).

The combination of these technologies—robotics, big data and IoT—forms the backbone for the development of smart cities in Latvia. Riga has set itself the goal of becoming one of the leading smart cities in Europe. To this end, innovative ICT solutions are being implemented to improve the quality of life of citizens while increasing resource efficiency (Riga City Council, 2023). An important aspect in the development of these technologies is data security and privacy. Latvian experts are working intensively on solutions to ensure the security of connected systems and increase citizens' trust in smart city technologies (Kalnina and Vitols 2024).

The Latvian government supports the development of these future technologies through targeted funding programmes and investments in research and development. The aim is to position Latvia as an attractive location for ICT companies and start-ups and to drive the country's digital transformation (Ministry of Economics of the Republic of Latvia 2024).

3 Technologies in Relation to Smart City Focus Fields

Information and communication technologies play a crucial role in the development of smart cities, especially in the areas of smart energy, smart mobility and smart health. Latvia is thus positioning itself as an innovative location for future technologies and offers attractive opportunities for investors and companies (Invest in Latvia 2025).

In the field of smart energy, Latvia relies on smart grids and renewable energies. The integration of IoT technologies enables more efficient energy distribution and use (European Commission 2024). By using smart meters, consumers can monitor and optimize their energy consumption in real-time, resulting in a reduction in overall energy consumption (Latvian Ministry of Economics 2023).

The development of smart mobility solutions is another focus in Latvia's Smart City strategy. Innovative traffic management systems and the promotion of electric mobility contribute to improving urban mobility (Riga City Council 2024). Car-sharing platforms and intelligent parking guidance systems optimize the use of public space and reduce traffic in city centers (Baltic Times 2023).

In the health sector, Latvia is focusing on the digitization and networking of health services. Telemedicine and e-health solutions improve access to medical care, especially in rural areas (Latvian Ministry of Health 2024). The implementation of electronic health records and AI-supported diagnostic systems increases the efficiency and quality of healthcare (European Health Forum Gastein 2023).

The integration of these smart technologies into Latvia's cities not only creates an improved quality of life for citizens, but also opens up new business opportunities for companies in the ICT sector. The Latvian government is actively promoting investment in these areas through various incentive programs and advanced digital infrastructure (Latvian Investment and Development Agency 2025).

4 Conclusion

In conclusion, Latvia has made significant progress in the field of telecommunications technology, cloud computing, and AI. The integration of these technologies into smart city concepts shows the potential for innovative solutions that can improve citizens' lives and promote economic development. Latvia's investments in robotics, big data and IoT show promising progress and form the basis for the development of smart cities and make a significant contribution to the country's modernization and digitalization. The combination of government support, technological expertise and a growing startup scene makes the country an attractive location for investment and development in these forward-looking technology areas. With its forward-looking strategy and commitment to digital innovation, Latvia is well positioned to play a leading role in Europe's digital economy.

References

Baltic Institute of Advanced Technology. (2024). *Future of ICT in Latvia: Trends and opportunities*. Riga: BIAT Press.

Baltic Times. (2023). Latvia's smart mobility initiatives gain momentum. The Baltic Times.

Berzins, A., Liepa, E., & Ozols, J. (2022). Big Data applications in Latvian smart city initiatives. Journal of Baltic Studies, 53(2), 178-195.

Cabinet of Ministers of the Republic of Latvia. (2022). *National artificial intelligence strategy 2022-2027*. Riga: Government Publications.

European Commission. (2023). *Digital economy and society index (DESI) 2023 - Latvia*. Brussels: Publications Office of the European Union.

European Commission. (2024). Smart grids and meters in Latvia. Digital Single Market.

European Health Forum Gastein. (2023). E-health innovations in the Baltic states. EHFG Conference Proceedings.

Invest in Latvia. (2025). Information and communication technologies - Smart City. https://investinlatvia.de/industry/informations-und-kommunikationstechnologien-smart-city/

Kalnina, L., & Vitols, G. (2024). Cybersecurity challenges in IoT-driven smart cities: A case study of Riga. In Proceedings of the International Conference on Smart Technologies (pp. 87–102). IEEE.

Klavins, M., & Zariņš, A. (2021). Implementation of IoT solutions for urban infrastructure management in Riga. Smart Cities, 4(3), 412-428.

Latvian Investment and Development Agency. (2025). ICT sector in Latvia: Opportunities and incentives. LIAA Publications.

Latvian Ministry of Economics. (2023). Smart energy solutions for sustainable development. Ministry Reports.
Latvian Ministry of Health. (2024). Digital health strategy 2024–2030. Government Publications.
Ministry of Economics of the Republic of Latvia. (2024). Digital transformation strategy 2021-2027: Progress report and future outlook. Riga: Government Publication Office.
Ministry of Environmental Protection and Regional Development of Latvia. (2023). *Cloud computing adoption in public sector: Progress report*. Riga: Government Publications.
OECD. (2024). *OECD digital economy outlook 2024*. Paris: OECD Publishing.
Ozolina-Ozola, I. (2020). The impact of robotics on Latvia's manufacturing sector. Baltic Journal of Management, 15(1), 65-83.
Riga City Council. (2023). Riga Smart City Strategy 2030. Retrieved from https://www.riga.lv/en/smart-city-strategy
Riga City Council. (2024). *Smart Riga: Transforming the capital through technology*. Riga: City Council Press.
World Bank. (2023). *Digital skills and labor market challenges in emerging economies: A case study of Latvia*. Washington, D.C.: World Bank Group.

Irland—Smart Cities Forum

Christian Schachtner

1 Introduction

A key feature of a smart city is the provision of better, more efficient and more accessible public services to citizens. By using modern technologies and data-driven approaches, cities can optimize their services and better tailor them to the needs of their residents (Brown et al. 2024). This can include, for example, the implementation of digital citizen portals, intelligent transport systems or automated administrative processes.

Another essential aspect of a smart city is the ability to develop new and innovative solutions to existing and future challenges that affect citizens, businesses and visitors. This often requires interdisciplinary approaches and the integration of different technologies such as the Internet of Things (IoT), artificial intelligence or big data analytics (Garcia and Martinez 2023). Examples of this can be intelligent energy management systems, predictive maintenance of infrastructure or data-driven urban planning.

A smart city is characterized by the creation of effective internal and external collaborations and partnerships. This includes cooperation between local authorities, other public institutions, universities, external agencies, businesses, citizens and international partners. Through this networking, resources can be pooled, experiences exchanged and joint solutions developed (Thompson and Lee 2024).

The benefits of a smart city are many, ranging from validating new technologies to putting research results into practice and improving the quality of life of citizens. Smart cities act as reference locations for testing and validating new technologies in an urban context. This makes it possible to test and optimize innovative approaches in real-world conditions before implementing them on a large scale (Roberts and Kim 2023).

C. Schachtner (✉)
RheinMain University of Applied Sciences, Wiesbaden, Germany
e-mail: Christian.schachtner@hs-rm.de

C. Schachtner (ed.), *The European Smart City Movement*,
https://doi.org/10.1007/978-3-031-96700-9_11

Another important aspect is the transfer of research results into reality. Smart cities provide a platform where theoretical concepts and research results can be translated into practical applications. This not only promotes innovation, but also accelerates the transfer of knowledge between science and practice (Chen et al. 2024).

By creating a collaborative framework to solve urban challenges, smart cities can respond more effectively to complex problems. This collaborative approach makes it possible to pool different perspectives and expertise and develop holistic solutions (White and Green 2023).

Increasing the efficiency of urban services is another key advantage of smart cities. By using technologies such as IoT and data-driven decision-making processes, resources can be used more optimally and processes can be improved. This not only leads to cost savings, but also to a higher quality of the services offered (Johnson and Brown 2024).

A particularly important aspect is the increased involvement of citizens and service users. Smart cities use digital platforms and technologies to improve communication between administration and citizens and promote participation. This can lead to higher citizen satisfaction and greater identification with their city (Miller and Davis 2023). Ultimately, the concept of the smart city aims to improve the quality of life of residents. By integrating smart solutions in areas such as mobility, energy, environment and security, smart cities can create a more sustainable and livable environment for their citizens (Taylor and Wilson 2024).

2 The Irish Smart City-Movement

The Smart City initiative in Ireland promises numerous benefits for cities and their residents. As project manager, I would like to explain the essential aspects of this innovative development and emphasize its importance for the urban future.

Smart cities in Ireland will be used as a reference location for the validation of smart city technologies. This makes it possible to test and optimize new solutions under real conditions. Through practical application, potential challenges can be identified and overcome at an early stage, which increases the efficiency and effectiveness of the implemented technologies (Angelidou 2014).

Another important vision point in Ireland is the creation of a cooperation framework to solve urban challenges is another key benefit[1]. Smart cities foster collaboration between different actors, including city administrations, businesses, research institutions, and citizens. This integrative approach makes it possible to look at complex urban problems from different perspectives and to develop holistic solutions (Caragliu et al. 2011).

Providing more efficient urban services is another core aspect of smart city projects there. By using innovative technologies and data-driven decision-making processes, cities can make better use of their resources and tailor services to meet their needs. This not only leads to cost savings, but also to an improved quality of life for citizens (Zanella et al. 2014).

A particularly important aspect is the increased involvement of citizens and service users. Smart cities use digital platforms and technologies to improve communication between city administration and citizens. This promotes transparency, allows for more direct citizen participation in decision-making processes, and increases trust in local governance (Cardullo and Kitchin 2019).

Finally, the implementation of smart city concepts in Ireland aims to improve the quality of life. By integrating smart technologies into areas such as transportation, energy, waste management, and public safety, urban operations are streamlined and made more efficient. This results in a cleaner environment, improved mobility solutions, and an overall livable city (Yigitcanlar et al. 2018).

3 The Project „Ireland Smart Cities Forum"

The All Ireland Smart Cities Forum plays a crucial role in realising these benefits. As a community of practitioners focused on the advancement of cities in both the Republic of Ireland and Northern Ireland, the Forum promotes knowledge sharing and collaboration between cities. This cross-border initiative, involving representatives from cities such as Cork, Dublin, Limerick, Galway, Waterford, Belfast, Derry and Newry, makes it possible to explore and address common challenges in the implementation of smart city policies and projects (All Ireland Smart Cities Forum, 2025). The support from Maynooth University highlights the importance of academic expertise in this process. Dr. Niall Connolly and Caroline Creamer from Maynooth University bring valuable expertise to the forum and strengthen the link between research and practice.

The All Ireland Smart Cities Forum is also a major initiative to promote smart urban development in Ireland. As a community of practice, the forum focuses on the advancement of cities in both the Republic of Ireland and Northern Ireland through the use and value creation of smart city programs. The forum is made up of city officials who have extensive knowledge and experience related to the evolving smart city agenda. This expertise enables them to identify and capitalize on the opportunities for future growth and investment. The primary objective of the Forum is to create a collaborative platform for the promotion of smart city agendas across all partner cities (Cork City Council 2024).

A key aspect of the forum is to address common challenges in the implementation of smart city policies and projects. By sharing experiences and best practices, participating cities can learn from each other and develop innovative solutions to urban problems. The Forum defines a smart city as a place that develops better services for citizens, promotes economic activity, drives innovative solutions, and increases collaboration and engagement between different actors. This holistic view highlights the importance of networking and cooperation for the success of smart city initiatives (Fingal County Council 2024.

The concrete goals of the forum include establishing cities as reference locations for the validation of smart city technologies, facilitating the transition from research to reality, and creating a collaborative framework to solve urban challenges. In addition, the Forum aims to deliver more efficient urban services, increase engagement with citizens and service users, and ultimately improve the quality of life in cities.

The Forum also attaches great importance to improving economic activity in cities. It supports regional economic development by helping to create an ecosystem that attracts and provides opportunities for entrepreneurs, investors, and businesses. This economic dimension underlines the importance of smart city initiatives for the sustainable development of urban areas.

4 Conclusion

In summary, the Smart City Initiative in Ireland represents a promising model for future urban development. By leveraging innovative technologies, fostering collaboration, and focusing on citizen-centric solutions, smart cities lay the foundation for more sustainable, efficient, and livable urban spaces. The benefits range from improving urban services to promoting economic activity and improving the quality of life of citizens.

With the All Ireland Smart Cities Forum, Ireland has created a platform that enables the exchange of best practices and co-tackling challenges, paving the way for a smarter and more connected future for Irish cities. It combines technological innovations with social and economic aspects to make cities more efficient, sustainable and livable. However, the implementation of this concept requires close cooperation between different actors and continuous adaptation to changing needs and technological possibilities. The forum thus represents an important platform to jointly address the challenges of urbanisation and digitalisation and to develop sustainable solutions for Irish cities.

References

Angelidou, M. (2014). Smart city policies: A spatial approach. Cities, 41, S3-S11.

Brown, A., Smith, C., & Jones, E. (2024). Smart city technologies and their impact on urban services. *Journal of Urban Innovation*, 15(2), 123–145.

Caragliu, A., Del Bo, C., & Nijkamp, P. (2011). Smart cities in Europe. Journal of Urban Technology, 18(2), 65-82.

Cardullo, P., & Kitchin, R. (2019). Being a 'citizen' in the smart city: Up and down the scaffold of smart citizen participation in Dublin, Ireland. GeoJournal, 84(1), 1-13.

Chen, H., Wang, L., & Zhang, Y. (2024). From research to reality: Implementing smart city concepts. *Smart Cities Review*, 8(1), 45–62.

Cork City Council. (2024). All Ireland Smart Cities Forum. https://www.corkcity.ie/en/ng-business-in-cork/research-and-development/smart-city/smart-city-networks/all-ireland-smart-cities-forum/

Fingal County Council. (2024). All Ireland Smart Cities Forum. https://www.fingal.ie/all-ireland-smart-cities-forum

Garcia, R., & Martinez, S. (2023). Innovative solutions in smart urban environments. *International Journal of Urban Technology*, 28(3), 301–320.
Johnson, K., & Brown, M. (2024). Efficiency gains through smart city initiatives. *Urban Management Quarterly*, 19(4), 210–228.
Miller, P., & Davis, R. (2023). Citizen engagement in smart cities: A participatory approach. *Journal of Digital Governance*, 12(2), 178–195.
Roberts, J., & Kim, S. (2023). Smart cities as living laboratories: Testing urban innovations. *Urban Studies and Technology*, 17(3), 289–310.
Smith, A., & Johnson, B. (2023). Defining the smart city: A comprehensive approach. *Journal of Urban Planning and Development*, 149(1), 78–95.
Taylor, E., & Wilson, F. (2024). Quality of life improvements in smart urban environments. *Sustainable Cities and Society*, 70, 103–120.
Thompson, R., & Lee, S. (2024). Collaborative networks in smart city development. *Urban Governance Journal*, 22(1), 55–72.
White, D., & Green, M. (2023). Solving urban challenges through smart city frameworks. *City Planning Review*, 41(4), 401–418.
Wilson, C., & Taylor, D. (2022). Economic ecosystems in smart cities: Attracting investment and talent. *Journal of Urban Economics*, 130, 103–121.
Yigitcanlar, T., Kamruzzaman, M., Buys, L., Ioppolo, G., Sabatini-Marques, J., da Costa, E. M., & Yun, J. J. (2018). Understanding 'smart cities': Intertwining development drivers with desired outcomes in a multidimensional framework. Cities, 81, 145-160.
Zanella, A., Bui, N., Castellani, A., Vangelista, L., & Zorzi, M. (2014). Internet of things for smart cities. IEEE Internet of Things Journal, 1(1), 22-32.

Smart Cities in Switzerland: Theory and Practice

Brian Fabrègue

1 Introduction

The Swiss understanding of a smart city (SC) can be summed up as an urban area that employs information and communication technologies (ICT) to enhance the quality of life for its citizens, optimize the efficiency of city services and operations, and promote sustainability (Musiolik et al. 2020). The general goal of a smart city is to create a more liveable, efficient (emphasis on the efficient), and sustainable environment for its residents. The ultimate aim is to make cities and municipalities more liveable, sustainable, and attractive. This is not by itself a substantially different approach from other scenarios (Ahmad et al. 2022).

Switzerland is recognized globally for its high standard of living, strong economy, and commitment to sustainability. As urbanization continues to increase, Swiss cities face challenges such as population growth, environmental pressures, and the need for efficient resource management. Two-thirds of Switzerland's population resides in urban areas, and this percentage continues to grow, topping the charts years over years.

The Smart City concept, with its comprehensive approach to creating innovative, resource-efficient, and liveable cities, has gained traction in Switzerland in the past 10 years. In Switzerland, the choice of smart urbanisation exceeds often the city itself, implication larger but still local authorities. For instance, all smart cities in the Canton Ticino are managed by a common program, for large scale tool and material acquisition. A Swiss Smart City integrates various fields, bringing together different stakeholders, involving residents in decision-making processes, and positioning itself as a catalyst for urban innovation.

B. Fabrègue (✉)
Faculty of Law, University of Zurich, Zurich, Switzerland
e-mail: brian@qienda.com

C. Schachtner (ed.), *The European Smart City Movement*,
https://doi.org/10.1007/978-3-031-96700-9_12

Smart city activities are on the rise in Swiss cities, and very fast, driven by a combination of technological advancements and strategic initiatives by local governments. Swiss cities are increasingly investing in smart infrastructure and technology to improve urban living conditions and ensure sustainable growth. Most large cities (including all the top 10 urban centres) have initiated projects and developed strategies to transition towards becoming Smart Cities. Some many medium-sized cities face significant challenges in following this path due to limited human and financial resources, and often, opportunity costs.

Swiss smart cities are recognized for their excellence and leadership in urban innovation, as evidenced by their excellence in the Smart City Index Report by the International Institute for Management Development (IMD) (IMD 2024). Zurich has led the ranking for the past several years, securing the first position globally, showcasing its comprehensive and effective implementation of smart technologies and policies. Geneva follows closely in fourth place and Lausanne ranks at seventh, demonstrating its commitment to smart city initiatives and sustainable urban development.

2 The Concept of Smart City in Switzerland

The concept of a smart city has arisen as a modern method to tackling intricate urban development difficulties. Globally, there is a noticeable trend among SCs to move away from plans that prioritize infrastructure and instead prioritize improving the quality of life for citizens and promoting sustainability (Myeong et al. 2022). The objective of transforming urban systems into smart systems is to enhance the quality of life, living conditions, environment, economy, transportation, and government. The importance is stressed that innovation should not be undertaken solely for the sake of it, but rather should be closely linked to and attentive to the needs of the community. Zhu et al. define a smart city by the presence of accessible and secure information and ICT infrastructures, dependable physical infrastructures, a thriving economy, an inclusive society, a sustainable environment, and transparent governance (Zhu et al. 2022). Research suggests that the fundamental basis of intelligent cities is the combination of human resources, social resources, and information and communication technology infrastructure (Capdevila/Zarlenga 2015). This combination works together to promote economic growth and improve the overall quality of life (Wang/Zhou 2023). Therefore, concepts such as human-centric cities, friendly urban spaces and participative smart citizens are commonly employed (Portal/Fabrègue 2022). The primary goal of smart cities, as widely debated, is to enhance the quality of life, liveability, and welfare of its citizens. Nevertheless, this is often an overlooked aspect of SC development by policymakers, while still presenting issues to be concretely measured (Fabrègue/Bogoni 2024). It has been noted that smart cities have sometimes fallen short of their goals due to neglecting the needs of their citizens (Dameri/Rosenthal-Sabroux 2014). Additionally, a systematic review of smart cities by Myeong et al. indicated that studies focusing on human aspects were the least prioritized, accounting for only 8.1% of the research Myeong et al.

2022). As we will see in Sect. 3, this is not the case of Switzerland, where human aspects are a key component of the municipalities' projects.

2.1 The Background of the Swiss Smart City Concept

In Switzerland, the concept of smart cities sparked from the focus on energy efficiency in an urban context. The "Smart City Switzerland" ("Smart City Schweiz") initiative (EnergieSchweiz 2024a, b), for instance, launched in 2012, represents a progressive and visionary approach for Swiss cities and municipalities with advanced energy policies aspiring to transform into smart cities. This initiative has since become a vital element of the "SwissEnergy for Municipalities" ("EnergieSchweiz für Gemeinden") program, building upon the foundational principles and experiences derived from the "EnergyCity" process. These were distinguished Swiss municipalities recognized with the "EnergyCity" label for their exemplary achievements in municipal energy and environmental policy. These cities prioritize the efficient use of resources, the promotion of renewable energies, and environmentally friendly mobility, positioning them as role models (EnergieSchweiz 2024a, b).

The "EnergyCity" ("Energiestadt") programme (Trägerverein Energiestadt 2024a, b), established in 1991, has been a fundamental initiative in promoting sustainable energy and climate policies at the local level in Switzerland. Known for its robust management system, the programme offers a way to assess and compare the energy efforts of municipalities, awarding the "EnergyCity" label to exemplary cities for nearly 30 years. Initiated in 1988 by WWF and the Swiss Energy Foundation (SES), the project aimed to empower municipalities and cities to actively pursue sustainable energy policies. In 1991, Schaffhausen became the first officially recognized energy city, and the "EnergyCity Association" ("Trägerverein Energiestadt") was established, responsible for awarding the "EnergyCity" label and ensuring quality assurance.

Concurrently, the federal "SwissEnergy 2000" ("EnergieSchweiz 2000") program was launched, incorporating the "EnergyCity" project. Key political frameworks introduced during this period significantly contributed to this rapid growth, including the launch of the Energy2000 investment program, the passing of the Energy Act by Parliament, Switzerland's signing of the Kyoto Climate Protection Protocol (1998), and the enactment of the CO_2 Act (1999). Since then, the "SwissEnergy for Municipalities" program has provided extensive support to cities and municipalities, offering information, campaigns, experience exchange, advisory services, and assistance from "EnergyCity" consultants (EnergieSchweiz 2024a, b). Notable milestones include Regensdorf as the 300th in 2012 and Camorino as the 400th in 2016. As of July 2024, the Association has around 650 member municipalities, all committed to energy efficiency, has assigned the "EnergyCity" label to 480 cities and 99 of these have the golden "EnergyCity" rank. Moreover, the Swiss "EnergyCity" label prompted the development of the European Energy Award (EEA) (Trägerverein Energiestadt

2024a, b), which oversees a quality management and certification system for municipalities. This Swiss model has been adopted internationally, promoting high standards for sustainable energy practices across Europe.

As mentioned before, the thematic areas of the "EnergyCity" process were largely adopted and further developed to align with smart city requirements. This holistic approach mandates that smart projects be cross-thematic and cross-departmental, involving various stakeholders. The smart city concept emphasizes the increased networking of individual subject areas, such as urban energy master planning, smart buildings, smart grids and energy supply, intelligent mobility, municipal management, and stakeholder processes.

The "Smart City Switzerland" concept serves as a comprehensive guide for cities and municipalities, facilitating their transition into smart cities by providing an overview of steps, methods, strategies, and options. This concept underscores the importance of involving all stakeholders in the development process to ensure solutions meet genuine needs and are widely accepted. Through this initiative, Switzerland aims to create more liveable, sustainable, and attractive urban environments, setting a benchmark for smart city development globally.

The blending that results from a prioritisation on energy, upon which the rest of the Swiss Smart City is built, is evident a few definitions. P. Richner (Swiss Federal Laboratories for Materials Testing and Research) proposed a definition of smart city as part of the Smart City programme "SwissEnergy for Municipalities": "*A smart city offers its residents maximum quality of life with minimum resource consumption thanks to the intelligent linking of infrastructure systems (transport, energy, communication,* etc.) *at different hierarchical levels (building, neighbourhood, city). In this context, 'intelligent' is not automatically synonymous with 'IT'. With similar performance, passive or self-regulating mechanisms are preferable to actively regulated approaches.*" And "*smart city projects should include multi-stakeholder partnerships, in particular involving city decision-makers, commercial and industrial service providers, citizens, researchers and other relevant stakeholders*" (Lobsiger-Kägi et al. 2016).

A more comprehensive definition, such as the one chosen by the Smart City Winterthur (SCW) project, placed greater emphasis on a holistic approach to solutions that also includes social issues in particular: "*A holistic approach that requires a cross-thematic and cross-organisational approach is important for increasing the quality of life and minimising the consumption of resources. Intelligent and innovative solutions for a smart city encompass technical as well as social, ecological and economic developments*" (Lobsiger-Kägi et al. 2016).

Switzerland, despite moving beyond a purely technical focus on energy efficiency and IT in urban spaces, has always been rooted in a holistic perception of smart cities. These perspectives illustrate that Switzerland has been committed to an integrative approach to smart cities from the outset.

2.2 *Contemporary Conceptualization of Smart Cities in Switzerland*

In Switzerland, the core concept of smart cities has always been seen as the interplay of different human-centred dimensions of SCs, which must be holistically developed to address citizens' needs (Eschenauer et al. 2017). This approach is clearly illustrated in the Smart City Wheel models described in Figs. 1 and 2, which represent the same concept differently. These models emphasize the integration of various dimensions: Smart Living, Smart Mobility, Smart People, Smart Economy, Smart Energy and Environment, and Smart Governance. The holistic and human-centric approach ensures that smart cities in Switzerland are not just technologically advanced but also socially inclusive and environmentally sustainable, aiming to create liveable, resilient, and dynamic urban spaces. Particularly, *SwissEnergy* highlights the multifaceted nature of smart city development, emphasizing that all aspects of urban life are interconnected and must be addressed collectively to create a truly smart city. Whereas *Smart City Hub Switzerland* takes this concept further by placing human needs and interactions at the centre of SC initiatives, it stresses that the ultimate goal of smart cities is to create environments where technology and innovation serve the people, enhancing their daily lives and ensuring that urban development is both sustainable and inclusive. The subcategories within these figures are notably stable within Swiss literature, reflecting a consistent approach to smart city development. Examining SC categories in Table 2 and SC definitions provided in Table 3, they reveal both convergences and distinctions in their approaches. These tables were built by categorizing the definitions from various Swiss cities according to specific criteria, including purpose and goals, technological focus, scope and integration, stakeholder involvement, and governance and policy. Furthermore, these are mostly choices that can be made locally, allowing individual cities to tailor their smart city initiatives to their specific needs and contexts (Tables 1, 2 and 3).

While all cities—probably as a marketing stunt—prioritize improving the quality of life and sustainability, other primary objectives differ between cities. For instance, Zurich, Geneva, and Basel emphasizing enhancements in living standards and resource-efficient development. Technological integration is a shared focus, with Geneva highlighting ICT, while Basel and Winterthur stress digital technologies and intelligent networking.

Differences are also notable in stakeholder involvement and scope. Zurich and St. Gallen emphasize citizen participation and networking among people, organizations, and infrastructures. Bern and Winterthur adopt holistic approaches, integrating social, economic, ecological, and cultural aspects. Geneva focuses on a broad vision encompassing sustainability, innovation, and quality of life, while Lucerne and Basel stress operational focus and policy-driven approaches, respectively.

Table 1 Six smart city dimensions developed by SwissEnergy and smart city hub Switzerland compared

Dimension	SwissEnergy	Smart city hub Switzerland
Smart governance/Smart government	City infrastructure, Smart government, Digital services and administration	Open government, Provision of technological infrastructure, Provision of online services through the administration
Smart environment/smart energy and environment	Smart public spaces, Resource management and digital Infrastructure, Circular and sharing economy	Sustainable urban development and planning, Resource conservation and renewable energy, Smart and sustainable buildings
Smart economy	Open innovation and collaboration, Smart work and workspace, Start-up ecosystem and green growth	Entrepreneurial and innovative spirit and acting, Productivity, Local and global networking
Smart living	Smart home, Culture, Creativity and community, Tourism	Quality of life and cultural offers, Safety and security, Health
Smart mobility	Logistics and transportation, International mobility, Clean and sustainable mobility	Access to intermodal mobility, Prioritising efficient means of transport, Integrating IT in transportation and infrastructure
Smart people	Inclusive and sustainable society, Social participation, Learning and education, Research and development	Inclusive and participatory society, Modern educational offers at universities, Openness for creativity

Source Author's own compilation

3 Outlook of Swiss Smart Cities

In a previous study, we evaluated smart projects in Switzerland, focusing on privacy considerations (Fabrègue/Bogoni 2023). The study examined whether specific privacy initiatives were undertaken and whether privacy policies specific to smart city initiatives were adopted. Interest in smart mobility was found to be high across Switzerland, with successful integration between local public transport and national railways. Moreover, Zurich stands out with specific privacy policies for its mobility projects, demonstrating a commitment to protecting citizens' privacy. On the contrary, other cities like Geneva and Lausanne, despite being recognized as smart cities, were found to lack sector-specific privacy policies. Moving from this, the next Sections aim to analyse Swiss smart cities from a broad perspective: the first Section will see how they rank at the international level, while the second and third Section will evaluate their state of the art using data from the Swiss Smart City Survey.

Table 2 Classifications of smart city concepts by category

Category	Sub-category	Description
Purpose and goals	Quality of life	Emphasizes improvements in living standards and overall well-being
	Sustainability	Focuses on environmental stewardship and efficient resource use
	Innovation	Highlights the promotion of technological and social advancements
Technological focus	ICT infrastructure	Stresses the importance of information and communication technologies
	Smart infrastructure	Includes enhancements to both physical and digital infrastructure
Scope and integration	Holistic approaches	Integrates multiple sectors such as energy, transport, and governance
	Specific domains	Concentrates on particular areas like mobility, energy, or governance
Stakeholder involvement	Citizen-centric	Emphasizes citizen participation and engagement in smart city initiatives
	Multi-stakeholder partnerships	Includes collaborations among public, private, and academic sectors
Governance and policy	Policy-driven	Rooted in policy frameworks and strategic planning
	Operational focus	Concentrates on the practical implementation and management of smart city projects

Source Author's own compilation

3.1 Swiss Smart Cities in Global Rankings

International smart city rankings serve as a tool for cities worldwide, providing benchmarks, attracting investment, guiding policy and encouraging innovation. Swiss cities, particularly Zurich and Geneva, excel in international rankings for digitalization and smart city development. Swiss Smart Cities in general consistently rank well in international evaluations. These rankings underscore the country's commitment to leveraging technology and innovation to enhance urban living. Zurich, Geneva, Lausanne, Basel, and Bern are frequently highlighted for their exemplary smart city initiatives. Table 4 presents some of the most recognized city rankings that assess these dimensions globally, illustrating how Swiss cities compare on the international stage.

Table 3 Smart city concepts of the ten most populous cities in Switzerland, categorised

City	SC concept (translated by the Author)	Source	Category
Zürich	*"For the city of Zurich, 'smart' means networking people, organizations and infrastructures in such a way that social, ecological and economic added value is created. The networking of data, sensors and applications allows new and more efficient solutions for users of urban infrastructures as well as for their operators. The good networking with the population strengthens participation opportunities and contact with the administration"*	Strategie Smart City Zürich	Quality of life, sustainability, smart infrastructure, citizen-centric
Geneva	*"A Smart City is a city that uses information and communication technologies to improve the quality of its urban services and reduce its operating costs. Geneva's smart city goes far beyond a purely technological vision, basing itself on three fundamental pillars: sustainability, innovation and quality of life"*	La smart city à Genève	Quality of life, ICT infrastructure
Basel	*"Smart City Basel uses digital technologies and data to secure the quality of life, to enable innovation and to foster a resource-efficient and sustainable development. Smart City is an approach rather than a goal to itself: Technologies and data serve a sustainable development and are to be used in the interest of the people in Basel. As a smart city, Basel is creating tomorrow together."*	Strategie smart city basel	Quality of life, sustainability, policy-driven
Lausanne	N/A	N/A	N/A
Bern	*"Smart cities use digitalisation to shape their development in a socially responsible, fair, energy- and resource-efficient way. They manage their digital transformation strategically and with the involvement of citizens. The goal is an intelligently functioning, future-orientated community that respects the needs of the population as well as climate protection and the sustainable use of the environment, energy and mobility"*	Was ist eine Smart city?—Smart city Verein Bern	Sustainability, holistic approaches, citizen-centric

(continued)

Table 3 (continued)

City	SC concept (translated by the Author)	Source	Category
Winterthur	*"The Smart City Winterthur is an innovative, progressive and networked city that places people and the environment at the centre. It is characterised by a high quality of life and efficient use of resources. This is achieved through the intelligent networking of infrastructure with modern technologies and the involvement of relevant stakeholder groups. Social, economic, ecological and cultural requirements are taken into account"*	Strategie smart city Winterthur	Quality of life, sustainability, holistic approaches, multi-stakeholder partnerships
Lucerne	*"A smart municipality, city or region uses technology, innovation and data to improve people's quality of life and the attractiveness of the location for companies, while utilising infrastructure more efficiently and conserving resources. People as residents, employees and visitors are always at the centre of all considerations and help to shape them. Smart City builds on various networked dimensions"*	Digitalstrategie und smart city Luzern	ICT infrastructure, smart infrastructure, operational focus
St. Gallen	*"In the Smart City of St.Gallen, technologies and data are linked in order to improve the quality of life for residents and the quality of the location for companies while using fewer resources. Residents are an essential part of the developments and can contribute, test and co-determine solutions for everyday life. They are therefore important partners for future developments"*	Smart St.Gallen—Der Mensch als Mittelpunkt der "Smarten Stadt" Strategie	Quality of life, sustainability, citizen-centric
Lugano	N/A	N/A	N/A
Biel/Bienne	N/A	N/A	N/A

Source *Author's own compilation based on official documents and websites*

While acknowledging the limitations of the calculations included in Table 4—such as the different scale and parameters employed in each ranking (Chai Keong 2022)—it can be useful to note the performance of Swiss cities in several prominent international city rankings. In fact, cities such as Zurich, Geneva and Lausanne consistently showcase their strengths in sustainability, digitalization, and smart city initiatives. Swiss cities are frequently included in major global rankings. The IESE Cities in

Table 4 Swiss smart cities in international rankings

Smart city index	Total cities examined	Basel		Bern		Geneva		Lausanne		Zurich		Swiss Average %
		N°	%	N°	%	N°	%	N°	%	N°	%	
IESE Cities in Motion Index 2022	183	29	16%	28	15%	47	26%			14	8%	16%
IESE Cities in Motion Index 2024	183	24	13%	30	16%	40	22%			12	7%	14%
IMD Smart City Index 2019*	102					7	7%			1	1%	4%
IMD Smart City Index 2020*	109					8	7%			1	1%	4%
IMD Smart City Index 2021*	118					6	5%	4	3%	1	1%	3%
IMD Smart City Index 2023	141					9	6%	5	4%	1	1%	4%
IMD Smart City Index 2024	142					4	3%	7	5%	1	1%	3%
The Smart Centres Index 8 (2023)	77					18	23%			4	5%	14%
The Smart Centres Index 9 (2024)	79					9	11%			2	3%	7%
Cities of the Future 2021 (600 k-3 m pop.)	50									38	76%	76%
Cities of the Future 2021(50 k-600 k pop.)	50					12	24%	16	32%			28%
Arcadis Sustainable Cities Index 2022	100					47	47%			11	11%	29%

(continued)

Table 4 (continued)

Smart city index	Total cities examined	Basel		Bern		Geneva		Lausanne		Zurich		Swiss Average
		N°	%	N°	%	N°	%	N°	%	N°	%	%
Arcadis Sustainable Cities Index 2024	100					37	37%			21	21%	29%
Kearney Global Cities Outlook 2023	156					13	8%			18	12%	10%
Kearney Global Cities Outlook 2022	156					16	10%			19	12%	11%
Kearney Global Cities Outlook 2021	156					16	10%			13	8%	9%
Kearney Global Cities Outlook 2020	156					16	10%			22	14%	12%
Kearney Global Cities Outlook 2019	156					12	8%			15	10%	9%
Kearney Global Cities Outlook 2018	156					16	10%			13	8%	9%
Count and Average	19	2	14%	2	16%	18	15%	4	11%	18	11%	13%

Source: Author's own compilation and calculation

* IMD Smart City Indexes 2019, 2020 and 2021 are considered with the 2023 methodology

Motion Index and the IMD Smart City Index are two such examples where multiple Swiss cities feature prominently. Particularly, Zurich and Geneva are consistently listed among the top cities globally, reflecting their advanced urban infrastructure and smart city strategies.

The average percentage position of Swiss cities in these rankings indicates their strong performance. In fact, Swiss cities average in the top 13% of rankings. Zurich stands out, averaging in the top 11% of rankings, while Geneva follows in the top 15% of the selected indexes.

3.2 *The Swiss Smart City Survey*

To provide a general depiction of the Swiss Smart City situation, we can refer to the 2020 and 2022 Swiss Smart City Survey (SSCS) (ZHAW 2022), organized by ZHAW School of Engineering. In 2022, 51% of the 170 cities/municipalities contacted, which amounts to 87, participated. In the first survey performed in 2020, 84 out of 171 cities/municipalities (49%) participated. The regular survey enables the monitoring of advancements and patterns in the smart city industry over an extended duration, facilitating the documentation of any modifications. Furthermore, the SSCS also assesses the maturity of SC projects.

The SSCS developed a specific Smart City Index for Switzerland, reflecting various aspects and developments of smart cities over time. Based on the Smart City Wheels developed by SwissEnergy and Smart City Hub Switzerland (Table 1), eight key dimensions were defined, with an additional ninth dimension focusing on organizational and structural elements within city administrations. The nine dimensions include *Smart Mobility, Smart Environment, Smart Economy, Smart People, Smart Governance, Smart Living, Smart Data, Smart Infrastructure, and Enabler* (see Table 5 for a more detailed overview of the 9 dimensions of the SSCS).

Each dimension comprises four to ten indicators and has a maximum score of 100 points. It weighted by a panel of independent experts through a three-phase Delphi procedure, ensuring a balanced and comprehensive assessment. The SSCS is conducted every two years (the next being finalized in 2024) and collects data via an online survey targeting all cities and urban municipalities classified by the Federal Statistical Office (Swiss Smart City Survey 2024). In addition to the nine dimensions, participating cities and municipalities were asked to identify up to 10 of their most significant smart city projects and categorize them according to one of the six smart city categories developed by *SwissEnergy* (see Table 1) further analysed in Sect. 3.

The expert team synthetized their findings in a short report, both for 2022 (Sütterlin et al. 2022) and for 2020. An interactive dashboard was also developed (ZHAW 2024), through which it is possible to generate charts according to parameters such as size of the municipality, region, strategy, language spoken and group.

As showcased by the chart (Fig. 3), the maturity phases of smart city projects in Switzerland, comparing data from 2020 and 2022, suggests a growing commitment among Swiss cities to advance through the maturity phases of SC projects. In 2020,

Table 5 The nine dimensions of the 2022 Swiss Smart City Survey

Dimension	Description
Smart mobility	Creating clean mobility and logistics, promoting efficient means of transport, intermodality and sharing concepts
Smart environment	Resource- and environmentally friendly development of the urban environment (buildings, public spaces, infrastructure systems), promotion of renewable energies and use of synergies
Smart economy	Establishment of an innovative, resource-saving and open economic system that relies on networking, cooperation, circular economy and flexible working models
Smart people	Use and promotion of resources of the inhabitants as well as ensuring lifelong learning, participation, social integration and openness to creativity
Smart governance	Intelligent, needs-oriented and transparent control of urban administrative processes and infrastructure as well as interaction between residents and the administration
Smart living	Ensuring a barrier-free, communal, safe and healthy life based on equal opportunities
Smart data	Collection, processing and use of data in real time (internet of things, artificial intelligence, virtual reality, etc.) and ensuring cyber security
Smart infrastructure	Establishment and maintenance of a modern ICT infrastructure as well as urban services and networks (health, education, emergency organizations, structural measures in the area of transport and traffic, etc.)
Enabler	Formal anchoring of the smart city in the administrative organization, corresponding strategic goals, political mandates or designated budgets, which serve as drivers in the development of a smart city

Source Swiss Smart City Survey

approximately 35 cities were not engaged in any smart city phase, but this number decreased in 2022, showing that more cities have begun to engage in smart city projects. There is a notable increase in the number of cities in the pilot project phase, growing from around 25 cities in 2020 to nearly 45 cities in 2022, indicating a robust shift towards experimentation and initial implementation of smart city solutions. The number of cities in the institutionalization phase remains relatively stable, with a slight increase, signifying a steady movement of cities adopting and integrating smart city initiatives into their regular operations. The establishment phase sees a modest increase, reflecting that a few cities have successfully implemented and institutionalized smart city initiatives into their urban infrastructure.

The positive trend in the development of smart city initiatives from 2020 to 2022 is furtherly analysed in Fig. 2, showing substantial strides made in the dimensions of Governance, Data and Enabler. The 2022 Index (depicted in blue) shows notable improvements across most dimensions compared to the 2020 Index (shown in gray). Key observations include significant marked progress in Governance and Data, suggesting better urban administrative processes and citizen engagement. A growth in the Enabler dimension, reflects the increasing adoption of the SC concept within municipalities, usually in the form of a smart city strategy.

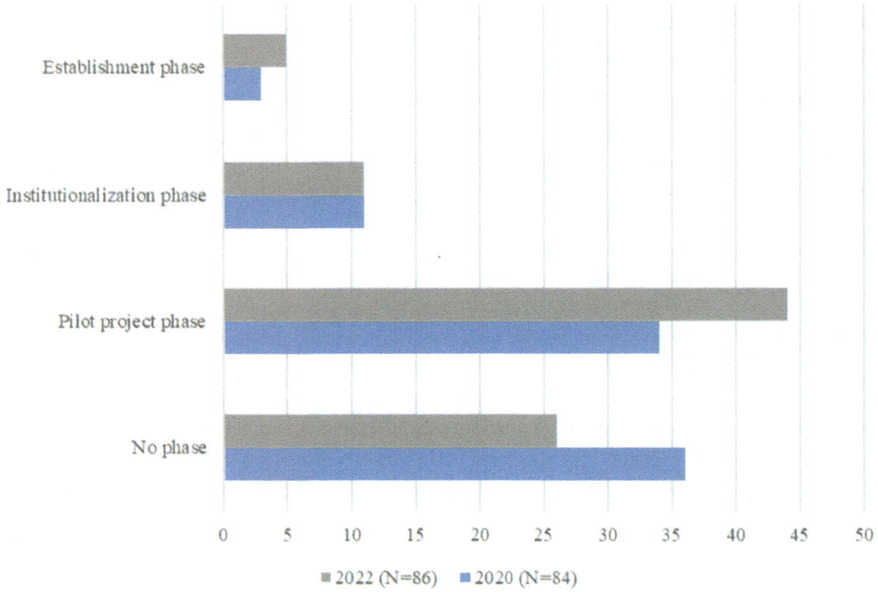

Fig. 1 Maturity phase of smart city projects in Switzerland. *Source* Author's own compilation based on SSCS data

Further analysing the indexes for the 9 SSCS dimensions, the chart depicted in Fig. 3 compares the 2020 smart city indexes for municipalities with more than 100,000 inhabitants (shown in blue) against the Swiss average index (shown in gray) seen before. Municipalities with populations over 100,000 generally perform better than the Swiss average in several dimensions. Notably, in 2020 larger cities excelled in every dimension more than the average Swiss smart city, especially within the Enabler dimension, which includes the formal anchoring of the smart city in the administrative organization. This suggests that bigger cities are more proactive in institutionalizing smart city initiatives and securing the necessary political and financial support to drive their development. Conversely, the Governance dimension is almost the same in both large municipalities and the Swiss average, indicating that urban administrative processes and citizen engagement practices are relatively uniform across cities of different sizes.

The radar chart below in Fig. 4 compares the 2022 smart city indexes for municipalities with populations over 100,000 (shown in blue) against the Swiss average index (shown in gray). Just as in 2020, larger cities outperform smaller ones in every dimension. For instance, the Economy index of larger cities improved from 2020 to 2022, suggesting a much more vibrant and innovative economic environment. It has thus become the dimension where small cities lag behind larger ones the most. Enabler still shows a great divergence between the two indexes but we can see small cities "catching up".

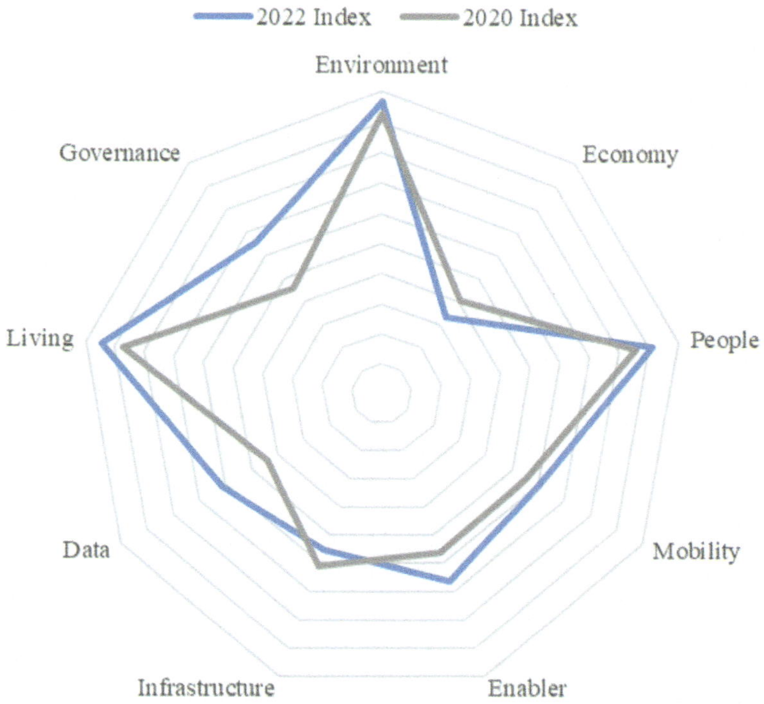

Fig. 2 Comparison of smart city indexes for the 9 dimensions in 2020 and 2022. *Source* Author's own compilation based on SSCS data

The difference between small and large cities in Environment, People and Living has remained relatively similar to 2022. In the People dimension, which includes social inclusion and education, the consistent emphasis on high-quality education and social programs across all urban areas ensures steady improvement in both small and large cities. Moreover, quality of life factors, such as healthcare, safety, and housing, are uniformly high due to national standards, resulting in similar advancements in the Living dimension across city sizes. However, the gap in Data, Infrastructure, Mobility, and Governance has widened significantly. This suggests that municipalities benefit from economies of scale, allowing them to invest heavily in urban projects, public transport networks, and robust digital infrastructure, while smaller cities struggle with limited budgets. Mobility solutions in larger cities are also more advanced, featuring smart traffic management and extensive public transit, driven by their complex transportation needs and greater financial and technological resources.

To better understand the main priorities of smart city development for Swiss municipalities, the next Section will analyse the 6 SC sub-dimensions of the SSCS.

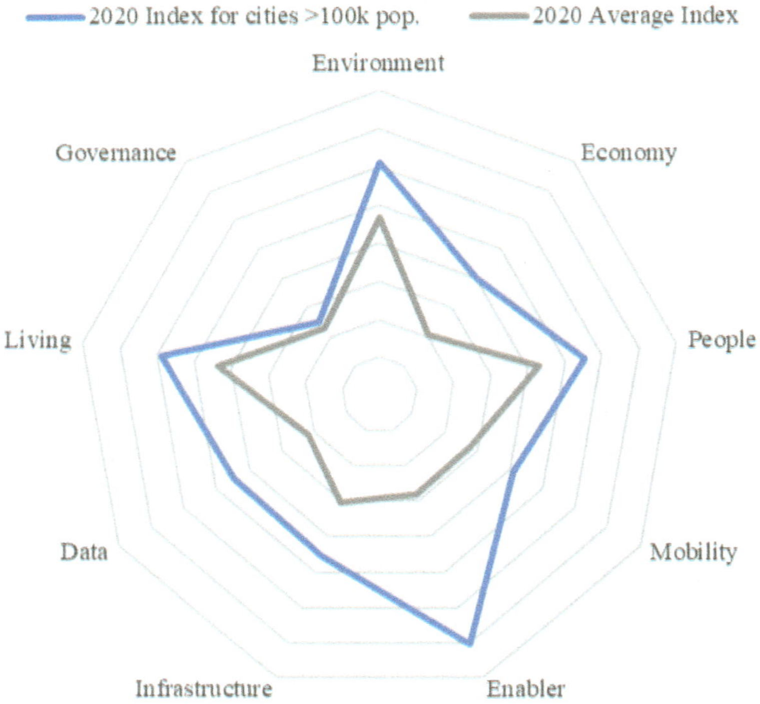

Fig. 3 Comparison of 2020 smart city indexes for the 9 dimensions of municipalities with more than 100,000 inhabitants and the Swiss average. *Source* Author's own compilation based on SSCS data

3.3 Priorities of Smart City Development

As mentioned before, more than 300 projects were categorized by municipalities into the 6 sub-dimensions developed by *SwissEnergy* (Table 1) both in 2020 and in 2022. The distribution of smart city projects across the six sub-dimensions highlights the varying levels of focus and investment of municipalities and cities in Switzerland: taking data from the SSCS, Fig. 5 showcases the percentages of projects per SC sub-dimension.

The dimension with the highest number of projects both in 2020 and 2022 is Smart Governance, accounting for 29.90% and 30.30% respectively, reflecting a strong emphasis on enhancing administrative processes and public services. In 2022 Smart Mobility, Smart Energy and Environment follow with 22.40% and 15.50% respectively, showing a strong interest to increasing the efficiency of urban systems. On the other hand, Smart People at 14.20% and Smart Living at 11.40% show a slightly lower priority to issues generally related to citizen participation and quality of life. Nevertheless, we can note that while Smart Energy and Environment and Smart Mobility are still ranking second and third respectively, they decreased their percentage in

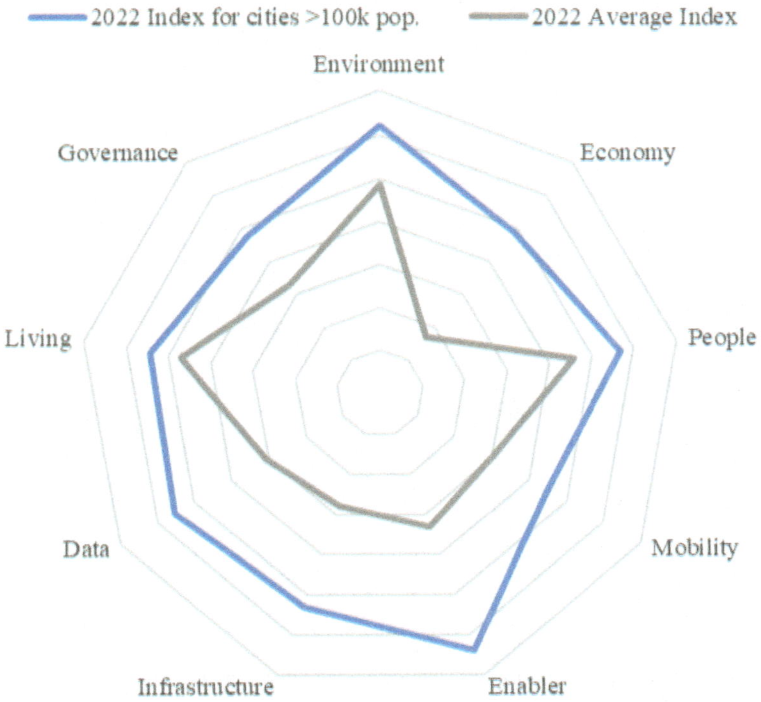

Fig. 4 Comparison of 2022 smart city indexes for the 9 dimensions of municipalities with more than 100,000 inhabitants and the Swiss average. *Source* Author's own compilation based on SSCS data

favour of both Smart Living and Smart People. This signals a shift towards participative and quality-of-life-oriented SC practices. Lastly, while having increased, the Economy is still the lowest priority of SC project development (Table 6).

Smart Governance. In the area of Smart Governance, 30.30% of projects were categorized. Many initiatives focus on administrative reform through digitization, with cities digitizing internal processes like traffic technology and energy data

Table 6 Project distribution among smart city dimensions

Smart dimension	Projects 2022 (%)	Projects 2020 (%)
Smart living	11.4	10.30
Smart mobility	15.50	20.50
Smart people	14.20	12.10
Smart economy	6.30	4.20
Smart energy and environment	22.40	23.00
Smart governance	30.30	29.90

Source Author's own compilation based on SSCS data

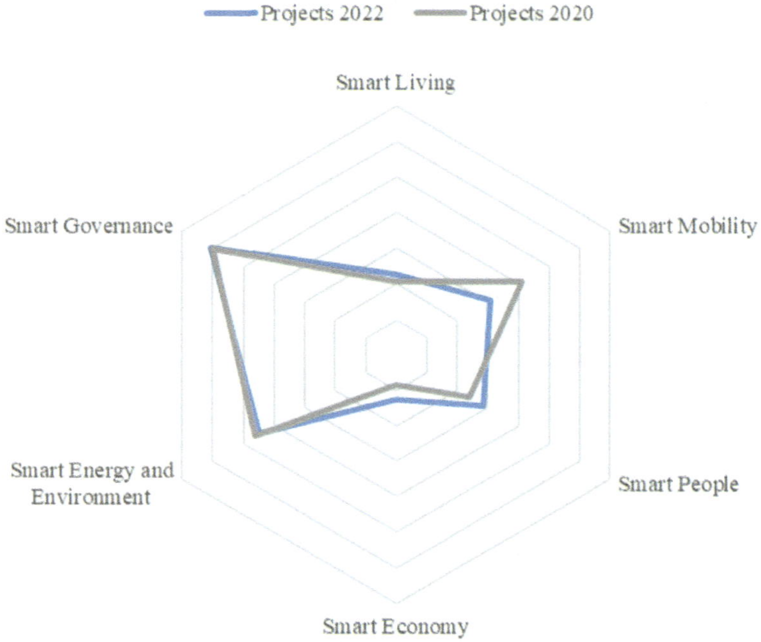

Fig. 5 Graphic illustration of project distribution among Smart City Dimensions. *Source* Author's own compilation based on SSCS data

management. Numerous projects aim to enhance digital services, such as revamping websites or developing city apps. Offerings include online counters, city notifiers, chatbots, eBau for building permits, and eUmzug for relocation notifications. Additionally, several projects promote paperless meetings and digital transformations within administrations. Finally, open governance initiatives include creating open data platforms and various participation opportunities, intersecting with the smart people dimension.

Smart Energy and Environment. In 2022, 22.40% of projects were identified under Smart Energy and Environment, a slight change from 2020 (23%) when the "EnergyCity" label certification was frequently mentioned. Many initiatives focus on promoting renewable energy, particularly photovoltaics, including funding and installation on public buildings. Resource conservation projects include smart metering, smart lighting, and heating networks. Urban planning projects often involve green space development, such as tree planting, and measures to prevent flooding. Additionally, sustainable building initiatives and projects optimizing urban green space irrigation and cleanliness measurement are also prevalent.

Smart Mobility. With 15.5%, Smart Mobility is the third the most focused areas. The survey reveals that numerous cities are developing new mobility strategies to enhance sustainability through more efficient systems. Many projects emphasize intermodal transport, expanding public transport infrastructures like real-time displays and bus routes, extending bicycle lanes, and incorporating bike-sharing systems. Smart parking and parking management are also significant, typically involving parking sensors and digital payment options. Despite this, it is also the sub-dimension that saw the highest shrinking, with 5 percentage points less than 2020. This could be due to the fact that municipalities have already implemented necessary projects related to Mobility and in 2022 shifted their priorities to other areas such as Smart People, Living and Economy.

Smart People. Within the domain of Smart People, 14.20% of initiatives were identified in 2022, with the majority of them incorporating diverse avenues for engagement such as eParticipation platforms, citizen surveys, and workshops. Overall, it experienced a 2.1 percentage points increase from 2020, signalling a growing interest. The listed measures aim to establish a conducive atmosphere for participation. Additional engagement activities support the growth of local businesses by offering a marketplace platform. Furthermore, the list of projects included initiatives focused on education, such as the implementation of ICT in elementary schools, as well as programs, campaigns, and games aimed at promoting a lifestyle that is both environmentally friendly and energy conscious.

Smart Living. In 2022, 11.40% of projects were classified in the area of Smart Living, which primarily focuses on enhancing the quality of life. As Smart People did, this sub-dimension grew from 2020 to 2022 by 1.1 percentage points, showing that Swiss municipalities are prioritising this area of the SC wheel. These projects promote smart and integrated neighborhood development, digital opportunities for interaction, and digital access to art. Several other initiatives are dedicated to elderly care, aiming for obstacle-free public spaces and better networking for older people. The expansion of the fiber-optic network is also a common theme among these projects.

Smart Economy. Lastly, Smart Economy, with 6.30%, encompasses key initiatives focus on establishing co-working spaces and innovation labs, such as the "WinLab Ko-Kreationsplattform" in Winterthur. Additionally, there are efforts to promote the circular economy and sharing economy. The relatively low percentage of Smart Economy projects, while showing an increase from 2020's 4.20%, can be also attributed to the nature of SCSS data, which is municipality-based. Many significant initiatives in the smart economy dimension, such as startup programs, innovation labs, and entrepreneurial initiatives, are led by universities and private partnerships. Therefore, while the municipal data highlights a broad range of smart city projects, it does not fully capture the extensive work being done in the smart economy sector by other key players.

It is advisable to coordinate integrated actions at the federal/cantonal level where municipalities and cities seem to lack the most. In fact, challenges such as climate change, economic innovation and the remodelling of energy and mobility infrastructure systems require us to rethink the current approach to urban development. Thanks to the SC concept, cities can address these challenges using an integrated and interdepartmental development approach, in a network of partners and with the support of digital technologies. As seen in Sect. 2, the Swiss SC concept is understood as being much more than internal e-government and digitalisation strategies. The overarching goal is to develop efficient and resource-friendly solutions while enhancing quality of life and site attractiveness. Moreover, as also seen in Sect. 3.2, small sized municipalities lack the capability—mostly in terms of budget—to address such challenges by themselves. In fact, they present difficulties especially in those areas that are capital-and skill-intensive.

For most small and mid-sized cities and municipalities in Switzerland, the introduction to the SC concept proves to be challenging. While pioneer cities have gained some experience of SC implementation in the past few years, most small and mid-sized towns need proper support. To aid administrators, the Swiss Federal Office of Energy and SwissEnergy produced a guide in 2020 (Musiolik et al. 2020), presenting and discussing various steps, tools, options and practical examples for the implementation of Smart Cities, from which interested towns can choose according to their needs. The challenges, steps, tools, and measures introduced in this guide are based on a three-phase development model consisting of (i) a pilot project, (ii) an institutionalisation, and (iii) a consolidation phase. During the pilot project phase, various departments gain experience by carrying out initial Smart City projects, with guidance provided to help cities develop, select, and implement project ideas. Following this, many cities transition to the institutionalisation phase, where they work to embed the Smart City concept by developing an overarching strategy and establishing an organizational structure. This phase includes creating a prioritised project portfolio, a Smart City organization, and a strategy. The final phase, consolidation, is gradual and characterized by the operational integration of the Smart City concept into the administrative procedures and routines of municipal services. This phase involves investments in digital infrastructure and the implementation of measures in communication, education, training, and data security.

4 Case Studies of Swiss Smart Cities

The complexity of public administration reforms varies by country, influenced by their institutional and political systems. In Switzerland, ties to historical roots and the nation's founding principles are notably strong. Switzerland is often described as a "Sonderfall" (unique case), building its prosperity on political neutrality and consensus decision-making. The Swiss policy-making process is shaped by three defining features: direct democracy, a consensual government model, and federalism, which collectively ensure stability and rigidity in political institutions (Mach/Trampusch 2011).

Swiss decision-making has historically been slow and resistant to major changes, with any adopted policies typically resulting in modest adjustments (Sciarini 2015). The consensus-based political structure tends to favour compromise solutions that maintain the status quo, sometimes limiting innovation. Moreover, the Swiss federal structure grants significant autonomy to its 26 cantons and 2202 municipalities.

The principle of subsidiarity dictates that policy decisions are made at the lowest competent level of government (Schenkel/Serdült 2004). Despite the strong local autonomy, the Swiss Federal Government also engages in smart city initiatives through consensus and collaboration (Kriesi/Trechsel 2008). Notable examples include the "eGovernment Strategy Switzerland," developed jointly by federal, cantonal, and municipal governments (eGovernment Suisse 2020). Most cantons have their own eGovernment policies, and the national eGovernment study regularly assesses progress. The aforementioned eGovernment Strategy Switzerland and the Smart City Switzerland initiative provide strategic direction for digital transformation and smart city development across federal, cantonal, and municipal levels.

Municipalities are, nevertheless, at the centre of smart city development. From Sects. 4.1 to 4.5 we will briefly analyse case studies of Swiss smart cities, showcasing selected SC projects that stand out related to the key characteristics of each city.

4.1 Zurich

Zurich, Switzerland's largest city and economic hub, is positioning itself as a leader in smart city initiatives. With a population expected to grow in the next years, Zurich faces the challenge of managing increased resource consumption while enhancing the quality of life for its residents and maintaining a conducive environment for businesses. Leveraging new technologies and digital transformation, Zurich aims to create a socially, ecologically, and economically sustainable urban environment.

The core of Zurich's smart city strategy is the seamless integration of people, organizations, and infrastructures to generate added value across social, ecological, and economic dimensions (Stadt Zürich 2024). This involves networking data, sensors, and applications to develop efficient solutions for users and operators of urban infrastructures. A significant component of this strategy is fostering dialogue between city administration and citizens, thereby enhancing participation and direct democracy.

Zurich's smart city initiatives are city-led, supported by an interdepartmental steering committee and a dedicated team within the urban development department. This structure ensures coordinated efforts across different sectors and projects, fostering innovation and efficient implementation of smart solutions.

Key elements of Zurich's digital infrastructure can be seen in Table 7. Moreover, selected smart city initiatives are featured in Table 8.

4.2 Geneva

Geneva, a global city renowned for its role in international diplomacy and high quality of life, is actively transitioning into a smart city. With a focus on sustainability, innovation, and enhancing the quality of life, Geneva's smart city initiatives

Table 7 Key elements of Zurich's digital infrastructure

Name	Description
Central data center infrastructure	A centralized hub for data management and processing
Fibre optic network (Zürinet)	Providing high-speed internet connectivity across the city
Open data platform	Facilitating data sharing and transparency
ICT for schools	Enhancing digital literacy and education
Citywide LoRa WAN	Supporting Internet of Things (IoT) applications with a city-wide wireless network

Source Author's own compilation

Table 8 Selected SC initiatives of Zurich categorised by their dimension

Name	Dimension	Description
Züri Mobil	Smart mobility	An integrated mobility platform that offers comprehensive travel options within Zurich. This app includes various transportation modes and presents them clearly to users, providing mobility as a service without additional costs for providers
Mein Konto	Smart governance	A digital gateway to the city's online services, consolidating multiple services under a single login. It simplifies interactions between residents, businesses, and the city administration, reducing the need for in-person visits and manual data entry
Stadtidee	Smart people	A participatory budgeting initiative launched in summer 2021, inviting residents and visitors to propose ideas for neighbourhood improvements. The focus is on projects related to climate, environment, and youth
StadtBox	Smart economy	The Smart City Zurich program promotes internal innovation through the "StadtBox" program, seed funding for pilot projects, and innovation fellowships. The Smart City Lab serves as an open space for developing and testing new ideas, providing a foundation for informed decision-making

Source Author's own compilation

aim to integrate technology into urban development to address the challenges of urbanization and climate change.

The "Smart Geneva" initiative, launched by the canton of Geneva (République et canton de Genève 2024), is a key component of the city's strategy to become a Smart City 3.0. This initiative, inspired by Scandinavian models, emphasizes a participatory approach to create an ecosystem that fosters innovation and sustainability through networking and project promotion. It also involves both public and private sectors, as well as academic institutions.

Geneva's smart city initiatives are also deeply intertwined with its sustainability goals. The city's approach includes using digital technologies to monitor and reduce energy consumption, optimize waste management, and promote sustainable transportation. One of the long-term objectives of the Smart Geneva initiative is to integrate the French border area into its smart city strategy. By 2030, Geneva aims to be

Table 9 Key elements of Geneva's digital infrastructure

Name	Description
Fully developed FTTH network	Ensuring high-speed internet access
4G Infrastructure	Providing extensive mobile connectivity
LoRaWAN Infrastructure	Used for specific applications in the city, supporting various IoT initiatives

Source *Author's own compilation*

Table 10 Selected SC initiatives of Geneva categorised by their dimension

Name	Dimension	Description
Open data platform	Smart governance	Geneva promotes transparency and innovation by making selected administrative records accessible through its open data platform
Participez.ge	Smart people	The city's e-participation platform aims to involve residents in future-oriented issues, fostering a culture of participation regardless of age, origin, or civic status
Genevecultive	Smart energy and environment	As part of the Smart Geneva initiative, an urban agriculture project and a crowdfunding platform have been launched to promote sustainable local food production and community-funded projects
Telemedicine platform	Smart living	At the HUG University Hospital, a telemedicine platform has been developed to improve healthcare access and efficiency

Source *Author's own compilation*

a fully-fledged smart city, setting an example for sustainable and innovative urban development that meets the needs of its residents while addressing global challenges.

Key elements of Geneva's digital infrastructure can be seen in Table 9. Moreover, selected smart city initiatives are featured in Table 10.

4.3 Basel

Basel, often referred to as the "small metropolis," is strategically located at the confluence of Switzerland, France, and Germany. This unique position has fostered a strong tradition of cross-border collaboration in areas such as transport, environment, education, and culture. Basel's smart city initiatives are designed to leverage its dynamic economy, high quality of life, and cultural richness to create a sustainable and innovative urban environment. Culturally, Basel is known as Switzerland's "cultural capital," boasting a rich array of museums, galleries, and cultural institutions with international acclaim. Basel's economy is one of the most dynamic in

Table 11 Selected SC initiatives of Basel categorised by their dimension

Name	Dimension	Description
EnergieStadt	Smart energy and environment	Basel is recognized for its excellence in energy management, holding the "EnergyCity" GOLD award and being a pioneer in the 2000-Watt Society since 2014
Tech park Basel	Smart economy	Tech Park Basel offers early-stage tech start-ups a collaborative, business-driven environment with office and lab spaces, essential services, and networking opportunities, fostering innovation and growth in the rapidly growing Stücki Park
Switzerland innovation park	Smart economy	Switzerland Innovation Park supports innovative ventures in therapeutics, digital health, and industrial transformation by providing collaborative workspaces, mentorship, funding, and networking opportunities
Institut für Jungunternehmen	Smart economy	Institut für Jungunternehmen (IFJ) supports Swiss start-ups by providing resources for company formation, business planning, accounting, and administration, along with free intensive courses and networking events, fostering entrepreneurship and innovation

Source Author's own compilation

Switzerland, driven by its capacity for innovation and its strong industrial base. The city is home to major global players in the life sciences sector, including Novartis, Roche, Lonza, Bayer, and Syngenta. Additionally, the finance sector, logistics, and trade also play crucial roles in Basel's economic landscape.

Basel's smart city developments and projects relate to its key characteristics (Kanton Basel-Stadt 2024), with a particular focus on Smart Economy. Table 11 showcases selected initiatives.

4.4 Bern

Bern, the capital city of Switzerland, serves as the country's political centre. Known for its scenic beauty, historical significance, and vibrant cultural scene, Bern is leveraging its unique position to become a leading smart city. With a population of over 142,000 and expected growth to 160,000 by 2050, Bern is proactively implementing smart city initiatives to enhance quality of life, foster sustainable development, and promote innovation. As Switzerland's political heart, Bern is also a cultural and economic hub. The city boasts a rich array of cultural institutions, including theatres, museums, and festivals, contributing to a dynamic urban atmosphere. Economically, Bern is home to various sectors, including government services, education, and healthcare, providing a stable foundation for growth and development.

Bern's smart city strategy is coordinated by the Konferenz Digital Stadt Bern (KDSB), an interdepartmental steering committee (Stadt Bern 2020). The initiative is then supported by the Fachstelle Digitale Entwicklung, a dedicated team within the city administration. Their efforts focus on enhancing digital infrastructure, promoting sustainable urban development, and fostering citizen engagement. Additionally, the city publishes open government data, encouraging innovative companies and individuals to develop new public services. Bern is also recognized for its sustainable practices, earning the European Energy Award Gold in 2019. Selected smart city initiatives are featured in Table 12.

Table 12 Selected SC initiatives of Bern categorised by their dimension

Name	Dimension	Description
BärnBoost	Smart people	A web-based platform aimed at increasing community participation and engagement in Bern's neighbourhoods. Residents can initiate social events and seek support from fellow citizens, fostering a strong sense of community
Base4kids	Smart living	An initiative to modernize the digital learning environment in Bern's schools
Bikesharing	Smart mobility	Bern offers a comprehensive bikesharing program with more than 1500 bicycles and e-bikes available at around 200 stations across the city
Bildungsplattform	Smart people	A new educational platform designed to encourage continuous learning among employees, equipping them with skills needed for the digital age
E-Bern	Smart governance	An initiative to expand digital access to urban services, allowing citizens and businesses to submit requests online while ensuring accessibility and data security
KU	Smart living	A mobile guidance system enabling blind and visually impaired individuals to experience art and culture in public spaces
Matte-Schnägg	Smart mobility	A self-driving minibus
Grünflächenmanagement	Smart energy and environment	A public green space management system overseeing over 800 parks and green areas. This system supports efficient urban planning and development decisions

Source Author's own compilation

4.5 Winterthur

Winterthur, the sixth largest city in Switzerland, is known for its cultural diversity, affordable living spaces, and high quality of life. With a population of 116,000 in 2021, expected to grow to 129,000 by 2030, Winterthur is embracing smart city initiatives to manage this growth sustainably and enhance the urban experience for its residents. Winterthur's smart city strategy, approved in 2018, is driven by the innovation team under the program management of the Smart City Unit (Fachstelle Smart City) (Stadt Winterthur 2024). The strategy promotes a "bottom-up" approach where individual city departments determine how to align projects with strategic goals. Winterthur is also recognized for its sustainability and innovation efforts; the city earned the "EnergyCity" Gold label and the Grünstadt Schweiz (Silber) for its green city initiatives.

A notable mention is the Winterthur WinLab (WinLab Ko-Kreationsplattform), a co-creation platform and innovation hub focused on sustainable energy and mobility solutions in Winterthur. It provides a collaborative environment and invites Public–Private-People Partnerships (PPPP) to test social and technological innovations, supporting sustainable smart city development. WinLab supports projects related to energy efficiency, renewable energy, and smart mobility by offering access to advanced infrastructure, expertise, and networking opportunities. Other selected SC initiatives in Winterthur are featured in Table 13.

Table 13 Selected SC initiatives of Winterthur categorised by their dimension

Name	Dimension	Description
Electric waste collection	Smart energy and environment	Winterthur has introduced electric waste collection vehicles
Naturfundbüro	Smart living	At the Natural History Museum Winterthur (Naturmuseum Winterthur) visitors can identify natural objects at various stations through APIs connected to a database
Lokstadt climate simulation	Smart energy and environment	This project uses computer-modeled climate simulations to test temperature-reducing measures in the densely built Lokstadt area
Winterthur-App	Smart people/smart governance	The city app offers information and participation opportunities for residents. Features like "Stadtmelder" allow citizens to report infrastructure issues, while the "Quartierleben" module promotes interaction and community engagement

Source *Author's own compilation*

5 Conclusion

The exploration of smart cities in Switzerland, as detailed in this paper, reveals the rather typical multifaceted approach to urban innovation and sustainability. The Swiss model emphasizes a holistic integration of technology, sustainability, and quality of life improvements, guided by robust frameworks integrated within administrative bodies and driven by local (municipal and cantonal) and federal initiatives. The "SwissEnergy for Municipalities" program, for instance, has been pivotal in promoting sustainable energy practices and smart city development. These programs provide extensive support to cities and municipalities, including information campaigns, advisory services, and financial assistance.

In general, the Swiss smart city aims to create an efficient, sustainable, and liveable environment. The country's high standard of living and strong commitment to sustainability provides a fertile ground for smart city initiatives, addressing challenges such as urbanization, environmental pressures, and resource management. The country's historical emphasis on energy efficiency has evolved into a comprehensive smart city conceptualisation that aligns with global best practices while addressing local needs. This is one of the reasons why Swiss smart cities score well in international benchmarks, averaging the top 13% of the rankings.

The biennial Swiss Smart City Survey (SSCS) provides valuable insights into the development and maturity of smart city projects across Switzerland. The survey highlights a growing commitment among Swiss cities to advance through the maturity phases of smart city projects, with notable progress in governance, data management, and the formal anchoring of smart city initiatives within municipal administrations. Larger cities tend to perform better across most dimensions, benefiting from economies of scale and more substantial investments in urban projects and infrastructure.

Additionally, as outlined in Sect. 3.3, the distribution of smart city projects reveals varying levels of focus among Swiss municipalities. Smart Governance, Mobility, and Energy and Environment are prominent areas of investment, reflecting a strong emphasis on administrative efficiency, sustainable urban systems, and environmental stewardship. There is also a growing interest in Smart People and Smart Living, indicating a shift towards participatory and quality-of-life-oriented practices. However, there are areas for improvement, particularly in the disparity between larger and smaller cities in terms of smart city development. This underscores the need for tailored support to address specific challenges faced by smaller municipalities.

Detailed case studies of cities such as Zurich, Geneva, Basel, Bern, and Winterthur illustrate diverse approaches and successes. Zurich leads in integrating urban infrastructures and citizen participation. Geneva focuses on sustainability and innovation, while Basel leverages its strong economic base and cross-border collaborations. Bern emphasizes digital governance and sustainable development, and Winterthur showcases innovation in sustainable energy and mobility solutions. All of these case studies suggest that smart cities developed with a holistic approach adhere to the specific characteristics of the living space.

The success of Swiss smart cities can be attributed to several factors: strong political will, effective federal and cantonal budgeting, active citizen participation, and a clear focus on sustainability. Swiss cities have demonstrated that a holistic approach, involving multiple stakeholders and emphasizing quality of life, can lead to significant advancements in urban living conditions. Looking ahead, the continuous evolution of SC strategies in Switzerland will likely focus on further integrating stakeholders and key actors—both public and private, both vertically and horizontally—to allow even more urban centres to adopt the smart city concept. Moreover, it will be interesting to see in 2024's SCSS if the trend towards SC dimensions related to participation (Smart People) and quality of life (Smart Living) will continue and to what extent.

References

Ahmad, Kashif, Majdi Maabreh, Mohamed Ghaly, Khalil Khan, Junaid Qadir, and Ala Al-Fuqaha. 'Developing Future Human-Centered Smart Cities: Critical Analysis of Smart City Security, Data Management, and Ethical Challenges'. *Computer Science Review* 43 (1 February 2022): 100452. https://doi.org/10.1016/j.cosrev.2021.100452.

Capdevila, Ignasi, and Matías I. Zarlenga. 'Smart City or Smart Citizens? The Barcelona Case'. Edited by Amel Attour and Thierry Burger-Helmchen. *Journal of Strategy and Management* 8, no. 3 (1 January 2015): 266–82. https://doi.org/10.1108/JSMA-03-2015-0030.

Dameri, Renata Paola, and Camille Rosenthal-Sabroux, eds. *Smart City: How to Create Public and Economic Value with High Technology in Urban Space*. Switzerland: Springer, 2014. https://link.springer.com/book/https://doi.org/10.1007/978-3-319-06160-3.

eGovernment Suisse. 'eGovernment Strategy Switzerland', 2020. https://www.digital-public-services-switzerland.ch/en/media/landingpage-egovernment.

EnergieSchweiz. 'Nachhaltig Unterwegs Mit Energiestadt'. Bern, April 2015. https://www.local-energy.swiss/dam/jcr:4aac7461-74e0-4693-8c6c-0edc4edda060/energiestadt_broschuere_d_150423.pdf.

EnergieSchweiz. 'EnergieSchweiz für Gemeinden: Fördergelder und Beratung'. EnergieSchweiz. Accessed 31 July 2024. https://www.energieschweiz.ch/energieschweiz/gemeinden/.

EnergieSchweiz. 'Smart City Schweiz'. Local Energy. Accessed 31 July 2024. https://www.local-energy.swiss/programme/smart-city.

Eschenauer, Ursula, Lukas Braunreiter, Tobias Kuehn, Onur Yildirim, Evelyn Lobsiger-Kägi, Harry Spiess, Vicente Carabias-Hütter, and Adrian W. Müller. 'Smart Cities in Theorie und Praxis : Szenarien, Strategien und Umsetzungsbeispiele', March 2017. https://doi.org/10.21256/zhaw-1237.

Fabrègue, Brian F. G., and Andrea Bogoni 'Privacy and Security Concerns in the Smart City'. *Smart Cities* 6, no. 1 (February 2023): 586–613. https://doi.org/10.3390/smartcities6010027.

Fabrègue, Brian F. G., and Andrea Bogoni. 'Assessing Quality of Life in German, French, Italian and Polish Smart Cities: Identifying the Need for Further Development'. *Smart Cities and Regional Development (SCRD) Journal* 8, no. 1 (20 January 2024): 131–52. https://doi.org/10.25019/yccjvz56.

Gassmann, Oliver, Jonas Böhm, and Maximilian Palmié. *Smart Cities: Introducing Digital Innovation to Cities*. Emerald Publishing Limited, 2019. https://doi.org/10.1108/9781787696136.

Hunter, Trevor, Peter Worthy, Stephen Viller, and Ben Matthews. 'Designing for Citizen-Centric Smart Cities: Supporting People's Needs of a Community Garden'. *Proceedings of the 30th*

Australian Conference on Computer-Human Interaction, 4 December 2018, 162–66. https://doi.org/10.1145/3292147.3292217.

IMD Smart City Observatory. 'Smart City Index 2024', 8 April 2024. https://www.imd.org/smart-city-observatory/home/.

Kanton Basel-Stadt. 'Welcome to Smart City Basel'. Accessed 6 August 2024. https://www.smartcity.bs.ch/en.html.

Kriesi, Hanspeter, and Alexander H. Trechsel. *The Politics of Switzerland : Continuity and Change in a Consensus Democracy*. Cambridge University Press, 2008. https://cadmus.eui.eu/handle/1814/34905.

Lobsiger-Kägi, Evelyn, Theresia Weiss Sampietro, Ursula Eschenauer, Vicente Carabias-Hütter, Lukas Braunreiter, and Adrian W. Müller. 'Treiber und Barrieren auf dem Weg zu einer Smart City : Erkenntnisse aus Theorie und Praxis', February 2016. https://doi.org/10.21256/zhaw-1052.

Mach, André, and Christine Trampusch. 'The Swiss Political Economy in Comparative Perspective'. In *Switzerland in Europe. Continuity and Change in the Swiss Political Economy*, 9–26. Routledge, 2011. https://www.researchgate.net/publication/317169247_The_Swiss_Political_Economy_in_Comparative_Perspective.

Jörg Musiolik, Anna Kohler, Pascal Vögeli, Evelyn Lobsiger-Kägi, Leticia Müller, and Vicente Carabias-Hütter. 'Smart City: Guide to the Implementation of Smart City Initiatives in Switzerland'. Bern: Swiss Federal Office of Energy, 2020.

Myeong, Seunghwan, Jaehyun Park, and Minhyung Lee. 'Research Models and Methodologies on the Smart City: A Systematic Literature Review'. *Sustainability* 14, no. 3 (January 2022): 1687. https://doi.org/10.3390/su14031687.

Pascal Sciarini. 'Introduction'. In *Political Decision-Making in Switzerland. The Consensus Model Under Pressure*, 1–23. Palgrave Macmillan, 2015. https://link.springer.com/chapter/https://doi.org/10.1057/9781137508607_1.

Portal, Léo, and Brian Fabregue. 'Establishing Participative Smart Cities: Theory and Practice'. *Smart Cities and Regional Development (SCRD) Journal* 6, no. 2 (13 April 2022): 43–62. https://doi.org/10.25019/scrd.v6i2.128.

République et canton de Genève. 'Smart City'. ge.ch. Accessed 6 August 2024. https://www.ge.ch/taxonomy/term/5416.

Walter Schenkel and Uwe Serdült. 'Intergovernmental Relations'. In *Handbook of Swiss Politics*, 393–427, 2004. https://www.researchgate.net/publication/302018230_Handbook_of_Swiss_Politics.

Stadt Bern. 'Smart City Verein Bern', 19 May 2020. https://www.smartcity-bern.ch/.

Stadt Winterthur. 'Smart City und Nachhaltigkeit'. Stadt Winterthur. Accessed 6 August 2024. https://stadt.winterthur.ch/themen/die-stadt/smart-city-winterthur.

Stadt Zürich. 'Smart City Zurich'. Accessed 6 August 2024. https://www.stadt-zuerich.ch/prd/en/index/urban-development/smart-city.html.

Sütterlin, Bernadette, Jasmin Oberkalmsteiner, Jonas Giger, Marisa Kappeler, and Vicente Carabias. 'Swiss Smart City Survey 2022: Final Report'. Winterthur: ZHAW School of Engineering, 27 March 2023. https://smartcity-survey.ch/wp-content/uploads/2023/05/2_Final-Report_EN.pdf.

Swiss Smart City Survey. 'Research Design and Methodology'. SSCS. Accessed 28 February 2024. https://smartcity-survey.ch/en/research-design/.

Toh, Chai Keong. 'Smart City Indexes, Criteria, Indicators and Rankings: An in-Depth Investigation and Analysis'. *IET Smart Cities* 4, no. 3 (2022): 211–28. https://doi.org/10.1049/smc2.12036.

Trägerverein Energiestadt. 'Association European Energy Award AISBL'. Energiestadt. Accessed 31 July 2024. https://www.energiestadt.ch/de/ueber-energiestadt/association-european-energy-award-24.html.

Trägerverein Energiestadt. 'Energiestadt'. Energiestadt. Accessed 31 July 2024. https://www.energiestadt.ch/.

Wang, Mengmeng, and Tao Zhou. 'Does Smart City Implementation Improve the Subjective Quality of Life? Evidence from China'. *Technology in Society* 72 (1 February 2023): 102161. https://doi.org/10.1016/j.techsoc.2022.102161.

ZHAW School of Engineering. 'Swiss Smart City Survey 2022', 2022. https://smartcity-survey.ch/en/.

ZHAW School of Engineering. 'Swiss Smart City Survey Dashboard'. Accessed 1 August 2024. https://dashboard.smartcity-survey.ch/en/.

Zhu, Huiying, Liyin Shen, and Yitian Ren. 'How Can Smart City Shape a Happier Life? The Mechanism for Developing a Happiness Driven Smart City'. *Sustainable Cities and Society* 80 (1 May 2022): 103791. https://doi.org/10.1016/j.scs.2022.103791.

AI in Public Administration: The Need for Training for the Successful Smart City Transformation

Serhan Ünalan

1 Introduction

In 1863, when kings and emperors ruled over ancient civilizations, Jules Verne revealed how he sees the future from his perspective in his work "Paris in the 20th Century" (Verne 1996). In this work, Jules Verne predicted that in the 1960s, Paris would have skyscrapers with windows, air conditioners, televisions, elevators, express trains, vehicles, fax machines, and even Internet connections. This book is also considered the first known work on "Smart Cities" in the history of mankind.

If we think this way, how can we judge the level of development of cities and civilizations? What criteria can be used to measure their progress? Is it enough to use global energy consumption or energy production? Or does the amount of data generated show how developed a city or civilization is? Both are important indicators, but they alone are not enough. And what about pollution? After all, the increasing demand for energy and the growing amount of data are closely related to the burden on our environment.

Moving from Jules Verne's visionary account to a modern definition of "smart" cities and advanced civilizations, the question arises as to which characteristics and criteria are actually decisive. While technical innovation and coping with growing energy and data demands play an important role, environmental aspects such as climate change and sustainability must also be taken into account. Against this background, the central challenge arises: How can a city be "smart" and at the same time remain safe, sustainable and liveable?

At the beginning of the nineteenth century, only 2% of the world's population lived in cities. At the beginning of the twentieth century, this proportion rose to 13% and reached 47% in the early 2000s. The United Nations estimates that this figure will rise to 80% by 2050—that would be about 7.5 billion people. Urbanization is thus

S. Ünalan (✉)
Berlin School of Business and Innovation, Berlin, Germany
e-mail: serhan.uenalan@berlinsbi.com

C. Schachtner (ed.), *The European Smart City Movement*,
https://doi.org/10.1007/978-3-031-96700-9_13

progressing three times faster than global population growth. With the development of humanity, global urbanization has become one of the greatest challenges of the twenty-first century. The cities of the future must meet this challenge while offering a high quality of life without radically changing the fundamental structures of the city.

If you superimpose the connections of the members of Facebook, the world's largest social network, with NASA's night view of the Earth, you get an impressive image that shows how urbanized and interconnected the world has become. In this illustration, cities on the east–west axis such as Los Angeles, New York, Amsterdam, Beijing, Seoul and Tokyo are characterized by their high light density, which are also called "smart cities".

This depiction of the networked and urbanized world not only refers to technological progress and the emergence of "smart cities", but also reflects the dynamic shifts in the global economic landscape. Similar to how light density is concentrated on the east–west axis, the world's economic power center has also moved along this axis over the centuries. This is confirmed by studies by Angus Maddison of the University of Groningen, which show that the global economic axis is indeed on the "east–west–east" axis (Maddison Project 2020). In fact, 1000 years ago, the global economic center was in the Middle East, where two-thirds of the world's trade took place. With industrialization, the center shifted to Europe. After the Second World War, especially due to the rise of the USA, it shifted further west. Today, the economic center is moving back east, driven by the rise of Japan and China, also known as the "Asian Tigers." They play a crucial role in this development.

2 Smart City Model and Fields of Action

The Smart City Model provides strategic guidelines for cities to drive the transformation towards a smart city. Cities can use the model to assess their current capabilities in key areas such as data, sensors, energy and infrastructure and develop a tailored strategy. The model also makes it possible to analyze the degree of integration of various smart platform technologies. One example of this is Istanbul, which uses big data, AI, and cloud computing to improve everyday urban life, manage resources efficiently, relieve traffic, and provide scalable data storage solutions.

The model also recommends the introduction of so-called "quick-win" solutions to demonstrate short-term benefits and gain early support from citizens and other stakeholders. This approach makes it easier for cities to balance the long-term goals of their smart city transformation with the short-term needs of the population. It also helps navigate regulatory requirements, optimize resource allocations, and align technology investments with sustainability goals. In this way, a successful transformation towards a sustainable, efficient and livable city is made possible.

Cities, the center of human life, should be based on smart platform technologies, and in particular AI, to create integrated infrastructures that enable solutions and services for smart cities. With the help of sensors, AI algorithms and communication networks, cities become observable, measurable and can be improved in a targeted manner. The availability of data, combined with AI, leads to the digitization of processes, their automation and greater networking. By analyzing and linking this data with the help of AI, innovative solutions and services are created that meet the needs of modern cities. As shown in the upper layer of the Smart City Model (see Fig. 1), there are numerous fields of action in which cities benefit significantly from the use of smart platform technologies, especially artificial intelligence (AI) (Table 1).

The middle layer of the Smart City Model emphasizes the importance of convergence, in which smart platform technologies such as AI, big data, blockchain and the Internet of Things (IoT) intertwine to improve, for example, traffic control, energy management or urban safety. As a result, urban processes can not only be optimized, but also better aligned to future challenges, which significantly drives the transformation towards a smart city.

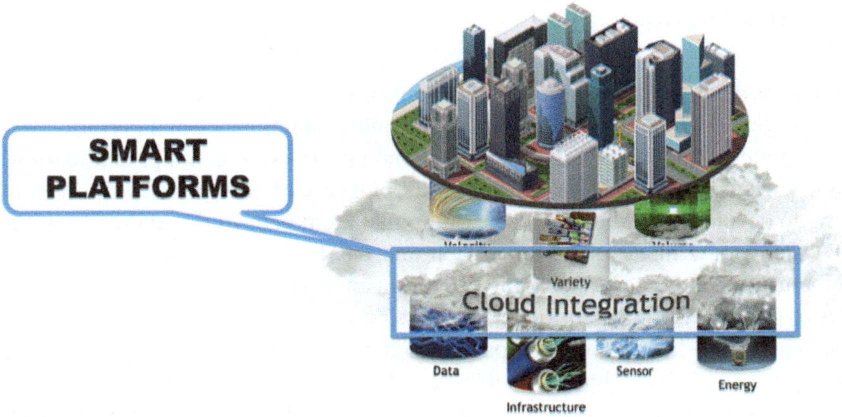

Fig. 1 Smart city model (Ünalan 2016)

Table 1 Fields of action of smart city

Field of action	Description
Life	Optimize resource allocation through AI-based population analytics
Mobility	Intelligent traffic control and autonomous transport solutions
Energy	Efficient energy management through AI-supported forecasts
Environment	Monitoring and improving air and water quality through smart sensors and AI
Security	Crime prevention through AI-powered analytics and real-time monitoring
Health	Use of AI in telemedicine and health monitoring
Administration	More efficient administrative processes through automated AI solutions
Education	Personalized learning platforms and AI-powered teaching methods

In the lower layer of Fig. 1, the megatrends that influence urban development as global changes are discussed. These include advances in data, sensing, infrastructure and energy, based on fundamental principles such as Moore's Principle (for data), Metcalfe's Principle (for network effects), Swanson's Principle (for energy efficiency) and Weiser's Principle (for sensing). These principles thus form the basis for "Smart Cities" to be understood as an integrated platform that ensures safe, sustainable and efficient urban development. AI in particular plays a central role here, as it makes it possible to efficiently analyze the data collected by sensors, recognize patterns and make well-founded predictions. For example, the use of AI in combination with IoT devices enables real-time monitoring and efficient management of urban services. AI is thus becoming a central driver of the transformation towards a smart city.

3 Artificial Intelligence: The Key to Smart City Transformation

AI is one of the most powerful platform technologies shaping the future of cities. It refers to systems or machines that can mimic human intelligence and perform tasks such as pattern recognition, speech understanding, decision-making, and learning from experience. John McCarthy, one of the pioneers of AI, defined it in 1995 as "trying to develop machines that can exhibit intelligent behavior" (Andresen 2002). This vision of AI has evolved since then and has become a central pillar in the development of intelligent and connected urban environments.

The concept of AI as a platform technology is now at the heart of economic growth and technological innovation. Companies such as Google, Microsoft, and Amazon are using advanced AI technologies to power their platforms and offer services such as cloud computing, natural language processing, and predictive analytics. By integrating AI into such platforms, industries have been revolutionized, processes have been made more efficient, innovative solutions have been developed, and the user experience has been significantly improved.

AI not only enables the analysis of large amounts of data, but also acts as the central intelligence of these platforms. In addition to AI, these platforms integrate additional technologies and solutions such as mobile devices, social media, internet connectivity, sensors, and smart metering solutions to create a seamless ecosystem. The integration of third-party solutions and AI makes it easier to reuse and adapt services so that they are accessible to users at all times. AI acts as the driving force behind these developments by enabling efficiency gains, improving data-based decisions and laying the foundation for forward-looking technological advances.

AI also plays a central role in public administration and offers solutions that can significantly increase efficiency and effectiveness. AI-powered systems, for example, analyze large amounts of citizen data to optimize public services such as waste management, energy distribution, or emergency planning. By leveraging data collected through IoT devices and public service platforms, AI ensures that these services are delivered exactly where they are needed most. This not only reduces costs and increases efficiency, but also increases citizen satisfaction.

In addition, AI supports public administration in policy-making through predictive analytics. It recognizes patterns, for example in urban traffic, and provides usable insights that can help improve public transport infrastructure. Anomalies in resource consumption can also be identified to uncover optimization potential. AI also enables dynamic adjustments, for example in real-time traffic management or in the load distribution of energy grids.

The exponential integration of AI into public administration can be explained particularly well by the GPT (General-Purpose Technology) theory, as it highlights the link between technology and civic engagement (Jovanovic and Rousseau 2005). GPT theory describes AI as a cross-sectional technology with the ability to produce transformative effects and provide a platform for innovation. In public administration, AI should not only be seen as a tool for specific tasks, but as a key technology that fundamentally revolutionizes processes and opens up new possibilities.

The GPT theory also highlights that the true value of AI lies in its ability to serve as the basis for further technological advancements. In this way, AI enables the development of new business models, improves efficiency in production and supports the solution of complex societal challenges, such as climate change or urbanization. At the same time, the transformative nature of AI requires a high degree of adaptability in organizations and societies to fully realize its potential.

The GPT theory also emphasizes that AI provides a platform for continuous innovation by serving as a foundation for technological advancements, thus driving the development of other areas such as education, mobility, energy, and health. This theory highlights that the successful integration of AI depends heavily on the ability of executives to strategically use this technology and embed it into existing processes.

Another important dimension of GPT theory is the emphasis on the interdependence between technology and the frameworks in which it is applied. For public administration, this means that the success of AI depends not only on its technological performance, but also on the ability of institutions to adapt their structures and processes accordingly. This includes investing in infrastructure, fostering innovation, and training professionals to ensure that AI is implemented in a way that is both effective and ethical.

Obviously, the smart city transformation requires not only technological adaptations, but also a fundamental change in the mindset and skills of executives and decision-makers.

4 Training Strategies for AI in Public Administration

In recent years, artificial intelligence (AI) has established itself as a key technology that not only automates processes, but also forms the basis for innovations in almost all areas of public administration and business. Their ability to analyze large amounts of data, support decisions and enable new approaches to solutions makes them an indispensable part of modern administrative and economic systems. To unlock the full potential of AI, targeted education and training of employees, especially in public administration, is crucial. Without the appropriate knowledge and competence to use AI efficiently and responsibly, the benefits of this technology cannot be realized.

However, the successful integration of AI in public administration is hampered by numerous barriers. Training programs offer a decisive opportunity to address these challenges in a targeted manner. The following aspects should be taken into account:

1. Cultural barriers: The adoption of new technologies such as AI is often hampered by cultural resistance within organizations. Risk aversion and a lack of acceptance of change prevent a proactive approach to AI. This reluctance is often based on a lack of knowledge or the fear of job loss due to automation. Training can reduce these fears through education and show the opportunities of AI, for example by improving efficiency and making administrative tasks easier.
2. Regulatory and governance barriers: Unclear legal frameworks and fragmented legislation make it difficult to implement AI. Data protection, data security, and ethical issues such as algorithmic fairness are common stumbling blocks. Targeted training can enable decision-makers to understand regulatory requirements and use AI safely and in compliance with the law.
3. Technology barriers: Technologies such as AI require specialized knowledge, especially with regard to their scalability and integration into existing systems. However, many decision-makers have limited technical expertise. Training must close this knowledge gap to ensure that decision-makers understand the technical possibilities and limitations of AI and can effectively control its application.
4. Economic barriers: The adoption of AI technologies is often associated with high costs, and the lack of proven use cases can inhibit investment. Training can help decision-makers analyze the economic benefits of AI initiatives, make informed investment decisions, and convincingly demonstrate the added value of these technologies.
5. Knowledge and competence barriers: Many employees in public administration do not have the necessary knowledge to use AI effectively. Knowledge silos and a lack of exchange between departments exacerbate the problem. Training programmes should be designed to be practical and provide technical and strategic knowledge to ensure effective use of AI.
6. Trust and legitimacy barriers: Distrust of AI as well as unrealistic expectations can hinder the adoption of this technology. Training that addresses ethical issues such as privacy and algorithmic fairness can help build trust and provide a realistic picture of AI.
7. Network and collaboration barriers: A lack of well-connected ecosystems and low stakeholder involvement hinder AI adoption. Training should emphasise the importance of cooperation and promote the development of networked innovation systems by providing a framework for exchange between different actors.

Incorporating AI into public administration training programs is critical to harnessing its transformative impact. Such programs should not only impart technical knowledge, but also address strategic, regulatory and ethical aspects. Decision-makers must be empowered to understand the possibilities of AI, to use it in a targeted manner and at the same time to take into account legal and social requirements. Only through comprehensive and targeted training can AI be used effectively to overcome the challenges of the smart city transformation and create sustainable, innovative solutions.

5 Case Study: AI-Powered Smart Platforms for Urban Energy Optimization

In the course of urbanization and increasing energy demand, cities around the world are striving for greater sustainability. The integration of renewable energy sources into urban systems is a central approach. Smart cities, which enable citizens to be both consumers and producers of energy, play a key role in this. A concrete example is the planned implementation of a platform where citizens can sell surplus solar energy to electric taxi fleets. Although this initiative has the potential to promote clean energy and reduce fossil fuels, its implementation could face significant challenges—especially due to a lack of AI expertise among public sector leaders.

The introduction of such a platform requires overcoming technical, strategic and regulatory hurdles. The main problems include:

1. Analysis of consumption patterns: Predicting peak energy demand and effectively managing excess energy requires the use of advanced AI models. AI-powered predictive algorithms can analyze consumption patterns and make predictions about when demand for energy is expected to be at its highest. This technology allows those in charge to proactively match supply and demand in the city while efficiently managing excess energy.
2. Development of the intelligent platforms: A functional platform requires the seamless integration of energy meters, fleet management systems and payment processing. AI plays a central role here by optimizing the integration of these systems, automating data collection, and making intelligent adjustments to maximize the efficiency of the entire infrastructure. Without the technical knowledge to design and implement such an AI-powered platform, there is a risk of inefficiency and failure.
3. Compliance with data protection laws: Real-time energy transactions involve the collection and processing of sensitive user data. AI can help ensure compliance with data protection regulations by implementing security protocols and privacy mechanisms, while increasing citizens' trust in the system. By using AI, anonymized data can be processed to maintain both privacy and transparency.
4. Promoting citizen participation: To motivate citizens to use the platform, effective communication strategies and user-friendly interfaces are required. AI-powered

engagement tools can improve engagement with the platform by providing personalized recommendations and providing citizens with real-time feedback on their energy consumption. Such tools not only support user engagement, but also promote active and informed participation in the smart city.

Targeted educational programs in the field of AI are crucial to empower those responsible in public administration to overcome these challenges. Such programmes should focus on:

1. Predictive analytics and demand management: Training in the use of AI-based algorithms to analyze historical consumption data and predict energy demand enables proactive decision-making, such as adjusting energy prices or reallocating resources during peak periods.
2. System design and integration: Through training on system architectures and the integration of AI, managers can design platforms that connect different components together and ensure seamless functionality.
3. Data protection and ethical AI: Courses on data management and ethical AI provide the knowledge to navigate complex legal frameworks, implement robust privacy measures, and increase public trust.
4. Stakeholder collaboration and citizen participation: Stakeholders can be trained to use AI tools to improve citizen participation, e.g. by creating dynamic dashboards that provide real-time updates on energy consumption and transactions.

This case study highlights the transformative impact of AI education in addressing urban challenges. By equipping public administrators with the necessary skills, cities can harness the full potential of AI to create integrated, efficient, and citizen-centered systems. Education not only closes knowledge gaps, but also empowers decision-makers to take a leading role in designing sustainable and smart urban environments.

References

Andresen, S. L. (2002). John McCarthy: Father of AI. Intelligent Systems, IEEE, 17(5), 84–85. https://doi.org/10.1109/MIS.2002.1039837

https://www.rug.nl/ggdc/historicaldevelopment/maddison/releases/maddison-project-database-2020?lang=en

Jovanovic, B., & Rousseau, P. L. (2005). General purpose technologies. In Handbook of economic growth (Vol. 1, pp. 1181–1224). Elsevier.

Maddison Project. (2020). Maddison project database 2020. Groningen Growth and Development Centre (GGDC).

Ünalan, S. (2016). Güvenli ve akıllı şehirlerde akıllı platformlar. Harvard Business Review Türkiye. https://hbrturkiye.com/dergi/guvenli-ve-akilli-sehirlerde-akilli-platformlar

Verne, J. (1996). Paris in the 20th century.

Integrating Climate Engineering and Artificial Intelligence: Strategies for Urban Resilience and Sustainable Development in the Face of Climate Change

Adam Jabłoński and Marek Jabłoński

1 Introduction

The importance of digital transformation cannot be overstated, as it has strengthened climate transformation and paved the way for modern solutions in automation, robotics, diagnostics, and artificial intelligence. These technologies, particularly AI, have guided the principles by which various organisations and communities are evolving to shift from negative polarisation to positively impacting the climate. The logic, methodology, and conceptual framework of climate engineering have incorporated artificial intelligence solutions, including for professional forecasting of climate changes in cities and regions, considering social, economic, and managerial aspects.

By framing climate data through the lens of climate engineering, we can delineate strategies across three temporal dimensions: short-term responses for crisis management, medium-term tactical adjustments to safeguard stakeholders, and long-term systemic adaptations incorporating advanced technological and organisational solutions. By examining the initiatives undertaken in Poland, we highlight the nation's commitment to leveraging advanced technologies in alignment with European Union regulations, ultimately contributing to enhanced climate resilience and sustainable development. In navigating the complexities of climate change, adopting a holistic perspective that balances environmental, economic, and social dimensions is essential, fostering resilient communities capable of thriving amidst climate adversities.

A. Jabłoński (✉) · M. Jabłoński
WSB Merito University, Poznań, Poland
e-mail: adam.jablonski@chorzow.merito.pl

C. Schachtner (ed.), *The European Smart City Movement*,
https://doi.org/10.1007/978-3-031-96700-9_14

2 Climate Change as a Platform for Generating Modern Climate Engineering Solutions

Climate change should be considered not only from the perspective of threats but also from the perspective of emerging opportunities for both society and the economy and business within urban and regional ecosystems. Timely information, data, and knowledge on climate change and related forecasts are crucial for decision-makers. Therefore, climate data, guided by climate engineering principles, should be focused on three temporal dimensions:

1. Short-term dimension—concerning rapid response within crisis management to address various climate threats.
2. Medium-term dimension—concerning immediate changes based on tactical-operational management mechanisms to protect various stakeholders from climate risks.
3. Long-term dimension—systematically introducing adaptation and resilience principles to climate change, using newly developed technological, digital, and organizational-management solutions.

In this approach, linking streams of climate-related changes with applicable climate engineering solutions becomes essential. These streams may specifically involve: Adaptive and resilience actions for various regions and cities, considering specific climate risks.

1. Adaptive and resilience actions for different economic sectors, considering their vulnerability to climate change as well as their attractiveness, stability, competitiveness, and maturity in the context of climate threats
2. Adaptive and resilience actions for various areas of human existence at the local community level, incorporating elements of public management impacted by climate change.

This substantively oriented approach can provide tangible and measurable climate benefits and value for various actors engaged in strategically pursuing a climate they find acceptable.

3 Adaptation, Resilience, and Regeneration in Response to Climate Change from the Perspective of Cities and Regions

Adaptation to climate change arises directly from the direct and indirect effects of climate change. According to the EN ISO 14090:2019 standard (EN ISO 14090:2019), climate change adaptation is adjusting to the actual and expected climate and its impacts. In this context, adaptive capacity becomes crucial. As defined by this standard, adaptive capacity includes the ability of systems, institutions, people, and other organisms to adjust to potential damages, take advantage of opportunities, or respond to consequences. Such adaptive capacity should be extrapolated to many areas of urban and regional functioning (Aaheim et al. 2008).

This is, among other things, implemented through the development of Urban Climate Change Adaptation Plans as part of city development programming. An urban adaptation strategy can serve as a roadmap for local authorities to increase cities' resilience to the impacts of climate change. It should be a document setting main and specific goals and action directions for adapting the city and its residents to current and expected climate threats. A priority should be to create conditions for stable socio-economic development in the face of climate change-related risks (Borek and Siwior 2021).

Investment-based solutions are applied in the context of climate change resilience in cities and regions, where specific mechanisms for ensuring climate security are implemented. Climate resilience is focused on actions related to climate neutrality, aimed at mitigating climate change, and infrastructure resilience to climate change through dynamic adaptation and adjustment. Regeneration, on the other hand, in the face of climate change, viewed from the perspective of cities and regions, is a dynamic process based on the implementation of a series of initiatives that effectively change the logic of understanding the climate, its threats, and potential growth and development opportunities. The concept of a regenerative economy stems from regenerative agriculture, which emerged when strategies to move away from intensive, industrial agriculture principles were implemented. It is important to note that in the case of cities and regions, applying the triad of climate change models for cities and regions is essential. This way, actions based on modernisation and strategic renewal of these areas will be comprehensive and systemic. In Figure 1, the discussed triad is presented.

When skillfully interpreted, this triad should generate streams of decision-making processes in climate change based on the principles of dynamic strategic equilibrium in the development and growth of cities and regions.

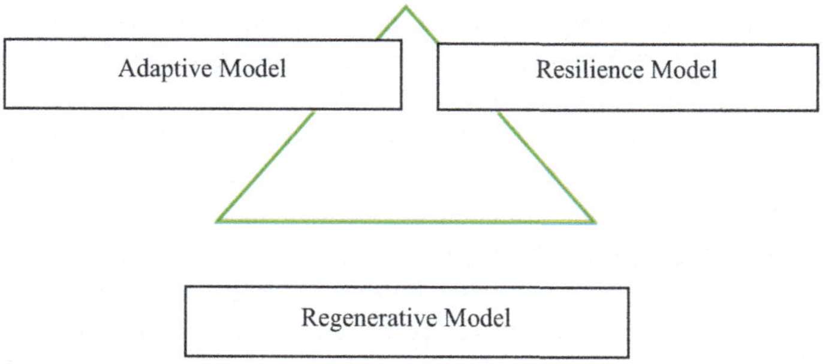

Fig. 1 Triad of climate change models for cities and regions

4 Climate Services as a Tool for Solving Climate Problems of Cities and Regions

Climate services result from an evolutionary process of understanding climate threats, seen spatially and across different dimensions of climate security management in cities and regions. The development of climate services should enhance their effectiveness and efficiency, and their quality should build a multi-dimensional trust in their credibility and utility for individual cities and regions. Climate services have been defined by the "WMO Global Framework for Climate Services" as "the provision of climate information in such a way as to assist decision-making" (Hewitt et al. 2012). Climate services include those functions that integrate them into business models (Larosa and Perrels 2017).

This requires a comprehensive understanding of the current socio-economic ecosystem open to generating positive climate changes that modify currently exploited processes, projects, strategies, and business models of various entities influencing the climate. Table 1 presents selected types of climate services applicable to cities and regions.

The presented set of climate services, capable of strongly shaping mechanisms for effective and efficient urban and regional management, is open. However, the diversity of these services clearly and transparently demonstrates their immense substantive potential. This can generate a range of positive actions by decision-makers to protect the climate and mitigate climate-related threats and potential disasters.

5 Climate Forecasting is Critical in Mitigating Climate Risks and Disasters

Climate forecasting based on prediction and proactive anticipation of climate risks and disasters determines the impact of climate change on cities and regions. Climate forecasting using artificial intelligence (AI) involves employing AI technologies such as machine learning and data analysis to predict future outcomes related to climate and its positive and negative changes based on available, extracted secondary historical data. By optimally utilising the potential of AI in climate change forecasting, decision-makers in cities and regions can generate precise climate forecasts, anticipate trends, better understand climate dysfunctions and anomalies, and thus make more accurate and logically justified organisational and management decisions. They can utilise various AI-based modelling solutions. The application of neural networks allows for broad recognition of complex patterns in climate data, which can result in highly accurate weather predictions. Implementing actions supported by support vector machines can enable precise classification and categorisation of climatic conditions and changes and comprehensive prediction of specific weather variables, such as temperature or rainfall. The random forest model facilitates the efficient management of large datasets containing numerous climatic variables, enabling the

Table 1 Selected types of climate services applicable to cities and regions

L.P.	Type of climate service
1	Maps and applications providing real-time climate status
2	Databases on climate change
3	Dedicated climate and agronomic models
4	Services generating climate data and forecasts
5	Expert services, including data analysis and model generation to develop contextual weather forecasts and related advice
6	We integrated advisory services for decision-making on climate change prediction and mitigation
7	Interdisciplinary services in climate management, climate engineering, and climate data management
8	Tools for creating climate projection charts
9	Climate risk modelling services focus on identifying and measuring climate-related hazards, opportunities, and risks
10	Decision-support systems for climate change processes
11	Integrated regional research and development programs on climate change
12	Joint systems and programs for climate change adaptation, creating climate knowledge
13	Support for local climate change adaptation through participatory GeoWeb systems
14	Workshops and training sessions raise awareness among society and businesses about climate change while promoting tools and solutions used in climate-supporting processes
15	Seasonal climate and agricultural forecasts tailored to specific locations
16	Organising dedicated events, festivals, and study meetings promoting climate services in traditional and online formats, increasing climate awareness and sensitivity, and fostering interdisciplinary networks and relationships in the field of climate change

creation of comprehensive and multi-criteria analytical models. The concept of scalability of climate change models becomes crucial here, as they enable real-time forecasting processes. For example, if climate-change and human-activity-related indicators are extracted from remote sensing images, with the power of AI methods, there would be possibilities for:

1. Creating rapid maps of land use and environmental resources on large scales,
2. Making predictions states of the extracted indicators,
3. Simulating hypothetical scenarios for disaster prevention,
4. Identifying abnormal situations (outlier identification),
5. Explaining the impact of the indicators with eXplainable AI (XAI) methods,
6. Identifying the relation between human activity, climate change and biodiversity-related changes,
7. Real-time predictions and actions in an urban environment based on sparse measurements from within the city (Sirmacek and Vinuesa 2022).

These solutions open the way, among other things, to a significant strategic change in decision-making processes regarding understanding and influencing climate change from the perspective of cities and regions.

6 Artificial Intelligence Solutions and the Principles of Digital, Climate, and Energy Transformation at the City and Regional Levels

Artificial intelligence (AI) solutions possess immense intellectual potential at the application level. The integrated implementation of digital, climate, and energy transformation principles at the city and regional levels should focus on the technological and organizational-managerial strength embedded in AI's logic and implementation-oriented nature. Figure 2 illustrates the triad of economic and societal transformation.

The presented economic and societal transformation triad principles can be implemented through business model concepts, especially as climate policy and climate change adaptation are now becoming strategic priorities in Europe (Dolsak and Prakash 2018; Orlove 2022). Decision-makers across various organisations strive to identify the best possible managerial solutions and optimal economic approaches, implementing them into different areas of the socio-economic ecosystem to improve human quality of life and enhance the effectiveness and efficiency of businesses, economies, and local communities. This ecosystem is supported by digitalisation solutions (Kohtamäki et al. 2019). In the context of climate change, a new taxonomy of business models is emerging. Among the business models shaping climate-related components are Sustainable business models (Jabłoński 2013a), Circular business models (Mentink 2014), Social business models (Yunus et al. 2010), Progressive business models (Rochlin 2021), Inclusive and adaptive business models (Rosenstock et al. 2020), Green business models (Jabłoński 2013b), and Technological business models (Baden-Fuller and Haefliger 2013; Jabłoński 2015). As a result, a new type of business model has emerged: climate-changing business models. These

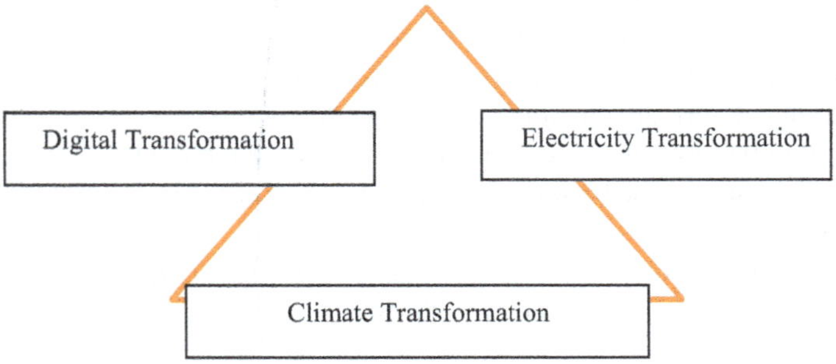

Fig. 2 The triad of transforming the economy and society

are closely linked to climate services (Larosa and Mysiak 2020). A climate-changing business model focuses on creating climate-related benefits and value for various groups within the societal, economic, and business ecosystem. The outcomes of such models are often climate-positive services that generate both environmental and economic benefits. The platform for creating such business models is the organisational culture shaped around the principles of climate responsibility in business (Jabłoński 2023). Climate strategies also have their conceptual basis in the scientific literature (Paul et al. 2017), supported by research-driven solutions related to digital transformation (Fischer et al. 2020).

7 Principles of Climate Adaptation for Cities and Regions Using the Climate Digital Twin Concept

Climate adaptation for cities and regions should be equipped with appropriate tools tailored to the specific solutions capable of generating the desired outcomes. The Climate Digital Twin concept is an essential solution within digital transformation for climate purposes. The Climate Digital Twin represents a new system for collecting and utilising climate-related information, data, and knowledge. It enables multidimensional assessments of the impact of climate change on various types of cities and regions, allowing for the individualised selection and implementation of diverse strategies for adaptation, resilience, and regeneration in response to dynamic climate changes. The origins of the Digital Twin concept are attributed to the work of Michael Grieves in collaboration with John Vickers from NASA, who introduced the idea during a lecture on product lifecycle management in 2003 (Grieves 2014). Digital Twin is a complete virtual description of a physical product that is accurate to both micro and macro levels. As highlighted in (Kazała et al. 2022; Jones et al. 2020), various types of Digital Twins correspond to different phases of a system's lifecycle:

1. Digital Twin Prototype (DTP) represents the design version of a product, including all its variants. DTPs should be developed for every manufactured product to capture the full scope of design possibilities.
2. Digital Twin Instance (DTI): a digital Twin of each individually produced artefact. DTIs are maintained only for products where retaining information about that specific product is crucial throughout its life. These instances operate within a Digital Twin Environment (DTE), enabling ongoing interaction and data exchange.
3. Digital Twin Aggregates (DTAs) are composites that aggregate all the DTIs. By compiling information from multiple DTIs, DTAs provide a holistic view, facilitating analysis and decision-making at a broader level.

The conceptual model of the digital twin DT was initially named the Mirrored Spaces Model (Grieves 2005) and later referred to as the Information Mirroring Model (Grieves 2006; Kazała et al. 2022). This evolution in terminology provides a broader perspective on applying the digital twin concept in monitoring and predicting climate change phenomena.

The concept of Climate DT encompasses several key components related to climate change phenomena, including:

1. a multi-model approach to global climate simulations,
2. both qualitative and quantitative assessments of climate data,
3. the development of climate and non-climate value carriers for cities and regions,
4. integrating diverse research and development management systems to address climate change.

This concept's key technological parameters include the principle of data consumer, understood as a person who uses data to make decisions, solve problems, or gain new insights. Data consumers encompass a wide range of analytical data, from individual data of specific entities to broadly interpreted collective data. They rely on data to make informed choices and find optimal economic, managerial, and technological solutions to previously defined problems. They use data to stay well-informed and make better decision-making processes. Another essential attribute of Climate DT is interactivity as an expression of the extent that in a given series of communication exchanges, any third (or later) transmission (or message) is related to the degree to which previous exchanges referred to even earlier transmissions (Rafaeli 1988). Another key attribute of Climate DT is scalability. Scalability refers to ensuring increasingly efficient performance as the number of components grows. It is, among other things, a feature of computer networks, referring to their ability to expand continuously. Scalability is sometimes defined as "the ease with which a system or component can be modified depending on the type of problem." A scalable system has three superficial characteristics:

1. The system can adapt to increased usage.
2. The system can accommodate more significant amounts of data.
3. The system is easy to maintain and operates with reasonable efficiency.

Scalability is not just about speed. The efficiency and scalability of a system differ and are correlated with each other. Efficiency measures how quickly and effectively a system can perform specific computational tasks, while scalability measures the trend of efficiency under increased. Daniel A. Menascé and Virgilio A.F. Almeida argue that a system is scalable if there is a "simple" way to update the system to handle increased traffic while maintaining proper efficiency. By "simple," they mean that no system architecture or software changes should be required to scale the system (Menascé and Almeida, 2000). The Universal Scalability Law (USL) in computer science is a model for predicting hardware and software scalability. It uses system performance as a load function to predict a system's scalability. The USL function creates a model from a formula and a data frame. The USL model results in two coefficients: sigma, which models the assertion, and kappa, which models the system's consistency delay. The Universal Scalability Law was formulated by Neil J. Gunther (Gunther 2007, 2008). The application of the Climate Digital Twin concept is becoming increasingly advanced, which should undoubtedly positively impact the quality of the climate data received and interpreted across cities and regions.

8 Strategic Recommendations for Cities and Regions Resulting from the Potential of Artificial Intelligence in Climate Change Forecasting

In the presented strategic context of socio-economic conditions within the process of ongoing dynamic climate changes, it becomes crucial to define a set of practical solutions dedicated to cities and regions to create positive climate-changing solutions derived from the potential of artificial intelligence. These solutions should result from a well-designed and implemented city and regional management model. Table 2 outlines key dedicated solutions that could have a robust application in the context of cities and regions.

The presented strategic recommendations can serve as a guiding compass for strategic forecasting, planning, and implementing dedicated development and growth strategies for cities and regions. They can also be an essential platform for what is known as endogenous regional development. This concept suggests that growth is a spatially cumulative process that can deepen regional development inequalities. According to this theory, well-developed countries or regions have a greater chance of further development than very underdeveloped countries. Earlier neoclassical models did not consider technological progress an endogenous development factor. The Solow-Swan model is a simple macroeconomic exogenous growth model that uses a production function, which relates the output level to the amount of production factors used (labor, capital, and the state of technology). The production function used can be, for example, the Cobb-Douglas function or the CES production function. In this context, technological progress related to the development of artificial intelligence could be a decisive factor driving endogenous regional development.

9 Report on the Application of Artificial Intelligence Solutions in Climate Change Forecasting in Cities and Regions in Poland

A key factor in applying artificial intelligence in climate change forecasting in cities and regions, in the context of creating social, economic, and managerial aspects, is the discovery of opportunities to apply these principles in the practical management of cities and regions. Therefore, an effort has been made to present solutions implemented in this area in Poland at various maturity levels. This was achieved by verifying different types of climate-related initiatives in Poland against the backdrop of technological development and the organisational and managerial framework of

Table 2 A set of strategic recommendations for cities and regions resulting from the possibility of using artificial intelligence in climate change forecasting

L.P.	Set of strategic recommendations for cities and regions resulting from the application of artificial intelligence in climate change forecasting
1	The climate digital twin concept (Climate Changing Digital Twin) application to develop effective and efficient weather risk management mechanisms
2	Construction of modern hydrological capacity management systems through analysing rainfall data concerning wastewater volumes in sewage systems
3	Development of analytical models and measurement streams in climate change forecasting using artificial intelligence solutions
4	Developing dedicated adaptation strategies to climate change using global simulation models (Multi-model global climate simulations)
5	Implementing climate value chain principles in urban resource management processes (Climate Services Value Chain)
6	Implementation of climate resilience principles in the effective management of urban resources
7	Development of measurement systems for urban resource management, energy consumption, and generation of energy surplus using intelligent diagnostic measurements
8	Development of analytical systems based on the frequency of climate change occurrences, meteorological data trends and their consistency with resource utilisation perspectives in a given region
9	Consideration of urban resource vulnerability (sensitivity and exposure) to climate change
10	Development of intelligent urban resource management systems
11	Conducting analyses to minimise uncertainties in climate forecasts related to a region's budget and energy balance and assessing its vulnerability to climate change
12	Development of analytical models based on the correlation between rising air temperatures and the consumption of urban resources in a given region
13	Development of technological solutions based on circular economy principles along with the use of renewable energy sources
14	Participation in electro-energy clusters to generate energy self-sufficiency for the region in a prosumer model
15	Development and implementation of technological solutions aimed at converting the use of fossil fuels into the creation of new renewable energy potential
16	Use of artificial intelligence for technological management of municipal enterprises
17	Linking new economic solutions with providing new services for the climate (circular economy, sharing economy, big data economy, Internet of Things economy, artificial intelligence economy)
18	Development of virtual reality as an accelerator for creating new analytical and measurement models for cities and regions
19	Development of digital platforms based on algorithms and Big Data environments dedicated explicitly to cities and regions as tools for creating new climate services
20	Transformation of linear business models in water and sewage enterprises to circular ones, using green business scalability mechanisms
21	Ensuring technological, energy, and cybersecurity for water and sewage enterprises through the use of digital business models

the country, as well as the applicable legal regulations, primarily those related to European Union legislation implemented into Polish conditions. Table 3 presents a set of selected artificial intelligence solutions applied to climate change forecasting in cities and regions in Poland.

Table 3 Set of selected artificial intelligence solutions applied in climate change forecasting in cities and regions in Poland

L.P.	Set of selected artificial intelligence solutions applied in climate change forecasting in cities and regions in Poland
1	Implementation of IT systems for calculating carbon footprints
2	Implementation of IT systems for calculating carbon footprints
3	Implementation of Google Cloud Platform solutions, which provide advanced AI tools, infrastructure, software solutions, and security systems
4	Implementation of energy management systems that monitor and optimise energy consumption in real time
5	The development of small modular reactors (SMR) supported by AI solutions offers new possibilities for a sustainable energy supply
6	Forecasting energy production using machine learning algorithms for predictive solar energy generation forecasting
7	Implement AI systems to increase wind energy by predicting wind patterns and optimising wind turbine performance
8	Implementation of predictive maintenance solutions in renewable energy systems
9	Implementation of AI systems for efficient energy storage, distribution, and consumption management
10	Use of predictive analytics solutions to forecast future levels of emissions into water, soil, and air based on real-time data and secondary analysis trends, focusing on proactive decision-making processes
11	Big Data analytics can be applied in climate change assessments by processing vast amounts of climate data to identify patterns and make highly accurate climate predictions
12	AI can be applied to monitor biodiversity and obtain early warnings about potential environmental degradation
13	Use image and sound analysis logic to identify and track animal species using AI algorithms
14	The application of quantum computers is due to their capability to process complex climate models much faster than traditional computers, generating more accurate results and forecasts
15	Implementing systems integrating AI solutions with IoT devices for more precise monitoring of ecosystems and energy management
16	Implement monitoring systems to ensure compliance with legal requirements and enhance the enforcement of environmental regulations
17	The application of AI in monitoring wildlife to prevent poaching and protect endangered species
18	We are employing AI solutions to forecast and optimise environmental conditions in agricultural resource management

The presented specific climate solutions reinforce the importance of this management area in Poland, mainly through the developmental cohesion of the entire European Union, supported by the generation of solid components of intellectual, social, relational, and innovative capital in Poland. This strategic approach aims to address immediate environmental challenges and fosters long-term sustainability and resilience within local communities. By leveraging advanced technologies and collaborative efforts, Poland can enhance its capacity to adapt to climate change while promoting economic growth and social equity.

10 Conclusion

Climate change's intricate and multifarious character mandates a comprehensive approach that considers both the perils and prospects it offers. Urban areas and regions can formulate effective strategies that bolster resilience and adaptability by applying climate engineering principles across short-, medium, and long-term frameworks. The capacity to acclimate to climate change's repercussions is paramount for urban and regional systems. The formulation of Urban Climate Change Adaptation Plans can function as critical frameworks for local governance, allowing them to delineate explicit objectives and actions to alleviate climate-related risks while fostering sustainable socio-economic advancement. Climate services constitute essential instruments in navigating climate-induced challenges. These services augment decision-making processes and cultivate stakeholder confidence by delivering timely and pertinent climate data. A broad spectrum of climate services can empower urban areas and regions to execute efficacious climate management strategies. Artificial intelligence (AI) deployment catalyses climate forecasting and presents substantial potential for enhancing predictive capabilities and decision-making methodologies. AI technologies can analyse intricate climate datasets, discern trends, and simulate various scenarios, thereby facilitating more precise and informed responses to climate-related hazards. The Climate Digital Twin paradigm signifies a transformative methodology towards climate adaptation, permitting real-time scrutiny and evaluation of climate effects. By amalgamating diverse data sources and enabling interactive decision-making, this paradigm can markedly elevate the efficacy of climate strategies. Strategic recommendations customised for urban areas, and regions can direct the practical utilisation of AI and other innovative methodologies in forecasting climate change. These recommendations underscore the importance of constructing resilient urban systems and endorsing sustainable practices congruent with climate objectives. The initiatives implemented in Poland exemplify the nation's dedication to harnessing advanced technologies for climate change forecasting and management. Poland can augment its climate resilience and contribute to overarching sustainability initiatives by aligning with European Union regulations and fostering innovation. Tackling climate change necessitates a comprehensive perspective that equilibrates environmental, economic, and social dimensions. Urban areas and regions can cultivate resilient communities capable of thriving amidst climate adversities by emphasising long-term sustainability and advocating for social equity.

References

Aaheim A., Berkhout F., McEvoy D., Mechler R., Neufeldt H., Patt A., Watkiss P., Wreford A., Kundzewicz, Z., Lavalle, C., Egenhofer, C. (2008). Adaptation to Climate Change: Why is it Needed and How Can it be Implemented? CEPS. Brussels. Belgium.

Baden-Fuller, Ch., Haefliger, S. (2013). Business Models and Technological Innovation, Long Range Planning 46, 419–426, https://doi.org/10.1016/j.lrp.2013.08.023

Borek, A., Siwior, P. (2021). Charakter prawny miejskich planów adaptacji do zmian klimatu w Polsce, w: Adaptacja do zmian klimatu w unijnej i polskiej polityce klimatycznej oraz prawie Klimatycznym, Wybrane zagadnienia, Wydawnictwo IOŚ-PIB, Warszawa.

Dolsak, N., Prakash, A. (2018). The Politics of Climate Change Adaptation, Annu. Rev. Environ. Resour. 43:317–41.

EN ISO 14090:2019, Adaptation to climate change—Principles, requirements and guidelines.

Fischer, M., Imgrund, F., Janiesch, Ch., Winkelmann, A. (2020). Strategy archetypes for digital transformation: Defining meta objectives using business process management, Information & Management 57, 103262, https://doi.org/10.1016/j.im.2019.103262

Grieves, M. (2005). Product Lifecycle Management: the new paradigm for enterprises. Int. J. Product Development, 2(Nos. 1/2), 71–84.

Grieves, M. (2014). Digital twin: manufacturing excellence through virtual factory replication. Google Scholar.

Gunther, N.J. (2007). Guerrilla Capacity Planning: A Tactical Approach to Planning for Highly Scalable Applications and Services. Springer, Heidelberg, Germany, 1st edition.

Gunther, N.J. (2008). A general theory of computational scalability based on rational functions. CoRR, abs/0808.1431.

Hewitt C., Mason S., Walland D. (2012), The global framework for climate services, "Nature Climate Change", Vol. 2, No. 12, pp. 831–832.

Jabłoński, A. (2013a), Modele zrównoważonego biznesu w budowie długoterminowej wartości przedsiębiorstw z uwzględnieniem ich społecznej odpowiedzialności, Difin, Warszawa.

Jabłoński, A. (2013b), Zielony wymiar budowy modeli biznesu i strategii przedsiębiorstw, [w:] Edytor: Urszula Wąsikiewicz-Rusnak (praca zbiorowa), Współczesne problemy rozwoju przedsiębiorstwa w koncepcji rozwoju zrównoważonego, WSB Dąbrowa Górnicza.

Jabłoński, A. (2015). Design and Operationalization of Technological Business Models, Acta Universitatis Agriculturae et Silviculturae Mendelianae Brunensis, Vol. 63, No. 3.

Jabłoński, A. (2023). Usługi klimatyczne w kształtowaniu modeli biznesu przedsiębiorstw, Przegląd Organizacji 4/2023.

Jones, D., Snider, Ch., Nassehi, A., Yon, J., Hicks, B. (2020). Characterising the Digital Twin: A systematic literature review, CIRP Journal of Manufacturing Science and Technology 29, 36–52, https://doi.org/10.1016/j.cirpj.2020.02.002

Kazała R, Luściński S, Strączyński P, Taneva A. An Enabling Open-Source Technology for Development and Prototyping of Production Systems by Applying Digital Twinning. Processes. 2022; 10(1):21. https://doi.org/10.3390/pr10010021

Kohtamäki, M., Parida, V., Oghazi, P., Gebauer, H., Baines, T. (2019). Digital servitization business models in ecosystems: A theory of the firm, Journal of Business Research 104, 380–392, https://doi.org/10.1016/j.jbusres.2019.06.027

Larosa, F., Mysiak, J. (2020). Business models for climate services: An analysis, Climate Services 17, 100111, https://doi.org/10.1016/j.cliser.2019.100111

Larosa F., Perrels A. (2017), Assessment of the existing resourcing and quality assurance of current climate services, Grant agreement 730500, EU-MACS European Market for Climate Services, Deliverable 1.2.

Menascé, D.A, Almeida, V. (2000). Challenges in scaling e-business sites, Proc. 2000 Computer Measurement Group Conference, Orlando, FL, December 10–15.

Mentink, B. (2014). Circular Business Model Innovation: A process framework and a tool for business model innovation in a circular economy. MSc thesis Industrial Ecology Master Thesis. Delft University of Technology in Delft (Holand) and Leiden University in Leiden (Holand).

Orlove, B. (2022). The Concept of Adaptation, Annu. Rev. Environ. Resour. 47:535–81.

Paul, A., Lang, J.W.B., Baumgartner, R.J. (2017). A multilevel approach for assessing business strategies on climate change, Journal of Cleaner Production 160, pp. 50–70, https://doi.org/10.1016/j.jclepro.2017.04.030

Rafaeli, S. (1988). 'Interactivity: From New Media to Communication', In: R.P. Hawkins, J.M. Wieman and S. Pingree (eds) Advancing Communication Science: Merging Mass and Interpersonal Processes, pp. 110–34. Newbury Park, CA: Sage.

Rochlin C. (2021). Climate change and the progressive business model, "The Electricity Journal", Vol. 34, No. 4, art. 106927.

Rosenstock, T.S., Lubberink, R., Gondwe, S., Manyise, T., Dentoni D. (2020). Inclusive and adaptive business models for climate-smart value creation, "Current Opinion in Environmental Sustainability", Vol. 42, pp. 76–81.

Sirmacek, B., Vinuesa, R. (2022). Remote sensing and AI for building climate adaptation applications, Results in Engineering 15, 100524, https://doi.org/10.1016/j.rineng.2022.100524

Yunus, M., Moingeon, B., Lehmann-Ortega, L. (2010). Building social business models: lessons from the Grameen experience, "Long-range Planning", Vol. 43, No. 2/3, pp. 308–325.

Smart Governance in Romanian Municipalities: Progress, Challenges, and Opportunities

Catalin Vrabie

1 Introduction

The COVID-19 pandemic has significantly altered various dimensions of our daily lives, including the functioning of cities and the services they offer to their residents. This global crisis has highlighted the critical need for public institutions to be resilient and adaptable, especially within the smart cities' context. A smart city leverages technology and data to enhance citizens' quality of life while ensuring sustainable development (Anthony and Andreas 2023; British Standards Institution 2014; Cisco IBSG 2012; Vrabie and Dumitrascu 2018).

The pandemic had a profound impact on the use of technology and its role in bolstering the resilience of public institutions in smart cities, driving a significant push towards the digital transformation of services, thereby substituting traditional in-person activities (Matthias et al. 2020; Kuhlmann et al. 2021; Müller-Török and Prosser 2021). The imperative for social distancing and remote communication during the pandemic has underscored the essential role of technology in maintaining the continuity of public services during crises. In this framework, smart cities are poised to play a crucial role in the provision of public services during emergencies. The ability of public institutions to adapt to technological changes and new conditions is essential for the successful implementation of smart city projects (Anthony and Andreas 2023; Schachtner 2021; Šiugždinienė et al. 2017; Nature Research 2022; WHO 2023).

This article seeks to examine the resilience of municipalities within the smart city paradigm in the aftermath of the COVID-19 pandemic. The focus is on how public institutions adapt to technological advancements and their capability to deliver services through remote methods (both online and mobile). The study aims to present

C. Vrabie (✉)
Faculty of Public Administration, National University of Political Studies and Public Administration, Bucharest, Romania
e-mail: catalin.vrabie@snspa.ro

C. Schachtner (ed.), *The European Smart City Movement*,
https://doi.org/10.1007/978-3-031-96700-9_15

new evidence on the durability of public institutions amidst technological changes and the importance of smart web platforms in this process. By analyzing the level of digitalization in public services across Romanian municipalities as of 2023 and comparing it with the progress of web initiatives from 2014 and 2019, this research aims to contribute to the ongoing discussion about the future of public services in the digital age, the role of public institutions in urban transformations, and the resilience of public institutions in times of crisis.

2 Literature Review

In the past decade, there has been significant advancement in the concepts and implementations of smart cities and digital governance. Smart cities leverage technology and data to enhance the quality of urban life, optimize resources, and ensure sustainable development. Recent research has focused on various dimensions of smart cities, including technological, social, economic, and environmental aspects.

One key area of focus has been Social and Economic Impact. Recent literature highlights the importance of citizen participation in the success of smart city projects. Digital platforms for citizen engagement, such as mobile apps and online portals, facilitate better communication between residents and municipal authorities (Albino et al. 2015). Additionally, the economic implications of smart cities are substantial, with improved infrastructure and services attracting investment and fostering innovation. Research indicates that smart city initiatives can lead to job creation and economic growth (Angelidou 2017; Hollands 2008).

Another dimension is Digital Governance Innovations. Digital governance involves the use of digital tools and technologies to improve public administration and service delivery. Recent advancements have significantly transformed how governments interact with citizens and manage public resources. A major development in this field is the evolution of E-Government Services. The transition from basic online services to more integrated and interactive platforms has been a key trend. The concept of e-government 3.0 emphasizes seamless service delivery across multiple channels, including web, mobile, and social media (Linders 2012; Anthopoulos 2017; Vrabie 2023; Schachtner 2022). Additionally, there is a growing focus on designing e-government services that are user-friendly and accessible. Research highlights the importance of involving users in the design process to ensure services meet their needs (Janssen et al. 2020; Vrabie 2022).

Transparency and Accountability have also seen significant enhancements. Governments are increasingly adopting open data policies to promote transparency and accountability. Studies show that open data can enhance public trust and enable citizens to hold authorities accountable (Janssen et al. 2017).

Another notable advancement is the application of AI and Machine Learning in Governance. AI and machine learning are being used to predict and respond to public needs more effectively. These technologies can analyze vast amounts of data to forecast trends and improve decision-making (Galetsi et al. 2020; Vrabie 2024a).

Additionally, AI is employed to automate routine administrative tasks, freeing up human resources for more complex duties. This can lead to more efficient public service delivery and reduced operational costs (Deng et al. 2021; Vrabie 2024b).

The concept of integrating web services was notably advanced by Tapscott (2008) in his seminal book "Grown Up Digital: How the Net Generation is Changing Your World," (Tapscott 2008; Vrabie 2015) and further explored by Homburg and Dijkshoorn (2011) in their examination of the "Diffusion of Personalized E-Government Services among Dutch Municipalities." These services are designed to facilitate citizen interaction with public administration, transforming the engagement into a personalized, one-to-one interaction through authentication processes and profile assignments.

The integration of Technological Innovations such as IoT and big data analytics has been a critical driver in smart city initiatives. These technologies enable real-time data collection and analysis, leading to more responsive and adaptive city management (Batty et al. 2012; Vrabie 2018). Furthermore, AI technologies are increasingly being utilized for predictive analytics, automated decision-making, and enhancing service delivery. Studies have shown AI's potential to improve traffic management, energy efficiency, and public safety (Kitchin 2013).

An exemplary framework for studying digitalization at the local level is presented by Marc Holzer, who conducted a comprehensive series of analyses on municipalities globally from 2005 to 2016, encapsulated in the work "Digital Governance in Municipalities Worldwide" (Holzer and Manoharan 2016). Additionally, the Journal of Web Semantics, published by Elsevier, offers a collection of articles (issues from 2016 to 2023) closely related to the design and content of websites.

A thorough review of research published in Government Information Quarterly (issues from 2020 to 2023) and the International Journal of Web Services Research (issues from 2020 to 2023) reveals that the bulk of scholarly attention is predominantly focused on e-government in a general context. There is a noticeable paucity of studies emphasizing its large-scale implementation, particularly at the national level, which aligns closely with the specific focus of our study.

Extensive literature on public management underscores the efficacy of web applications in addressing these challenges and fostering robust governance (Klievink and Janssen 2019; Deng et al. 2019; Cai and Zhu 2021; Alves and Gonçalves 2021; Nesti 2018). For instance, web applications can significantly enhance the quality and efficiency of public service delivery (Maiti et al. 2021; Thakhathi and Langa 2022), streamline administrative operations, and support decision-making processes, particularly when augmented by artificial intelligence applications (Schachtner 2021; Shafique et al. 2020). Furthermore, online platforms can improve transparency and accountability, while simultaneously increasing citizen participation and engagement (Etscheid 2019; Kolkman 2020).

In the European Smart Cities sphere, Estonia and Denmark stand out. Estonia is renowned for its e-government solutions, offering comprehensive digital services to its citizens. Denmark has also made significant strides in integrating digital technologies in public administration. Comparative studies can illustrate how different governance models and cultural contexts impact the success of smart city initiatives (Lember et al. 2018; Kattel et al. 2019).

From a Global Perspective, countries like Singapore and South Korea are at the forefront of smart city development. In North America, cities such as Toronto and New York are implementing innovative digital governance practices. Examining these examples can provide insights into diverse approaches to smart city challenges (Chourabi et al. 2012; Kim et al. 2017).

3 Romania's E-Gov Implementation Research Context

Romania is located in Central-Eastern Europe and is the twelfth-largest country on the continent. It is also the sixth most populous in the European Union. The country is divided into 41 counties, each with several municipalities, cities, and towns. Municipalities are the richest administrative units and have the most self-governing tasks, which is why this study focuses on them. Municipalities significantly influence the level of digitalization in their areas.

The country is classified as a decentralized state, which means that municipalities enjoy a high degree of autonomy relative to the central public administration (Iancu 2013). This autonomy extends to the design and management of electronic services (Baltac 2011; Vrabie 2009a, 2011). At the central level, digitalization initiatives are coordinated by the Ministry of Research, Innovation and Digitalization, often in collaboration with the Ministry of Development, Public Works and Administration. Unlike other European countries such as Estonia and Denmark, Romania does not have a Chief Information Officer (CIO) at both central and local levels to coordinate digitalization efforts aimed at achieving inter and intra-institutional collaboration.

The concept of digitalization, particularly its various "stages" or "maturity grade," is often assessed by the manner in which electronic public services are delivered via the Web. At one end of this spectrum is the static display of information, while at the other end is the delivery of services in a fully integrated and continuous form (Baltac 2011; Pardo 2000; Vrabie 2009b). The characteristics of the latter emphasize a seamless, comprehensive, and effortless provision of information within the administrative Web space.

The absence of a CIO in Romania may result in a lack of coherent strategy, inefficient use of resources, and slower progress in digitalization efforts (United Nation 2020; Ojo et al. 2007; Obi 2007).

4 Research Methodology

We collected data from the official websites of all 103 municipalities in Romania using the ParseHub API. This process was done twice within a week to ensure comprehensive and accurate data collection, even if some websites were temporarily down during the initial query.

We organized our analysis into five main categories:

1. Transparency: How well the municipality adheres to regulations concerning transparency.
2. E-Documents: Availability and functionality of electronic documents and forms.
3. Communication: How the municipality uses new media for communication.
4. Useful Content: Availability of useful online information, such as city maps and live cameras.
5. Generalities: Overall user-friendliness and design of the municipal websites.

Each category had specific indicators, and we rated the responses on a fixed-point scale.

1. Transparency Indicators: Employee declarations of fortune, organizational chart, meeting minutes, employee resumes, budget information, and legislation availability.
2. E-Documents Indicators: Online forms, application tracking, online petitions, and public announcements.
3. Communication Indicators: Contact options for the mayor's office, online suggestions, social media presence, and citizen login sections.
4. Useful Content Indicators: City maps, search functionality, multi-language options, city news, live cameras, and newsletter subscriptions.
5. Generalities Indicators: Pleasant design and easy navigation.

Each indicator was scored on a scale, with higher scores indicating better performance. For the Generalities category, we converted qualitative assessments (e.g., design quality) into points using specific formulas.

5 Results

All Romanian municipalities have active websites. The presence on social media platforms like Facebook, Twitter, and Instagram is also high (95.15%), showing active engagement with citizens.

Progress Over Time:

Table 1 shows the improvement in digital services from 2014 to 2023. Significant advancements were noted particularly after the COVID-19 pandemic, which accelerated the digital shift.

Table 1 Electronic public services available on Romanian cites web sites

E-service	2014 (%)	2019 (%)	2023 (%)
Active web site for the city hall	99.02	100.00	100.00
Official social media profile of the city hall	08.73	14.56	95.15
Sign in/log in section for citizens	22.33	28.16	66.02
The existence of electronic forms on the website	31.06	35.92	88.35
Online/mobile tracking of submitted applications	26.21	32.04	73.79
Online/mobile petitions	41.74	46.60	85.44
The citizen's possibility to subscribe to a newsletter	na	15.53	80.58

Data source authors own compiled data

Table 2 Aggregated view over Romanian cites web sites (relative scores)

Grade	2014 (%)	2019 (%)	2023 (%)
Very good	9.71	9.71	53.40
Good	32.04	35.92	24.27
Satisfactory	44.66	50.49	14.56
Poor	11.65	2.91	5.83
Very poor	1.94	0.97	1.94

Data source authors own compiled data

In 2014, 99.02% of municipalities had active websites. This increased to 100% by 2019 and 2023. The presence of online forms increased from 31.06% in 2014 to 88.35% in 2023. Online petitions rose from 41.74% in 2014 to 85.44% in 2023.

Table 2 illustrates the improvement in website quality from 2014 to 2023. In 2014, only 9.71% of websites were rated 'Very Good'. By 2023, this had increased to 53.40%. The percentage of 'Good' ratings remained relatively stable, while 'Satisfactory' ratings decreased significantly, showing overall improvement.

Table 3 breaks down the performance across different categories.

Transparency 2014–2023: Transparency scores have significantly improved from 42.72% in 2014 to 85.44% in 2023. This indicates a notable enhancement in how municipalities adhere to regulations concerning transparency, including the availability of employee declarations, organizational charts, meeting minutes, and budget information. The consistent improvement over the years reflects a growing commitment to making government operations more open and accessible to the public. However, this progress was relatively slow until 2019, when a more substantial leap in transparency efforts is observed, likely due to increased digital initiatives and pressure for greater accountability resulting from the pandemic.

E-Documents 2014–2023: The availability and functionality of electronic documents have seen a substantial increase, from 18.45% in 2014 to 69.90% in 2023. This remarkable growth signifies that more municipalities are now providing essential forms and documents online, streamlining administrative processes. The jump

Table 3 Relative scores obtained by the cities of Romania on each class of analysis

	Class of analysis					
Score	Year	Transparency (%)	E-documents (%)	Communication (%)	Useful content (%)	Generalities (%)
5	2023	85.44	69.90	61.17	62.14	85.44
	2019	44.66	21.36	17.48	22.33	44.66
	2014	42.72	18.45	15.53	20.39	42.72
4	2023	6.80	15.53	33.01	20.39	6.80
	2019	34.95	15.53	54.37	24.27	34.95
	2014	33.98	12.62	42.43	24.27	33.98
3	2023	0.00	0.00	0.00	13.59	0.00
	2019	14.56	25.24	0.00	26.21	14.56
	2014	16.50	26.21	0.00	30.10	16.50
2	2023	6.80	5.83	4.85	0.97	6.80
	2019	3.88	20.39	25.24	18.45	3.88
	2014	4.85	22.33	27.18	16.50	4.85
1	2023	0.97	8.74	0.97	2.91	0.97
	2019	1.94	17.48	2.91	8.74	1.94
	2014	1.94	20.39	4.85	8.74	1.94

Data source authors own compiled data

between 2019 (21.36%) and 2023 (69.90%) suggests a focused effort to digitize services, driven possibly by the need for remote accessibility during the COVID-19 pandemic.

Communication 2014–2023: Communication scores also improved from 15.53% in 2014 to 61.17% in 2023. This increase demonstrates enhanced use of new media for communication by municipalities, including social media presence, online suggestions, and contact options. The significant rise post-2019/COVID indicates a shift towards more interactive and responsive communication channels, aligning with modern expectations of citizen engagement.

Useful Content 2014–2023: Useful content scores have risen from 20.39% in 2014 to 62.14% in 2023. The availability of useful online information such as city maps, search functionality, and live cameras has improved markedly. This growth highlights the municipalities' efforts to provide more comprehensive and practical information online, enhancing the utility and user experience of their websites.

Generalities 2014–2023: Generalities, which cover the overall user-friendliness and design of the municipal websites, improved from 42.72% in 2014 to 85.44% in 2023. Although there is significant improvement, the 'Generalities' category had the least correlation with the overall outcomes at 0.295. This can be attributed to many websites being developed in-house, which might lack professional quality. The increase post-2019 suggests some municipalities began to prioritize better design and user experience, but there is still considerable room for growth in professional website development to ensure all critical features are included.

However, the analysis shows a clear trajectory of improvement across all categories, with transparency and e-documents leading the way. However, the 'Generalities' category highlights the need for more professional and comprehensive website development. The data underscores the importance of continued investment in digital infrastructure and professional development to maintain and further these advancements.

6 Study Limitations

While this study provides valuable insights into the digital transformation of municipal services in Romania, it is important to acknowledge its limitations and discuss their potential impact on the findings. Understanding these limitations can help refine future research and provide a clearer roadmap for further studies in this field.

One significant limitation is the absence of a centralized Chief Information Officer (CIO) role at both the central and local government levels in Romania. In many advanced digital governance models, a CIO is pivotal in coordinating digital transformation efforts, ensuring consistency in technology adoption, and driving strategic initiatives across various departments.

The study's reliance on data collected from municipal websites presents another limitation. Municipal websites vary significantly in terms of design, content, and functionality. This variability can affect the accuracy and comparability of the data collected. Some municipalities might have more comprehensive and updated websites, while others may lag behind, skewing the results.

Limited Indicators: The study focuses on a specific set of indicators to evaluate digital transformation as presented in the methodology. While these indicators provide a useful snapshot, they may not capture the full spectrum of digital services and citizen engagement activities. Additional indicators, such as mobile app usage, online payment systems, and digital literacy programs, could offer a more comprehensive assessment.

While the COVID-19 pandemic accelerated the adoption of digital services, it also introduced several challenges such as temporary measures; some digital initiatives implemented during the pandemic were be temporary measures rather than permanent solutions. Additionally, the rapid shift to digital services may have led to uneven adoption, with some municipalities adapting quickly and others struggling to keep up. This uneven adoption can create disparities in service availability and quality across different regions.

By addressing these limitations and exploring these areas for future research, we can gain a deeper understanding of the digital transformation process in Romanian municipalities and develop more effective strategies to enhance e-government services.

7 Findings

The concepts of smart governance signify the forefront of employing technologies to improve urban life and public services. When applied to municipal websites, these ideas translate into the implementation of various digital tools, solutions, and strategies that enhance communication between citizens and local governments and simplify the delivery of public services. What we found by this study are few key ideas that worth being shared with the community of smart city researchers.

- Improved User Experience: Municipal websites should prioritize a user-centered design, offering straightforward navigation, compelling content, and an easy-to-use interface that motivates citizens to engage in governance activities and access online services. This idea was, to a large extent, extracted from exploring the Romanian municipalities web sites.
- Unified Services: This aims for the smooth integration of public services across various departments and agencies, enabling citizens to access services more effectively. Municipal websites can serve as centralized portals, linking different departments and providing multiple services from one platform.
- Data-Informed Decision-Making: Smart cities depend on data to enhance decision-making and optimize resource allocation. Municipalities can incorporate data analytics tools and dashboards that provide insights into urban trends, patterns, and issues, helping local governments make well-informed decisions and develop focused policies.
- Transparency and Open Data: Municipalities should promote transparency and openness in governance. Municipal websites can function as platforms for disseminating open data, budget details, and legislative updates, fostering accountability and encouraging citizen involvement in the decision-making process.
- Digital Engagement and Collaboration: Municipal websites can support digital engagement by providing electronic referendums and public consultations. This approach allows citizens to directly influence governance processes and cultivates a sense of ownership and responsibility within the community.

In summary, the positive outlook of the present study can be mirrored in municipal websites. By integrating modern technologies, user-focused design, and innovative solutions that boost citizen involvement, cities have the opportunity to enhance transparency and simplify the delivery of public services.

8 Discussions and Conclusions

This study provides an insightful analysis of the state of digitalization in Romania by examining the websites of its municipalities. Observing the e-government evolution over time reveals a steady advancement toward maturity within the ICT sector. This ongoing progression sets the stage for the advent of e-government 3.0, which is expected to feature a broader integration of artificial intelligence (AI) tools in government operations.

The COVID-19 pandemic has undoubtedly served as a catalyst for the development and adoption of Information and Communication Technology (ICT) across various sectors, including e-government. As governments worldwide faced the challenges posed by the pandemic, the demand for efficient and contactless service delivery surged. The necessity for social distancing and remote work accelerated the integration of digital tools and platforms, speeding up the transition to a more advanced e-government infrastructure. This global crisis highlighted the critical role of ICT in maintaining the continuity of essential public services while ensuring the safety of citizens and government staff. Consequently, the pandemic has significantly driven the expansion and maturation of ICT, underscoring its importance in the resilience and adaptability of governments during crises, as demonstrated in this study.

References

V. Albino, U. Berardi and R. M. Dangelico, "Smart Cities: Definitions, Dimensions, Performance, and Initiatives," *Journal of Urban Technology*, 2015.

A. C. M. Alves and A. M. A. Gonçalves, "Public innovation and artificial intelligence: A systematic literature review," *Journal of Open Innovation: Technology, Market, and Complexity,* vol. 7, no. 1, 2021.

M. Angelidou, "The Role of Smart City Characteristics in the Plans of Fifteen Cities." *Journal of Urban Technology*, 2017.

L. Anthony and H. Andreas, Designing Smart and Resilient Cities for a Post-Pandemic World, London: Routledge, 2023.

L. G. Anthopoulos, Understanding Smart Cities: A Tool for Smart Government or an Industrial Trick?, Springer, 2017.

V. Baltac, Tehnologia informatiilor - notiuni de baza, Bucharest: Andreco Educational, 2011.

M. Batty, K. W. Axhausen, F. Giannotti, A. Pozdnoukhov, A. Bazzani, M. Wachowicz, G. Ouzounis and Y. Portugali, "Smart cities of the future," *The European Physical Journal Special Topics*, 2012.

British Standards Institution, "PD 8100 Smart city—Overview," British Standards Institution, London, 2014.

W. Cai and Y. Zhu, "Government innovation in the era of artificial intelligence: A systematic literature review," *Journal of Open Innovation: Technology, Market, and Complexity*, vol. 7, no. 1, 2021.

H. Chourabi, T. Nam, S. Walker, J. R. Gil-Garcia, S. Mellouli and K. Nahon, "Understanding Smart Cities: An Integrative Framework," in *45th Hawaii International Conference on System Sciences*, 2012.

Cisco, "Smart City Framework," Cisco IBSG, San Jose, 2012.

Y. Deng, H. Li, M. Huang and J. Li, "Can artificial intelligence improve public management? A case study of the civil service training system in China," *Sustainability*, vol. 11, no. 24, 2019.

T. Deng, K. Zhang and Z.-J. Shen, "A systematic review of a digital twin city: A new pattern of urban governance toward smart cities," *Journal of Management Science and Engineering*, p. 2021.

J. Etscheid, "Artificial Intelligence in Public Administration," in *Electronic Government. EGOV 2019. Lecture Notes in Computer Science*, Springer, 2019.

P. Galetsi, K. Katsaliaki and S. Kumar, "Big data analytics in health sector: Theoretical framework, techniques and prospects," *International Journal of Information Management*, 2020.

R. G. Hollands, "Will the real smart city please stand up? Intelligent, progressive or entrepreneurial?" *City*, 2008.

M. Holzer and Aroon P. Manoharan, "Digital Governance in Municipalities Worldwide," E-Governance Institute, Newark, 2016.

V. Homburg and Andres Dijkshoorn, "Diffusion of Personalized E-Government Services among Dutch Municipalities: An Empirical Investigation and Explanation," *International Journal of Electronic Government Research*, vol. 7, no. 4, pp. 21–37, 2011.

D. C. Iancu, "European compliance and politicization of public administration in Romania," *Innovative Issues and Approaches in Social Sciences,* vol. 6, no. 1, pp. 103–117, 2013.

M. Janssen, H. v. d. Voort and A. Wahyudi, "Factors influencing big data decision-making quality," *Journal of Business Research*, 2017.

M. Janssen, M. Hartog and G. Kuk, "Will Algorithms Blind People? The Effect of Explainable AI and Decision-Makers' Experience on AI-supported Decision-Making in Government," *Social Science Computer Review*, 2020.

R. Kattel, W. Drechsler and E. Karo, "UCL," 12 2019. [Online]. Available: https://www.ucl.ac.uk/bartlett/public-purpose/sites/public-purpose/files/final_innovation_bureaucracies_20_dec.pdf

T.-h. Kim, C. Ramos and S. Mohammed, "Smart City and IoT," *Future Generation Computer Systems*, 2017.

R. Kitchin, "The real-time city? Big data and smart urbanism," *GeoJournal*, 2013.

B. Klievink and M. Janssen, "Artificial intelligence and public values: A public values mapping exercise in the Dutch police domain," *Government Information Quarterly*, vol. 36, no. 4, 2019.

D. Kolkman, "The usefulness of algorithmic models in policy making," *Government Information Quarterly*, vol. 37, no. 3, 2020.

S. Kuhlmann, G. Bouckaert, D. Galli, R. Reiter and S. V. Hecke, "Opportunity management of the COVID-19 pandemic: testing the crisis from a global perspective," *International Review of Administrative Sciences*, vol. 87, no. 3, p. 497–517, 2021.

V. Lember, R. Kattel and P. Tõnurist, "Technological capacity in the public sector: the case of Estonia," *International Review of Administrative Sciences*, 2018.

D. Linders, "From e-government to we-government: Defining a typology for citizen coproduction in the age of social media," *Government Information Quarterly*, 2012.

S. Maiti, A. Dey and S. Saha, "A systematic review of artificial intelligence in public service delivery," *Journal of Open Innovation: Technology, Market, and Complexity*, vol. 7, no. 1, 2021.

D. Matthias, D. Axel, L. Abdulkader and R. Frauke, "Digital public services: How to achieve fast transformation at scale," McKinsey & Company, 15 07 2020. [Online]. Available: https://www.mckinsey.com/industries/public-and-social-sector/our-insights/digital-public-services-how-to-achieve-fast-transformation-at-scale. [Accessed 23 04 2023].

R. Müller-Török and A. Prosser, "The Corona Warning App of the German Federal Government—How perceived data protection issues hindered any effectiveness.," *Smart Cities and Regional Development (SCRD) Journal*, vol. 5, no. 2, pp. 23–31, 2021.

Nature, "Reining in the pandemic with smart city platforms," Nature Research, 2022. [Online]. Available: https://www.nature.com/articles/d42473-022-00042-z. [Accessed 23 04 2023].

G. Nesti, "Defining and assessing the transformational nature of smart city governance: insights from four European cases," *International Review of Administrative Sciences*, vol. 86, no. 1, pp. 20-37, 2018.

T. Obi, E-Governance—A Global Perspective on a New Paradigm, IOS Press Ebooks, 2007.
A. Ojo, Tomasz Janowski, Elsa Estevez and Irshad Kamal Khan, "Human Capacity Development for e-Government," United Nation University, 2007.
T. Pardo, "Realizing the promise of digital government: It's more than building a web site," *Albany, NY: Center for Technology in Government*, 2000.
C. Schachtner, "Smart government in local adoption—Authorities in strategic change through AI," *SCRD Journal*, vol. 5, no. 3, pp. 53–62, 2021.
C. Schachtner, "Wise Governance—Elements of the digital strategies of municipalities," *SCRD*, vol. 6, no. 2, pp. 23–29, 2022.
M. M. Shafique, S. Hassan and N. Ur Rehman, "A review of artificial intelligence applications in decision-making," *Expert Systems with Applications*, vol. 139, 2020.
J. Šiugždinienė, E. Gaulė and R. Rauleckas, "In search of smart public governance: the case of Lithuania," *International Review of Administrative Sciences*, vol. 85, no. 3, p. 587–606, 2017.
D. Tapscott, Grown Up Digital: How the Net Generation is Changing Your World, McGraw-Hill Education, 2008.
V. G. Thakhathi and L. R. D, "The role of smart cities to promote smart governance in municipalities," *SCRD Journal*, vol. 6, no. 2, pp. 9–22, 2022.
United Nation, "United Nation E-Government survey 2020. Digital government in the decade of action for sustainable development," United Nation, New York, 2020.
C. Vrabie, "Developing E-Government in Reforming Romania," in *NISPAcee Conference*, 2009a.
C. Vrabie, "Just Do It—Spreading Use of Digital Services," in *EGPA Conference*, 2009b.
C. Vrabie, "Digital Governance (in Romanian Municipalities). A Longitudinal Assessment of Municipal Web Sites in Romania," in *European Integration - Realities and Perspectives*, Galati, 2011.
C. Vrabie, "Book Review: 'Grown Up Digital: How the Net Generation is Changing Your World' by Don Tapscott," *HOLISTICA Journal of Business and Public Administration*, vol. 6, no. 1, 2015.
C. Vrabie, "IoT and Its Role in Developing Smart Cities," in *Information Systems: Research, Development, Applications, Education. SIGSAND/PLAIS 2018. Lecture Notes in Business Information Processing*, 2018.
C. Vrabie, "Artificial Intelligence Promises to Public Organizations and Smart Cities," *Digital Transformation. PLAIS EuroSymposium 2022. Lecture Notes in Business Information Processing*, vol. 465, pp. 3–14, 2022.
C. Vrabie, "E-Government 3.0: An AI Model to Use for Enhanced Local Democracies," *Sustainability*, 2023.
C. Vrabie, AI de la idee la implementare. Traseul sinuos al inteligentei artificiale catre maturitate, vol. 1, Pucharest: Pro Universitaria, 2024a.
C. Vrabie, ELEMENTE DE E-GUVERNARE [Elements of e-government] Ed. a II-a, Bucharest: Pro Universitaria, 2024b.
C. Vrabie and E. Dumitrascu, Smart cities: de la idee la implementare, sau, despre cum tehnologia poate da strălucire mediului urban, Bucharest: Universul Academic, 2018.
WHO, "Digital technology for COVID-19 response," WHO, 01 04 2023. [Online]. Available: https://www.who.int/news/item/03-04-2020-digital-technology-for-covid-19-response. [Accessed 23 04 2023].

Cyprus—Platform Framework is on the Way

Christian Schachtner

1 Introduction: The Project Smart City Platform Cyprus

In the following, contributions are presented based on project report evaluations, the reference of which relevant activities in Cyprus was based on a onsite-visit at the University of Nicosia after a notice about Smart City Activities in this regions by Professor Hans-Jürgen Kaufmann, University of Nicosia, among others.

The Cyprus Smart City Platform represents an innovative approach to the digitization and networking of urban spaces in Cyprus. The platform aims to integrate different urban systems and services to achieve more efficient administration and improved quality of life for citizens (Smith et al. 2023). A key aspect is the implementation of Internet of Things (IoT) technologies that make it possible to collect and analyze real-time data from the urban environment (Johnson and Brown 2024).

This data includes areas such as traffic flow, energy consumption, air quality, and public safety. A special focus is on the development of intelligent transport systems. This not only leads to a reduction in congestion and emissions, but also improves overall mobility in the cities of Cyprus. By integrating sensors and cameras at intersections and main arteries, traffic flow can be monitored and optimized in real time (Garcia et al. 2022).

The platform also includes an advanced energy management system. By connecting electricity grids, renewable energy sources and consumers, energy consumption can be managed more efficiently and the share of renewable energy can be maximised (Papadopoulos and Christodoulou 2024). This contributes significantly to the sustainability and environmental friendliness of Cypriot cities.

Another important aspect is citizen participation. The platform provides a user-friendly interface through which citizens can access city services, provide feedback, and participate in decision-making processes (Wilson 2023). This not only promotes

C. Schachtner (✉)
RheinMain University of Applied Sciences, Wiesbaden, Germany
e-mail: Christian.schachtner@hs-rm.de

C. Schachtner (ed.), *The European Smart City Movement*,
https://doi.org/10.1007/978-3-031-96700-9_16

transparency in the city administration, but also strengthens the sense of community and the active participation of citizens in urban life.

However, implementing the Cyprus Smart City Platform also presents challenges. Robust security protocols and encryption technologies are used to ensure the integrity and confidentiality of the data collected. Data protection and cybersecurity are critical aspects that need to be carefully addressed (Andreou et al. 2025). The interoperability of different systems and technologies is another challenge. To enable seamless integration, open standards and interfaces are used, allowing for flexible expansion and customization of the platform (Koutsouris and Mavromoustakos 2023). An important aspect of the project is scalability. The platform will be designed to be adaptable not only in the larger cities of Cyprus such as Nicosia or Limassol, but also in smaller municipalities (Michaelides 2024). This enables the widespread implementation of smart city solutions across the island. The economic impact of the Cyprus Smart City Platform is also significant. The creation of an innovative digital infrastructure is expected to create new business opportunities and increase Cyprus' attractiveness for technology companies and start-ups (Economou and Hadjigeorgiou 2024). This could lead to significant economic growth and the creation of new jobs in the technology sector.

2 Infrastructure Revolution

The Upgrading Infrastructure and Connectivity project is part of a broader digital strategy of Cyprus that aims to position the country as an innovative and technologically advanced location in the region (Digital Europe Programme 2024). Improving digital infrastructure and connectivity will lay the foundations for a dynamic digital economy that will create new jobs, foster innovation and strengthen Cyprus' international competitiveness. The Upgrading Infrastructure and Connectivity project is a central part of Cyprus' digital transformation, which aims to take the country's technological infrastructure and connectivity to a new level (European Commission 2023). As part of this initiative, extensive measures are being taken to modernise Cyprus' digital landscape and make it fit for the future (Ministry of Research, Innovation and Digital Policy 2024).

An important aspect of the project is the modernisation of public IT infrastructure (ECA 2023). This includes updating hardware and software in government and public institutions to provide more efficient and secure e-government services. Improving cybersecurity plays a key role in increasing citizens' trust in digital services and protecting critical infrastructure (ENISA 2024).

A key focus of the project is the expansion of the 5G network, which is intended to serve as the backbone for innovative technologies and services (GSMA 2022). The implementation of 5G will not only significantly increase the speed and capacity of mobile data transmission, but also open up new opportunities for the Internet of Things (IoT) and smart city applications (ITU 2023). In the area of connectivity, Cyprus' international connectivity will also be improved. New submarine cables are expected to connect the island to major internet hubs in Europe and the Middle East, increasing the redundancy and reliability of internet connections (TeleGeography 2023). This is particularly important for the development of Cyprus as a regional data hub and for the promotion of the digital economy sector. In parallel, the expansion of the fibre-optic infrastructure is being driven forward to ensure a nationwide high-speed internet connection (OECD 2024). This measure is crucial to bridge the digital divide between urban and rural areas and to provide equal access to digital services for all citizens and businesses (World Bank 2023).

The implementation of the project requires close cooperation between the public and private sectors (World Economic Forum 2024). Public–private partnerships are used to mobilise investment and contribute technical know-how. At the same time, it incentivizes telecoms companies to invest in underserved areas and offer innovative services (BEREC 2023). A special focus is on the sustainability and energy efficiency of the new infrastructure (European Environment Agency 2024). The use of modern technologies and intelligent network management solutions aims to optimize energy consumption and reduce the CO2 footprint of the digital sector. This is in line with the EU's climate goals and underlines Cyprus' commitment to a sustainable digital future.

The education and training of professionals is another important part of the project (Cedefop 2023). In order to be able to use the new technologies effectively, training programs are offered for IT specialists, but also for the general public. This is to ensure that the improved infrastructure is used optimally and contributes to increasing digital skills among the population.

The challenges in implementing the project are manifold, ranging from technical difficulties to regulatory issues (ETSI 2023). Cyprus' geographical location as an island nation places special demands on network planning and implementation. In addition, privacy and network security issues must be carefully addressed to ensure user trust and comply with international standards (European Data Protection Board 2024).

3 Smart Tourism: A Key Component

In the rapidly evolving landscape of smart city initiatives, Cyprus is leading the way by using innovative technologies to improve the tourism experience and make it more sustainable. The integration of information and communication technologies (ICT) into Cyprus' tourism industry has led to a significant improvement in the visitor experience (Gretzel et al. 2015). By implementing smart tourism solutions, travelers can now receive personalized recommendations based on their preferences and locations. This enables a more efficient use of tourist resources and contributes to increasing visitor satisfaction.

A core aspect of our Smart Tourism project is the development of a comprehensive mobile application that acts as a digital travel companion. This app not only provides information on attractions and cultural events, but also integrates real-time data on traffic, weather and visitor traffic (Buhalis and Amaranggana 2015). This allows tourists to plan their itineraries in the best possible way while avoiding crowded places, resulting in a more balanced distribution of visitor flows. A major focus of our project is the promotion of sustainable tourism. Through the use of IoT sensors and big data analytics, we can accurately monitor and control the ecological footprint of tourism activities (Femenia-Serra and Ivars-Baidal 2021). This data allows us to implement targeted measures to reduce energy and water consumption and minimize waste in tourist facilities.

Cyprus has also introduced an innovative reward system that incentivizes tourists for environmentally friendly behavior. By using public transport, participating in beach clean-ups or visiting eco-attractions, visitors can earn points that they can exchange for local products or services. This system not only promotes environmentally conscious travel, but also strengthens the local economy (Buonincontri and Micera 2016).

In collaboration with local historians and archaeologists, detailed 3D models and interactive content have been developed that convey the rich history of Cyprus in an innovative way. These AR experiences are not only educational, but also entertaining, helping to raise awareness about the protection of cultural heritage. A particularly fascinating aspect of our Smart Tourism project is the integration of augmented reality (AR) to enhance Cyprus' cultural heritage. Visitors can experience historical sites in their original state through their smartphones or AR glasses. This technology makes it possible to virtually reconstruct ancient ruins and bring historical events to life (tom Dieck and Jung 2018).

A central component of our Smart Tourism approach is the use of big data to optimize tourism planning and destination management. By analyzing visitor flows, booking patterns, and social media data, we can make accurate predictions about tourism trends and respond accordingly (Li et al. 2018). This data-driven approach allows us to allocate resources more efficiently, refine marketing strategies and continuously adapt the tourist offer to the changing needs and expectations of visitors. At the same time, data analysis helps us to identify potential congestion at an early stage and to take measures to guide visitors in order to minimise negative impacts on the local environment and community.

Despite the promising progress, we also face challenges in implementing smart tourism solutions. Privacy and cybersecurity are critical issues that we need to continuously address to ensure visitor trust (Gretzel et al. 2020). In addition, rapid technological development requires constant adaptation and training of all those involved in the tourism sector. In the future, we plan to integrate artificial intelligence to further personalize the visitor experience, as well as develop virtual reality offerings that will allow Cypriot attractions to be experienced from anywhere in the world. These innovations aim to establish Cyprus as a leading smart tourism destination while upholding the principles of sustainability and responsible tourism.

4 Policy Plans for Smart City Initiatives in Cyprus

The development of smart cities has gained global traction in recent years, and Cyprus is no exception in this trend. The Cyprus Forum Cities, which focuses on local governance and smart cities, highlights the importance of this issue for the island (Georgiou 2024). The Cypriot government has recognised that the implementation of smart city concepts is crucial for modernising and increasing the efficiency of urban spaces. A key aspect of the political agenda is the promotion of digital transformation in cities and municipalities (Papadopoulos 2023). The focus here is on the integration of information and communication technologies (ICT) into urban infrastructure in order to improve the quality of life of citizens and at the same time optimize the use of resources.

An essential part of the policy plans is the development of a comprehensive data strategy. The government aims to create a central platform that collects and analyzes data from different urban areas (Christodoulou 2024). This data should be accessible not only to the administration, but also to citizens and companies in order to promote transparency and enable innovative solutions.

In the field of mobility, policymakers are focusing on promoting sustainable transport concepts. These include the introduction of intelligent traffic management systems, the promotion of electric mobility and the improvement of public transport through digital solutions (Andreou 2023). These measures aim to optimize traffic flow and reduce environmental impact.

Another focus is on energy efficiency. Policy plans include implementing smart grid technologies and increasing the share of renewable energy in the energy mix (Kyriacou 2024). The integration of IoT devices into public buildings and infrastructures is intended to optimize energy consumption and reduce CO2 emissions.

The promotion of e-government services is another central aspect of the smart city strategy. The aim is to digitise administrative processes and provide citizens with easy and efficient access to public services (Nikolaou 2023). This is not only intended to increase the efficiency of the administration, but also to increase citizen participation and satisfaction.

To implement these ambitious plans, the Cypriot government is relying on public–private partnerships. Funding programmes have been launched to support companies and start-ups developing innovative solutions for smart cities (Demetriou 2024). At the same time, cooperation with universities and research institutions will be intensified in order to integrate the latest technological developments into urban planning. An important aspect of the political strategy is also the promotion of digital literacy among the population. Education programmes and training are designed to ensure that all citizens can reap the benefits of smart city initiatives (Stavrou 2023). Particular attention will be paid to older people and disadvantaged groups in order to avoid a digital divide.

However, the implementation of smart city concepts also brings challenges that policymakers must face. Data protection and cybersecurity are of central importance here. The government is working on a robust legal framework to protect citizens' privacy while maximizing the benefits of data use (Antoniou 2024).

5 Conclusion

In conclusion, the Smart City Initiatives in Cyprus is an ambitious and forward-looking project that has the potential to significantly improve the quality of life in Cypriot cities and position Cyprus as a pioneer in smart city development in the Eastern Mediterranean. Cyprus creates the technological conditions for an interconnected, innovative and competitive society and economy. Successful implementation requires close cooperation between government, the private sector, research institutions and citizens. Cyprus also takes a holistic approach that harmoniously combines technology, sustainability and cultural heritage. Through continuous innovation and the use of cutting-edge technologies, we strive to revolutionize the tourism experience in Cyprus while making a positive contribution to the local economy and environment. To shape a sustainable and smart urban future for Cyprus through continuous innovation and adaptation to the changing needs of cities and their inhabitants. The success of these projects will be instrumental in establishing Cyprus as a digital leader in the region and laying the foundation for sustainable growth in the digital era.

References

Andreou, A., Christodoulou, E., & Mavromoustakos, S. (2025). Cybersecurity challenges in smart city implementations: The case of Cyprus. Journal of Urban Technology, 32(1), 45–62.

Andreou, M. (2023). Sustainable urban mobility in Cyprus: Challenges and opportunities. Journal of Transport Policy, 45(2), 78–92.

Antoniou, G. (2024). Cybersecurity and data protection in smart cities: A case study of Cyprus. International Journal of Information Security, 18(3), 215–230.

BEREC. (2023). Report on the implementation of the Open Internet Regulation. Body of European Regulators for Electronic Communications.

Buhalis, D., & Amaranggana, A. (2015). Smart Tourism Destinations Enhancing Tourism Experience Through Personalisation of Services. In I. Tussyadiah & A. Inversini (Eds.), Information and Communication Technologies in Tourism 2015 (pp. 377–389). Springer

Buonincontri, P., & Micera, R. (2016). The experience co-creation in smart tourism destinations: a multiple case analysis of European destinations. Information Technology & Tourism, 16(3), 285–315.

Cedefop. (2023). Digital skills and competences in Europe: Trends and forecasts. European Centre for the Development of Vocational Training.

Christodoulou, A. (2024). Data-driven governance: The role of big data in smart city management. Urban Studies Review, 29(1), 45–60.

Demetriou, L. (2024). Public-private partnerships in smart city development: Lessons from Cyprus. Journal of Urban Technology, 31(2), 180–195.

Digital Europe Programme. (2024). Country profile: Cyprus—Digital transformation progress. European Commission.
Economou, G., & Hadjigeorgiou, M. (2024). Economic impacts of smart city initiatives in Cyprus. Mediterranean Journal of Economics, 18(3), 301–315.
ENISA. (2024). Threat Landscape Report: Cybersecurity challenges in the EU. European Union Agency for Cybersecurity.
ETSI. (2023). 5G deployment challenges and solutions. European Telecommunications Standards Institute.
Europäische Kommission. (2023). Digital Economy and Society Index (DESI)—Cyprus. Publications Office of the European Union.
Europäische Umweltagentur. (2024). Environmental impacts of the digital transition in Europe. EEA Report.
European Data Protection Board. (2024). Guidelines on data protection in 5G networks. EDPB.
European Court of Auditors. (2023). Special report: Digitalisation of public administration in the EU. Publications Office of the European Union.
Femenia-Serra, F., & Ivars-Baidal, J. A. (2021). Do smart tourism destinations really work? The case of Benidorm. Asia Pacific Journal of Tourism Research, 26(4), 365–384.
Garcia, R., Patel, S., & Nikolaou, K. (2022). Intelligent traffic management systems for smart cities: A comparative study. Transportation Research Part C: Emerging Technologies, 135, 103721.
Georgiou, N. (2024). Smart cities and local governance: Insights from the Cyprus Forum Cities. Mediterranean Politics, 30(1), 55–70.
Gretzel, U., Fuchs, M., Baggio, R., Hoepken, W., Law, R., Neidhardt, J., Pesonen, J., Zanker, M., & Xiang, Z. (2020). e-Tourism beyond COVID-19: a call for transformative research. Information Technology & Tourism, 22(2), 187–203.
Gretzel, U., Sigala, M., Xiang, Z., & Koo, C. (2015). Smart tourism: foundations and developments. Electronic Markets, 25(3), 179–188.
GSMA. (2022). The Mobile Economy Europe. GSM Association.
ITU. (2023). Measuring digital development: Facts and figures. International Telecommunication Union.
Johnson, L., & Brown, T. (2024). IoT applications in urban environments: Opportunities and challenges. IEEE Internet of Things Journal, 11(4), 3256–3270.
Koutsouris, D., & Mavromoustakos, S. (2023). Interoperability frameworks for smart city platforms: A review. Smart Cities, 6(2), 178–195.
Kyriacou, E. (2024). Smart grids and renewable energy integration in Cyprus: A roadmap for sustainable urban development. Energy Policy, 156, 112233.
Li, J., Xu, L., Tang, L., Wang, S., & Li, L. (2018). Big data in tourism research: A literature review. Tourism Management, 68, 301–323.
Michaelides, A. (2024). Scaling smart city solutions for smaller municipalities: Lessons from Cyprus. Journal of Regional Science, 64(2), 289–305.
Ministry of Research, Innovation and Digital Policy. (2024). Digital Strategy for Cyprus 2025–2030.
Nikolaou, P. (2023). E-government services in Cyprus: Adoption rates and user satisfaction. Electronic Government, an International Journal, 19(4), 389–405.
OECD. (2024). Broadband Portal. Organisation for Economic Co-operation and Development.
Papadopoulos, A. (2023). Digital transformation strategies for Cypriot municipalities. Local Government Studies, 49(3), 321–337.
Papadopoulos, N., & Christodoulou, A. (2024). Smart grid integration in the Cyprus Smart City Platform: Challenges and opportunities. Energy Policy, 165, 112946.
Smith, J., Davis, A., & Georgiou, C. (2023). Smart city initiatives in the Mediterranean: A comparative analysis. Urban Studies, 60(5), 1023–1042.
Stavrou, S. (2023). Bridging the digital divide: Promoting digital literacy in Cyprus. International Journal of Lifelong Education, 42(5), 478–493.
TeleGeography. (2023). Submarine Cable Map. PriMetrica, Inc.

tom Dieck, M. C., & Jung, T. H. (2018). A theoretical model of mobile augmented reality acceptance in urban heritage tourism. Current Issues in Tourism, 21(2), 154–174.

Weltbank. (2023). World Development Report: Data for Better Lives. The World Bank Group.

Wilson, E. (2023). Citizen engagement in smart cities: Strategies and best practices. Public Administration Review, 83(2), 312–326.

World Economic Forum. (2024). The Global Competitiveness Report: Special Edition 2024. WEF.

Insights From Great Britain—Smart Rural: The Open Data Gap

Gefion Thuermer, Johanna Walker, Elisavet Koutsiana, Elena Simperl, and Leslie A. Carr

1 Introduction

Digital inequality can be understood as the difference between individuals, companies, regions, and countries in their access to, use of and ability to benefit from the transformational power of information and communications technologies (ICT) (Halford and Savage 2010). The urban–rural divide is one of the most well-known foci of digital inequality (Vicente Cuervo and Lopez Menendez 2006). In comparison to their urban or suburban counterparts, rural populations tend to experience gaps in education, income, device availability, and mobile and internet access (Philip et al. 2017). The Eurostat Rural Development Statistics (Eurostat 2023) illuminate points of difference in underlying challenges in rural and urban areas across issues such as health, housing and education, which mean that the digital demands in each of these domains in rural areas differ from those experienced in the urban setting.

Recent approaches to addressing urban challenges have often taken the form of data-driven innovation to create digital products and services. This interaction of people, technology, data and services has produced the concept of the smart city. But what of rural challenges? In this paper, we posit that urban solutions are not simply transferable to the rural realm, and as such, require their own, rural data to shape them—data which is not easily available. We present three research questions:

*RQ*1—How available is rural open data and how is this changing over time?

*RQ*2—What are the challenges in collecting and publishing rural open data?

*RQ*3—How is rural open data used?

G. Thuermer (✉) · J. Walker · E. Koutsiana · E. Simperl
Department of Informatics, King's College London, London, Great Britain
e-mail: gefion.thuermer@kcl.ac.uk

L. A. Carr
University of Southampton, Southampton, Great Britain

C. Schachtner (ed.), *The European Smart City Movement*,
https://doi.org/10.1007/978-3-031-96700-9_17

This paper makes two main contributions. Firstly, we offer a definition of what rural data might consist of and identify some dimensions of its availability and use in Europe. Secondly, we engage with the smart city paradigm to explore its use for invigorating 'smart rural' open data.

2 Background

In this section we explore the size and location of the European rural population, and its access to information and communication technologies (ICT), including the impact of Covid. We explore a key ICT concept, smart cities, and the related concept of open data. We further outline key challenges for rural populations that can be addressed with data.

2.1 European Rural Population

Just over a quarter (25.2%) of EU citizens live in rural areas. The more densely populated EU member states have a very high proportion of urban population, but the opposite is true for many countries, especially in Eastern Europe. Over 40% of the Lithuanian, Slovenian and Slovakian populations live in rural areas as designated by Eurostat (2023). However, even in some Western member states, such as Ireland, France and Austria, over a third of the population are classified as living rurally. Although in Western and Northern member states rural living offers comparatively better housing and relatively available employment, in Eastern, Southern and Baltic member states, rural citizens tend to be among the poorest. According to Holmes and Burgess (2020), "[t]he link between poverty and digital exclusion is clear: if you are poor, you have less chance of being online." Being on the wrong side of the rural digital divide can therefore also be exacerbated by rural poverty.

2.2 ICT as an Economic Driver

In many areas, including the European Union, ICT are codified as the basis of a competitive, dynamic, and knowledge-based economy. It is intuitive to think that ICT are a way to reduce rural–urban digital inequalities, by lessening the barrier of physical distance, and being theoretically as available in a metropolis as a remote village. Salamink et al. (2017), in a systematic review of 157 articles on digital development in Western rural areas, conclude that, on the contrary, technological infrastructures in rural areas were mostly reactive and quickly obsolete, and that lower levels of education and computer skills presented a significant barrier to adoption amongst rural populations. They also found that marked differences between rural populations prevented previous policy in this area from meeting with real success. Particularly isolated or tight-knit rural communities, such as those studied in Chile by Pavez et al. (2017), may hold negative attitudes towards the internet in general, perceiving it as disruptive or associating it with loneliness and addiction.

2.3 Cities as a Focus of Economic Growth

Despite communication technologies being predicted to signal the end of the domination of cities as hubs of economic growth, "urbanisation continues to be a megatrend," (Braesemann et al. 2022). Information flows easily in close quarters, often supported by informal networking enabled by the cultural offer of a city, and young, mobile, educated people ('talent') flock to cities to enable the development of innovations based on that information (Zukin 2020). ICT were conceived of as creating a near substitute for this proximity of people and ideas. Nevertheless, cities continue to be a geographical and cultural contemporary locus of innovation, especially around digitally driven economic growth focused on 'platform capitalism' (Zukin 2020).

2.4 Cities and Data

Within this assemblage of digital innovation activities, cities and urban areas across both the Global North and South have become avid users and sources of data in a variety of ways. These often coalesce around, and are heavily incentivised by, data-driven innovations under the umbrella term 'smart cities'. There are many definitions of smart cities, but the term can generally be understood as an urban area that automatically collects data on various activities, and uses the insights gained from that data to efficiently manage assets, resources and services (Arroub et al. 2016). In fact, data is perceived to be so vital for cities and citizens that the Urban Data Platform (Centre and REGIO 2019) project has the aim of speeding up the adoption of common open urban data platforms and ensuring that "300 million European citizens are served by cities with competent urban data platforms, by 2025". This enables (amongst other things) deliberative democracy and the open innovation of services and applications useful to citizens (Salemink et al. 2017; Maja et al. 2020). Thus open data—data that is available for anyone to use for any legal purpose—is at the very heart of the smart cities concept (Ojo et al. 2015). In a circular fashion, smart city initiatives not only create vast amounts of data, but many of these data sets have also been released under open licences.

2.5 Rural Areas and Data

Conversely, the Rural Open Data Project*[1] states that, "few if any rural local governments provide open data, and little is known about how open data affects rural communities. If there is a benefit to communities from open data policies, it is likely that rural communities are benefiting less than urban ones, if at all." There is extensive research, both academic and driven by governments, into aspects of smart rurality such as smart specialisation (da Rosa Pires et al. 2014; Prause and Boevsky 2015; Talbot 2016). It is not just cities that need to make data-driven decisions (Maja et al. 2020). However, there is very little focus specifically on open data, even in contexts where open data is often seen as a natural part of the ecosystem, such as Mobility as a Service (Eckhardt et al. 2018). Studies—and recent coronavirus examples—show that the digital divide can cause political exclusion, social participation exclusion and economic exclusion (Barry 2013). This exclusion is now compounded by the potential for exclusion from the beneficial possibilities of open data.

2.6 The Impact of Covid on ICT Use

It is tempting to think that the increase in remote working and improvement in online collaboration tools created in response to the COVID-19 pandemic will, in the near future, render many of these issues null. In fact, the pandemic highlighted the digital vulnerability of many citizens. To work remotely, people are still required to have access to ICTs, be able to use them competently, and have skills and knowledge appropriate to the organisation utilising remote working. Even more severely than this, the pandemic caused a situation in which the access to the internet, to use the language of Esteban-Navarro et al. (2020), "seems to have become necessary even for survival, especially if it is related to health issues".

Furthermore, the pandemic emphasised the digital divide in rural areas as it removed the opportunity to access the internet in shared areas such as workplaces, libraries and schools. Across Europe, during the pandemic only 29% of workers living rurally accessed remote working compared to 44% of workers living in major cities and 35% of those living in the suburbs. The European Commission characterized this as a "new digital divide" (Sostero et al. 2020).

2.7 Rural Population Data and Challenges

Rural population data is data regarding or pertaining to those who live in rural areas. This includes, but is not limited to, demographic data. Demographic data is generally some of the most robust data across Western nations, however, in remote areas

[1] https://www.ruralopendata.ca.

with very little population, data may be subject to rounding or suppression to avoid re-identification (Pateman 2011). As noted above, there are key differences in the underlying challenges for rural populations. This means simply transferring solutions created with or on data from urban areas is unlikely to be successful. One of the most obvious examples is the intense urban focus on transport and mobility challenges regarding aspects such as traffic density, pedestrian areas, cycle paths and storage, heavy goods vehicle routing, transport mode interchanges and other issues that spring from the requirement of many forms and purposes of transport to coexist with each other. Whereas urban transport often involves addressing relatively short journeys, the average person living in the most rural areas of England, UK, travels 53% further per year than those living in urban areas, according to Action with Communities in Rural England.[2] Rural transport issues are more likely to concern affordability and availability, and necessitate solutions such as buses on-demand, or community rail partnerships. However, requirements also vary across such key areas as housing, health and education.

Whereas housing stock and availability is a major challenge for cities, 80% of the rural EU population lives in a house rather than apartment blocks. Health challenges also differ. Rural populations are often deterred from seeking medical treatments due to the distances involved, unlike urban populations, where the focus is on speeding up the diagnostic process to reduce waiting times. Across the EU, the level of those not in education, employment or training is higher in rural areas. In addition, the gender gap is wider in the rural labour market. Further, solutions developed for more densely populated areas—smart city solutions—may be technically inappropriate in a rural context.

2.8 Rural Economy Data and Challenges

This is data pertaining to the (predominantly) rural industries of agriculture, fisheries and forestry. Issues around food supply and environmental concerns such as CO_2 levels and water run-off are common to both urban and rural settings. However, significantly different challenges include accurate and verifiable forestry mapping and monitoring, livestock monitoring for animal welfare standards, prevention of disease, and supply chain monitoring. More transparent supply chains are needed to reduce the likelihood of events such as that which arose in 2013, when several European countries discovered that processed meat sold in a variety of locations as beef, in fact included traces of horsemeat (Premanandh 2013). Data is required not only for decisions at the producer level (for instance, geospatial and meteorological data is useful for agriculture practices), but also for state level production policy decisions and planning for farming and use of land.

Regulators of these industries also require this data.

[2] https://assets.publishing.service.gov.uk/media/5a79c15540f0b670a8025301/pb13741-fuel-cost-rural.pdf.

3 Methodology

To analyse the relationship between open data and rural economies and populations, we pursue a mixed methods approach, combining qualitative interviews with statistical analysis. We chose to focus on Europe (broadly defined, as opposed to just the EU) as it has mature statistics on its rural and urban populations and a supranational open data portal, data.europa.eu (formerly the European Data Portal).

3.1 Defining Rural

How to define 'rural' is virtually a research area of its own. As Wineman et al. (2019) show, there are a variety of ways to define rural, including by population, light pollution, impervious surfaces and 'subjective assessment'. Important key indicators, such as the rural poverty rate, will vary as measured by the different definitions. According to an article in The Washington Post, the US government alone has 15 separate definitions used by different departments for various purposes.[3] These are often recursive, as in the example of the US Census, which counts everything that is 'not urban' as 'rural'. We define the rural population as those living in the countryside as determined by the EU Degree of Urbanisation (DEGURBA) population statistics.[4] We define the rural economy as those economic activities denoted by EuroStat Rural Development as 'rurally thematic industries'(Eurostat 2023). These are primarily agriculture, forestry and fisheries.

3.2 Qualitative Data Collection and Analysis

For our qualitative analysis, we conducted interviews with open data experts from European countries with high levels of rural population (Romania, Slovenia and Ireland). We also spoke to experts from the rural data economy Global Fishing Watch, for fishery; the Swedish National Forest Inventory, for forestry and Regio FoodValley, for agriculture. We asked them about their views on the current status of rural open data, the challenges and potential use they saw, and their own and observed use of rural open data. The interviews were then thematically coded, both deductively, using the categories from the interview questions as a starting point, and inductively, with themes emerging from the conversations and observations of the interviewees themselves. Existing research was used to develop generalisations

---

[3] https://www.washingtonpost.com/politics/the-federal-definition-of-rural–times-15/2013/06/08/a39e46a8-cd4a-11e2-ac03-178510c9cc0a_story.html.

[4] https://ec.europa.eu/eurostat/web/gisco/geodata/reference-data/population-distribution-demography/degurba.

from the interviews. We also present examples taken from a webinar held in 2020 with representatives of Regio FoodValley and the Rural Services Network.

These interviews were carried out prior to the Covid pandemic. However, based on desk research, including reviewing the most recent versions of relevant publications, and attending relevant presentations and talks, it did not appear that the situation as presented by our interviewees had changed sufficiently to warrant further interviews.

3.3 *Quantitative Data Collation and Analysis*

For our statistical analysis, we collated data from several sources, as shown in Table 1. We tested several combinations of indicators for correlations, using Spearman rank coefficient, (Zar 2005) to compare the rurality of countries with their open data practices pertaining to agriculture, focusing on open data maturity, GDP, population in rural and urban areas, and number and proportion of published data sets. In all cases we sought to test data from the same, or where that was not possible due to publishing cycles, the nearest years.

4 Results

In this section we address each of our research questions in turn.

*RQ*1—*How available is rural open data and how is this changing over time?*

To answer this, we explored the open datasets available as shown in Table 1 to understand the overall availability of rural open data, and factors that might affect this, such as open data maturity and the importance of agriculture to GDP. To do this we tested 3 hypotheses.

Hypothesis 1: Countries with more rural populations are publishing more rural data.

Germany, France, Italy, and Spain presented the highest total and urban populations. France has the largest rural population. In addition, in 2022, 11 out of 29 countries possessed higher rural than urban populations. As a percentage of total population, Netherlands, Spain and Iceland have the greatest rural populations, which has remained consistent between 2018 and 2022, as shown in Figs. 1 and 2.

Figure 3 shows the number of total and rural datasets for the 35 countries which have data on the supranational portal data.europa.eu.

Countries with large populations such as Spain, Germany, Italy and France are leaders in publishing datasets, as shown in Fig. 3 and also have high rural populations. We can therefore hypothesize that they will publish a lot of rural datasets.

We found a correlation between the total number of datasets and the total population (0.77 for 2020, calculated as this was the most recent population data available). In contrast, we did not find a correlation with the total number of datasets and the size of the rural population (0.41 for 2020). This suggests that there is a relationship

Table 1 Datasets used in quantitative analysis of data availability

Feature	Source	Countries	# Countries	Years	Source link
Population	Eurostat	European Union member states, Switzerland, Iceland, Norway Albania, Serbia	29	2018–2022	• https://ec.europa.eu/eurostat/databrowser/view/demo_gind_custom_8259328/default/table?lang=en • https://ec.europa.eu/eurostat/databrowser/view/urt_pjangrp3_custom_8259145/default/table?lang=en
Numberof datasets	Data.europa.eu	European Union member states, Switzerland, Iceland, Liechtenstein, Norway, Moldova, Serbia, Ukraine, UK	35	2019, 2020, 2023	• https://data.europa.eu/catalogue-statistics/currentState/countries?locale=en • https://data.europa.eu/data/datasets?locale=en&categories=agri&page=1&query=Austria
OpenData Maturity	Data.europa.eu	European Union countries, Albania, Azerbaijan, Bosnia and Herzegovina, Georgia, Iceland, Liechtenstein, Moldova, Montenegro, Norway, Serbia, Switzerland, Ukraine, UK	40	2018–2023	• https://data.europa.eu/en/publications/open-data-maturity/2022
GDP	World Bank		40	2018–2022	• https://data.worldbank.org/indicator/NV.AGR.TOTL.ZS?end=2022&start=1960&view=chart

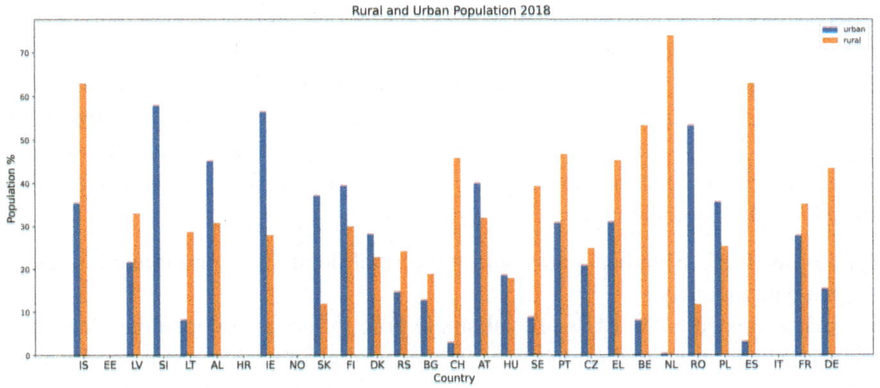

Fig. 1 Rural and urban population percentages 2018

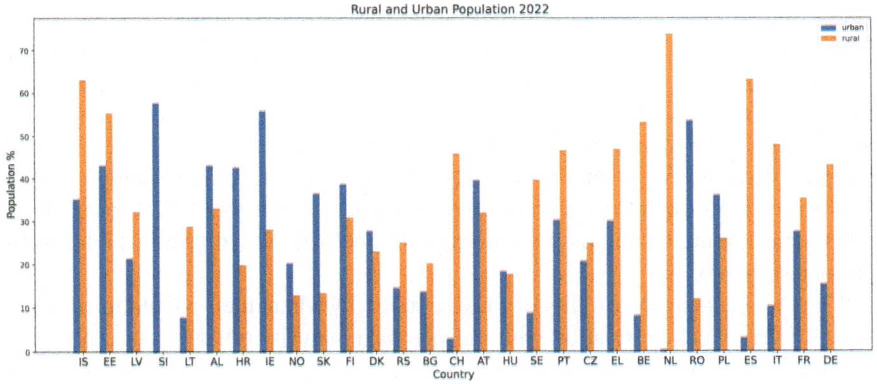

Fig. 2 Rural and urban population percentages 2022. Issues with Eurostat's allocation methods mean these do not always add up to 100%

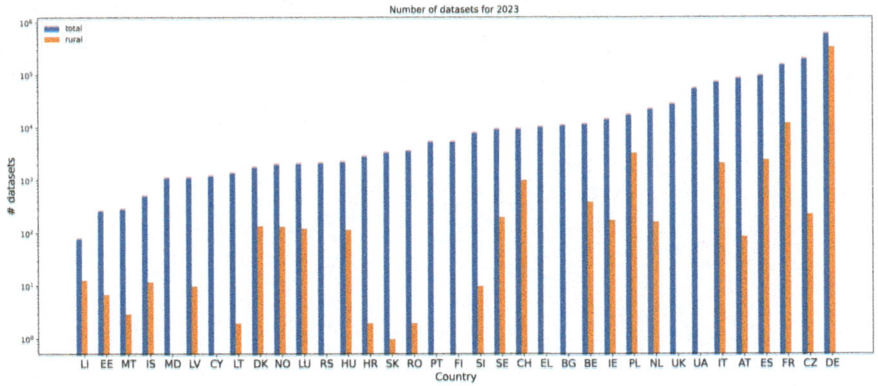

Fig. 3 Number of total and rural datasets published by country on data.europa.eu in 2023

between the total and urban population in a country, and a higher number of published datasets. However, rural population did not influence the published datasets.

Using the same correlation test we found that the number of total datasets has a weak correlation with the number of rural datasets (0.51 for 2019, 0.4 for 2020, and 0.46 for 2023). This indicates that a country that publishes many datasets may not necessarily publish a high number of rural datasets. We conclude there is a comparatively insufficient number of rural datasets.

Hypothesis 2: EU Member States with more rural populations are in earlier stages of open data maturity.

Data.europa.eu is the EU's official single point of access to open data from EU countries, institution, bodies and agencies, as well as other European and Europe-affiliated countries. Since 2018 it has has published a yearly report regarding the open data maturity (ODM) of each country (Open data in Europe 2018). The maturity of open data is calculated using evaluations for the quality (monitoring, completeness), impact (reuse of the dataset and strategic awareness,) portal (features, usage, sustainability), and policy (framework, governance, implementation) of these datasets. We hypothesised that, as our interviews suggested that rural data could be more challenging to collect and use, the EU Member States with less urban populations are in earlier stages of open data maturity. Figure 4 presents the ODM over the years for 40 European countries we found in those reports. Results varied significantly over the years with Ukraine and France presenting the highest values for 2022. Ukraine is an exceptional example with no ODM metrics before 2020 but with stable high results for the last three years.

When we measure the rurality of an EU country by where their population resides, there is a very weak correlation with its open data maturity (ODM) score. A higher percentage of both rural and urban population correlated with a higher ODM. As the effect is tiny and occurs for both population groups, this does not provide further insight. We investigated the correlation between the ODM of a country and the number of total and rural datasets published. We find a weak correlation between the ODM and the total (0.44 for 2019, 0.46 for 2020, 0.46 for 2023) and rural number of datasets (0.46 for 2019, 0.5 for 2020, 0.4 for 2023). This weak correlation could indeed be a function of how the ODM is calculated. However, the similarity for both

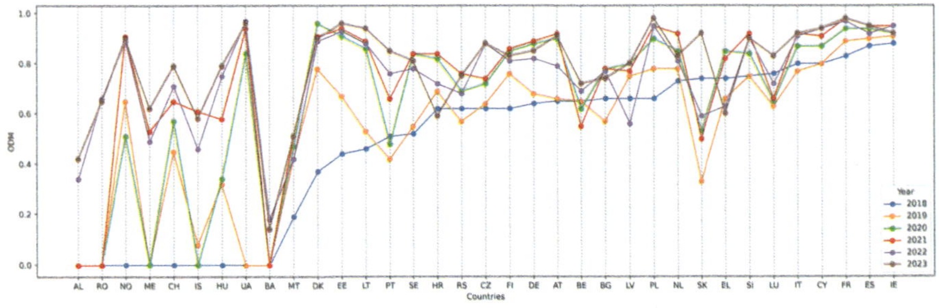

Fig. 4 Open data maturity trends 2018–2023

total datasets and rural datasets suggests whatever the cause, rural datasets are not necessarily related to lower open data maturity.

In terms of data availability for the rural economy we looked once again at the correlation between a country's Open Data Maturity and this time, the relevance of agriculture for their economy, approximated through the proportion of agriculture to their GDP. We find a significant correlation ($p = 0.000$, st. effect $= 0.871$), indicating that higher.

ODM is linked to a higher relevance of agriculture. Assuming that the less mature countries would have more agriculture is incorrect—indeed, the opposite is the case. We therefore reject the idea that rural open data challenges are simply caused by a lack of maturity.

Hypothesis 3—Countries with a higher percentage of their GDP derived from agriculture will publish more agricultural open data sets.

A third key finding from our statistical analysis is the lack of relationship between the importance of agriculture to a country's GDP, and the number of open data sets published in the category of 'agriculture' on the data.europa.eu (which includes fisheries, forestry and food). Agriculture covers 47% of the EU territory and represents around 40% of the EU budget; forestry accounts for another 1% of total EU GDP (Negre 2020), and is also important culturally. Fishery is a key investment area for developing sustainability and growth across the EU. Agriculture is the largest category on the EDP, but highly concentrated: Germany, Poland, and France provided over 94% of agriculture datasets in 2020. As Fig. 5 shows, this has changed over the years, with a number of countries, including Czechia, publishing substantial numbers of agricultural datasets, but Germany, Poland and France continue to publish the most.

If we use the percentage of agriculture of GDP in 2018 as a proxy for the importance of the rural economy sector, and correlate this with the total number or proportion of agricultural datasets published, as shown in Table 2, the results remain negative: there is no significant correlation between a country's economic reliance on agriculture and the total number of agricultural datasets, or the proportion of agriculture in their overall published datasets, with either the dataset count from 2019 nor 2020.For instance, France's agricultural sector was 1.91% of GDP in 2022 and it

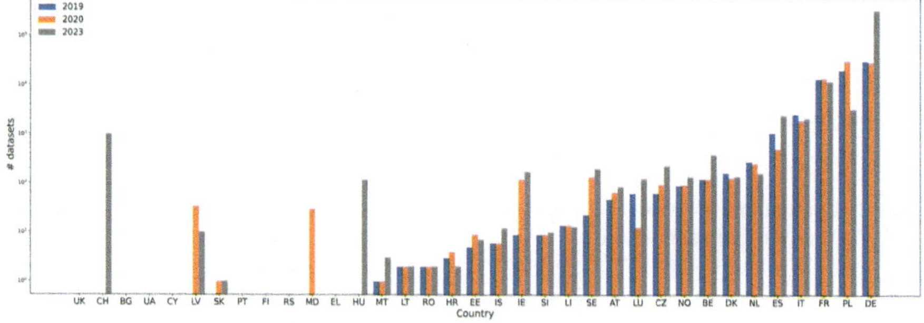

Fig. 5 Agriculture datasets published on data.europa.eu by year

published tens of thousands of agricultural datasets. Slovakia's agricultural contribution to GDP was similar at 2.22%, but it published tens of datasets and the statistical analysis shows these are not outliers. We did not replicate this in 2022/2023 as other trends suggested there would be no discernible change.

To evaluate the possibility that economic factors may influence the publication of rural datasets, we explored the impact of the gross domestic product (GDP). We examined the correlation between GDP for agriculture, forestry, and fishing, and the number of published total and rural datasets. We found no correlation between rural GDP and the published datasets. This suggests that economic factors did not influence the publication of datasets. But if rural economy is so important, why are the datasets not being published? The key finding here is the inconsistency between the relative value of agricultural data to these countries, and the availability of datasets; and also the inconsistency of data publication across the member states in general. This raises crucial questions. Why are so comparatively few member states publishing agricultural data? Are Germany, France and Poland simply opening agricultural datasets because they are easily accessible, or should other member states be opening agricultural data at the same levels?

Some of the reports from our respondents shed some light on these issues. They informed us about agricultural data that was only available to citizens or organisations on a request basis, and often shared only as a web page or PDF, and datasets published on the Ministry of Agriculture site, but not on the relevant national portal, all of which would prevent inclusion in the data.europa.eu catalogue. Many of the welldocumented barriers to opening data—data hoarding, lack of incentives, unclear licensing—still prevail. Regarding the usefulness of open data, many institutions and provinces are still operating at an ad hoc level of data publishing. Performing meaningful, timely research with such data is difficult. An illustration of the challenges created by not having available data was shown by the COVID-19 pandemic (Walker et al. 2020). Farms in the Netherlands faced a problem regarding harvesting on farms as workers could not travel across country borders. Without good data regarding the numbers of workers from abroad participating in seasonal agriculture they were unable to assess the size of the problem nor plan to mitigate it. Again, this is an exclusively rural problem but one which would eventually impact the entire country's food supply. We explore more of the challenges of data publication in the next section.

RQ2—What are the challenges in collecting and publishing rural data?

The challenges of collation and publication are those of digitalisation, awareness, IT capacity and capabilities, usefulness, and standardisation. Digitalisation, even in

Table 2 Correlations (p-values) between total and proportional agriculture datasets and countries % of GDP derived from agriculture (2018)

Year/%	2019	2020
Total	0.308	0.253
Proportion	0.166	0.293

highly urban countries, can often take the form of publishing PDF documents. In more rural areas of countries such as Romania, administrations still use analogue tools, although budgets and expenses are intended to be digitalized.

Awareness of open data in rural areas may be non-existent, even amongst public servants and decision makers. This is despite a decade of European legislation and guidance regarding the availability of Public Sector Information. In highly centralised countries, such as Slovenia, local authorities may view datasets as just another reporting obligation to the central government, instead of seeing them as a potential utilisable asset.

The lack of IT capacity in many rural administrations is a key challenge, not least because responsibility for open data initiatives, even in urban governments, often lies with the IT department. Ireland, a highly digital nation, has no IT capacity in many rural areas. The Roscommon County Council Open Data site[5] is a well-known example of a wide range of open data being published by a rural (and very small) council. However, this was largely driven by the enthusiasm and skill of one employee, although it has survived their leaving the council.

In terms of usefulness, rural data is not always prioritised for opening, especially when a ‘low hanging fruit’ strategy is used. As noted by Pateman (2011), small datasets are often amalgamated into larger units, where individual needs and circumstances are lost. This is sometimes for privacy reasons, but also for administrative purposes, or simply because of the fallacy of the small dataset being useless. An example is the island of Tiree’s electric meter usage data, which is amalgamated with neighbouring islands Coll and Mull, so that patterns specific to Tiree are hidden (Dix 2014). Onerous cultural and language requirements for the publication of relatively small datasets, likely to be of interest to a limited group such as speakers of a minority language, mean that some data is deemed too challenging to publish openly. Lastly, large, decentralised countries like Romania struggle with standardisation. Almost 3000 local authorities submit data for potential publication, but the format and content of this data is hugely varied, and requires substantial work to be made cohesive.

RQ3—How is rural open data used?

In this section we explore how rural open data is currently used, and also potential use cases for open data.

The uses of rural population data Largely, our interviewees agreed that the greatest potential of rural open data was to impact rural areas rather than have rural users. This theme—which we characterised as ‘urban drivers of rural data’—occurred across both our population and economy interviewees. This arises because the skills, knowledge and connections to ideas and innovation are most often found in urban areas with exposure to the quadruple helix of business, government, academia and citizenry. This mirrors other research on ‘smart and competitive rural areas’, such as that done by the European Network for Rural Development (Rural businesses 2016).

[5] http://www.roscommoncoco.ie/en/

One example of use which sits between urban and rural use is the work of the Rural Services Network (England), which is based in a town in a rural area. The Network uses official statistics to establish a fair position for rural service providers. By understanding the difference between urban and rural context, they have shown that rural residents are funding a higher proportion of government services through their council tax and receive less government support per head than their urban counterparts. Such data was also explored to understand the impact of COVID-19 on these areas.

The uses of rural economy data We interviewed several open data portal representatives from countries with large rural populations about the level of agricultural datasets requested to understand which datasets were being used. These were rarely amongst the most requested datasets, which in Slovenia, for example, concern vehicle licensing, public sector salaries, and building permits, with agricultural statistics on crops and the number of livestock further down the list. There are a number of plausible reasons for this, one of which might be fewer exemplar uses available. We found relatively few use cases (28 out of 548) concerning agriculture-adjacent topics on the European Data Portal.

However, there was no shortage of potential uses suggested by our respondents. Many of those centered around the quality and origin of products, driven by a need to control pests or ensure the food products were indeed what they claimed to be; by consumer demand for knowing who their food is produced by and where it is from; and the emergent circular economy. For example, Slovenia, is at a trading crossroads, neighboured by Austria, Croatia, Italy and Hungary. Fruit and vegetable provenance is hard to trace, leading to many more tons of produce being sold annually as 'produce of Slovenia' than are actually grown within the country. Open data would allow interested parties to monitor which produce truly was Slovenian. Risks of misuse of residual flows from the food chain (for instance, food byproducts that are not suitable for human consumption but can be fed to livestock) can be minimised with clear provenance.

When data is available and is actively used, this can create insights that are deeply specific to rural industry. For instance, the FoodValley region of the Netherlands contains many poultry farms. It also appeared to have an intense concentration of pollution, which could be seen from open data. This incentivised the region to explore what was happening. They discovered that, over the period 1995–2014 pollution by dust decreased, but that pollution by poultry increased ten fold. This was driven by the change from barndwelling hens to free range. This change was better for hen welfare and preferred by consumers, but demonstrates the importance of data in revealing such unexpected externalities (Walker et al. 2020).

Data also aids regional smart specialisation for competition. To participate in this regions need to know their strengths, and how well they are specialising. This reciprocally requires data on other regions to compare the specialisations and identify complementarities. As with population data, interviewees noted that in many cases these use cases were likely to be driven by urban-based initiatives or agribusiness, creating a gap between the potential audience and the potential creators. This gave rise to the fisherman in the boat image: while open data about fisheries might be of much value to individuals in the fishing industry, the majority are unlikely to have the skills or resources to realise this value. One organisation we spoke to produces maps of fishing effort and presence at very high resolution and publishes them as open data. Not all of their users are sufficiently well-funded or skilled to use the data. While they wish to engage with and encourage their users, delivering support and training is not always practical.

5 Discussion

Much of what we found regarding the lack of institutionalisation, in infrastructure, skills and champions, reflected earlier challenges of open data in urban settings. So it is unsurprising that there have been several attempts to import the tools and techniques of smart cities to rural settings. Two such projects have happened in Scotland, based on the idea that, "Rural [is] just a low density city" (coo). The Smart Fintry project focused on providing the village of Fintry with an open data portal and interactive energy dashboard, hailed as the United Kingdom's first "smart" village. However, as noted above, such transfers may fail to take into account the specific needs and requirements of rural areas.

A solution may be the emergence of 'smart regions'. Smart regions have extended the smart city to reflect the fact that urban areas are not completely independent of the rural areas that surround them. This creates the necessary links between urban and rural, while acknowledging the differences. Examples include Cork Smart Gateway; Smart Atlantic Way (Brolchain et al.. 2018) and Helsinki Smart Region. Understanding the key factors of smart regions allows for the identification of the key data that should be opened. The EU's 2018 Action for Smart Villages notes that, "Smart means thinking beyond the village itself. Some initiatives are taking place at village level, but many involve the surrounding countryside, groups of villages, small towns and links to cities" (EU Action For Smart Villages 2017). This kind of thinking can be used to help develop the smart region concept. However, it is crucial that this focuses on the part the rural area plays in the region. Even in 2023, 5 years after our initial analysis, the term'smart' in the context of data is synonymous with'urban', and rural data continues to be overlooked.

Approaches that link rural areas with nearby cities with particular technological strengths have been advised in work looking at tech clusters (McFarland and Grabowski 2022). These links with urban areas are also established through universities with complementary interests. There are already examples of this in the agricultural sector, for instance, the Wageningen University Masterclass Accelerator, which ran in 2017. This approach fosters new agricultural business models based on rural–urban linkages. It was managed by a partnership of two organisations: Wageningen Economic Research (WecR, Wageningen University) and the Municipality of Rotterdam's 'Food Cluster' (Agriculture, E. C. D. and Rural Development 2017). FoodValley, the leading agro-food centre in Europe, brings together eight municipalities to create knowledge and innovation in healthy food, and also uses

Wageningen's data. Such approaches could also help to strengthen the intentional selection of key agricultural, forestry and fishery datasets for publication by countries whose rural economy is important, to assist and empower local actors.

Identifying local actors is both challenging but key. As one of our interviewees said, "There are few fishermen sitting in a boat thinking of a way to use open data." Developing local champions—institutionalising the Roscommon example—enables this. The Romanian national data portal developed a new methodology for publishing at the national level. They attempted to disseminate this to the local level, but were successful only to county level. However, they experienced increased success, with one of the less developed regions starting to publish on the national data portal.

Farmers are rurally-located consumers and creators of a great deal of data. In many cases, the value of the data is created by combining with other farms, or other data types. This is the focus of the development of the Common European Agricultural Data Space. While there may be some commercial sensitivity issues, there are also strong arguments for publishing openly, particularly around identifying and combating the spread of disease in livestock. Poultry farmers in particular have experience in virus management due to the bird flu outbreaks over the past decade (Walker et al. 2020).

Another particular approach used in agriculture is 'high tech partnerships'. Shared platforms within agricultural specialisations link farmers with technical solutions. In this way, information can be made available between farmers, policy officers and technical suppliers. This requires a lot of data on agriculture and the technical solutions to be available to all the relevant parties. Where data cannot be made open for commercial reasons, efforts are still made to reveal as much as possible to key stakeholders.

6 Smart Regions and Data

The key insight is the potential value of the smart region. This term, more than the smart village, recognises that a city is not independent of its surroundings (especially in terms of transport, supply chains, food security etc.) while acknowledging rural areas benefit from metropolitan skills and capabilities. This addresses the clear problems of attempting to simply transpose urban smart city practices to rural areas, and redresses the imbalance of the concept of the smart city that leaks some smart practices at the margins into the surrounding countryside. To engage with this concept, we need to ask, How does the ecosystem of a smart region differ from other'smart' areas? What are the challenges? How is a 'smart region' successfully integrated with the larger 'smart city' concept?

In terms of opening data, it is clear that local champions—whether in local authorities or public sector institutions, or working with larger communities or industries such as farming—are crucial. Local champions should be supported to establish and institutionalise the processes and practices of open data in rural administrations, with close relationships to urban open data groups. Such groups might be the national portal, a city portal or a civic technology group. This could potentially be achieved by the development of training materials and incentives for local champions to participate in this training. This will also assist with improved digitisation and standardisation both within the rural administration and across other, related regions.

Sharing skills and fostering support can also be engendered by establishing links between complementary data owners, including Not for Profits, Non-Governmental Organisations, researchers and private organisations, that allow data to be opened and made available together. This could also be done via alternative models by which data can be shared, when open licensing is not an option (for instance, as exemplified by the JoinData rural data sharing platform in the Netherlands (Join Data 2024)) or the proposed Common Agricultural Data Space (Atik 2022).

Similarly, effort should be made to establish links from rural areas with urban areas, via such instruments as hubs or universities with complementary interests, as in Wageningen.

Finally, we suggest identifying and supporting a wider range of rural-specific challenges, including those mentioned by our respondents, such as seasonal workers, health solutions that address distance, and food provenance. These can/should still be run in urban settings, in order to ensure access to sufficient and varied technical skills. As well as assisting with the opening of key datasets, or financial resources, support might take the form of introducing urban teams with the right skills to rural groups with problems that can be addressed with data.

7 Conclusion

The need for institutionalisation of the commitment to, and appropriate skills for, open data, remains a key barrier to the publication and use of open data in both rural areas and sectors. More acknowledgement of the importance of data and data capacity in the urban rural digital divide literature will also give impetus to rural open data.

During the execution of the original research for this paper, we examined many reports and policies on open data, and on rural populations and industries. We were struck by the limited overlap between the two areas. Rural development reports frequently urged an increase in digital accessibility, hubs and businesses, (e.g. ENRD 2016). Few even contained the word 'data', far less 'open data', let alone views on how the data sharing economy would benefit the rural sectors. In subsequent research we located a report by Rural England, commissioned by Amazon, which featured a case study of a company that provides dairy cow behaviour monitoring data to farmers (Wilson et al. 2018). However, the technological possibility that was of interest to the researchers was the cloud computing via which the data was stored and shared, rather than the possibilities of the data itself. It is crucial that rural populations and industrial sectors are not (unintentionally or otherwise) excluded from the benefits of the data economy. If the focus of data collection and its subsequent re-use in technological artefacts in urban areas is prioritised, those who are not part of the data become invisible, and are subsequently ignored.

To do this, 'smart rural' must become something more real than a few isolated projects, but also something more than a faint facsimile of smart cities. The greatest opportunity for this lies in 'smart regions', that create the necessary links between urban and rural, while acknowledging the differences.

References

Agriculture, E. C. D. and Rural Development, E. N. f. R. D. (2017). Reimagining rural business opportunities.

Arroub, A., Zahi, B., Sabir, E., and Sadik, M. (2016). A literature review on smart cities: Paradigms, opportunities and open problems. In *2016 International conference on wireless networks and mobile communications (WINCOM)*, pages 180– 186. IEEE.

Atik, C. (2022). Towards comprehensive european agricultural data governance: Moving beyond the "data ownership" debate. *IIC-International Review of Intellectual Property and Competition Law*, 53(5):701–742.

Barry, J. J. (2013). Crossing the digital divide: Is access to the internet an economic right? *SSRN Electron. J.*

Braesemann, F., Lehdonvirta, V., and Kassi, O. (2022). ICTs¨ and the urban-rural divide: can online labour platforms bridge the gap? *Inf. Commun. Soc.*, 25(1):34–54.

Brolchain, N.´ O., Ojo, A., Porwol, L., Minton, D., and Barry,´ C. (2018). Examining the feasibility of a smart region approach in the north west atlantic and borders region of ireland. In *Proceedings of the 11th International Conference on Theory and Practice of Electronic Governance*, New York, NY, USA. ACM.

Centre, J. R. and REGIO, D. (2019). Urban data platform plus. Accessed: 13/02/2024.

[coo] Smart rural coop. https://smartrural.coop. Accessed: 13/02/2024.

Dix, A. (2014). Open data islands and communities white paper. https://tireetechwave.org/odic/. Accessed: 13/02/2024.

Eckhardt, J., Nykanen, L., Aapaoja, A., and Niemi, P. (2018).¨ Maas in rural areas-case finland. *Research in Transportation Business & Management*, 27:75–83.

Esteban-Navarro, M.-A., Garc´ ´ıa-Madurga, M.-A., Morte-´ Nadal, T., and Nogales-Bocio, A.-I. (2020). The rural digital divide in the face of the COVID-19 pandemic in europe— recommendations from a scoping review. *Informatics (MDPI)*, 7(4):54

EUAction For Smart Villages (2017). https://enrd.ec.europa.eu/news-events/news/eu-action-smart-villages_en. Accessed: 13/02/2024.

Eurostat (2023). Rural europe.

Halford, S. and Savage, M. (2010). RECONCEPTUALIZING DIGITAL SOCIAL INEQUALITY. *Inf. Commun. Soc.*, 13(7):937–955.

Holmes, H. and Burgess, G. (2020). *Pay the wifi or feed the children": Coronavirus has intensified the UK's digital divide. Cambridge Centre for Housing and Planning Research.*

Join data. https://join-data.nl/en/. Accessed 13 Feb 2024.

Lakomaa, E. and Kallberg, J. (2013). Open data as a foundation for innovation: The enabling effect of free public sector information for entrepreneurs. *IEEE Access*, 1:558–563.

Maja, P. W., Meyer, J., and Von Solms, S. (2020). Development of smart rural village indicators in line with industry 4.0. *IEEE Access*, 8:152017–152033.

McFarland, C. K. and Grabowski, E. H. (2022). Local employment impacts of connectivity to regional economies: The role of industry clusters in bridging the urban-rural divide. *Economic Development Quarterly*, 36(3):317–328.

Negre, F. (2020).' *Fact Sheets on the European Union: The European Union and forests*. European Parliament.

Ojo, A., Curry, E., and Zeleti, F. A. (2015). A tale of open data innovations in five smart cities. In *2015 48th Hawaii International Conference on System Sciences*. IEEE.

Open data in Europe 2018. https://www.europeandataportal.eu/en/dashboard/2018. Accessed 2023-10-26.

Pateman, T. (2011). Rural and urban areas: comparing lives using rural/urban classifications. *Reg. Trends*, 43(1):11–86.

Pavez, I., Correa, T., and Contreras, J. (2017). Meanings of (dis)connection: Exploring non-users in isolated rural communities with internet access infrastructure poetics 63. *Poetics*, 63.

Philip, L., Cottrill, C., Farrington, J., Williams, F., and Ashmore, F. (2017). The digital divide: Patterns, policy and scenarios for connecting the 'final few' in rural communities across great britain. *J. Rural Stud.*, 54:386–398.

Prause, G. and Boevsky, I. (2015). Smart rural development. *Agricultural Economics and Management*, 60(4):63–69.

Premanandh, J. (2013). Horse meat scandal – a wake-up call for regulatory authorities. *Food Control*, 34(2):568–569.

da Rosa Pires, A., Pertoldi, M., Edwards, J., Hegyi, F. B., et al. (2014). Smart specialisation and innovation in rural areas. *S3 Policy Brief Series*, 9:2014.

Ruijer, E., Dymanus, C., van Kesteren, E.-J., Boeschoten, L., and Meijer, A. (2024). Open data work for empowered deliberative democracy: Findings from a living lab study. *Government Information Quarterly*, 41(1):101902.

Rural Businesses (2016).https://enrd.ec.europa.eu/enrd-thematic-work/smart-and-competitive-rural-areas/ rural-businesses_en. Accessed: 2023-10-26.

Salemink, K., Strijker, D., and Bosworth, G. (2017). Rural development in the digital age: A systematic literature review on unequal ICT availability, adoption, and use in rural areas. *J. Rural Stud.*, 54:360–371.

Sostero, M., Milasi, S., Hurley, J., Fernandez-Mac´ ´ıas, E., and Bisello, M. (2020). Teleworkability and the COVID-19 crisis: A new digital divide? *European Commission.*

Talbot, S. (2016). Creating a smart rural economy through smart specialisation: The microsphere model. *Local Economy*, 31(8):892–919.

Vicente Cuervo, M. R. and Lopez Menendez, A. J. (2006). A multivariate framework for the analysis of the digital divide: Evidence for the european union-15. *Inf. Manag.*, 43(6):756– 766.

Walker, J., Thuermer, G., Berkvens, J., and Worth, D. (2020). *The Open Data Rural Gap (Webinar) European Data Portal.*

Wilson, B., Atterton, J., Hart, J., Spencer, M., and Thomson, S. (2018). *Unlocking the Digital Potential of Rural Areas Across the UK.Rural England/Scotland's Rural College.*

Wineman, A., Alia, D., and Anderson, C. (2019). *Definitions of 'Rural' and 'Urban' and Understandings of Transformation in Sub-Saharan Africa. Presentation at Data for Policy.*

Zar, J. H. (2005). Spearman rank correlation. *Encyclopedia of biostatistics*, 7.

Zukin, S. (2020). *The Innovation Complex: Cities, Tech and the New Economy*. Oxford University Press.

Germany: Procuring Smart City Solutions Through GovTech

Luca T. Bauer, Frederik R. Heinz, and Björn Niehaves

1 Introduction

Smart city solutions have gained significant traction over the past decade as municipalities strive to enhance innovation and citizen-centric governance (Dai et al. 2023). These solutions offer a pathway to modernization without necessarily requiring a complete overhaul of existing administrative infrastructure. By integrating digital tools and platforms, cities can deliver improved services while maintaining operational continuity (Bharosa 2022).

However, the adoption of smart city technologies has introduced new challenges. Chief among these is the technical complexity of these solutions, which often cannot be developed in-house (Collington 2022; Mazzucato and Kattel 2020). This dependency on external expertise creates potential vulnerabilities, including risks to digital sovereignty, as municipalities may rely heavily on proprietary systems controlled by private providers, often foreign companies and with non-local data storages (Neri 2022).

To address these challenges, a promising approach has emerged: procuring solutions from GovTechs, a term that blends "*government*" and "*technology*" (Niehaves and Klassen 2024). GovTechs are specialized companies that develop innovative technologies tailored to the unique needs of public sector entities (Commission of the European Union 2024). Their involvement represents a shift in how governments approach digitalization, switching over into a collaborative effort between public and private sectors.

The role of GovTechs in public sector digitalization has been steadily rising and is now seen as a cornerstone of administrative modernization (Bharosa 2022; Mergel

L. T. Bauer (✉) · B. Niehaves
University of Bremen, Bremen, Germany
e-mail: bauerl@uni-bremen.de

F. R. Heinz
Hochschule Hildesheim, Hildesheim, Germany

C. Schachtner (ed.), *The European Smart City Movement*,
https://doi.org/10.1007/978-3-031-96700-9_18

et al. 2022; Niehaves and Klassen 2024). This trend is evident across various levels of governance, from local municipalities to the European Union, where it has been highlighted in numerous strategic initiatives and policy documents (Hoekstra et al. 2023). This growing emphasis underscores the potential of GovTechs to reshape how governments manage and implement technological solutions.

In this paper, the term "*GovTech*" aligns with the European Commission's definition in the Interoperable Europe Act: "*technology-based cooperation between public and private sector actors supporting public sector digital transformation*" with a focus on small and medium sized enterprises (SMEs) and startups (Commission of the European Union 2024). This definition captures the essence of GovTechs as enablers of efficient, scalable, and interoperable solutions designed for the public sector's unique needs.

The following chapter uses Germany as a case study to explore how cities procure solutions from GovTechs, examining both the opportunities and challenges that arise in the process. Practical examples will further illustrate how optimizations in procurement processes—both organizationally and through the use of tailored procurement instruments—can unlock the potential of GovTechs for advancing smart city objectives (Klassen et al. 2024, 2023).

2 Procuring Smart City Solutions

One of the most pressing challenges faced by public administrations in their pursuit of smart city innovations is not merely the implementation or operation of these solutions but their procurement (Herdi ansyah 2023; Jarvers and Schachtner 2024; Kogan and Lee 2014). Navigating the procurement landscape involves reconciling diverse requirements such as technical compatibility, legal compliance, budgetary constraints, and alignment with long-term strategic goals (Jarvers and Schachtner 2024). This process can become a bottleneck, delaying the deployment of technologies and impeding progress toward creating more innovative and citizen-centric cities.

Historically, public administrations and national governments relied heavily on in-house development to address technological needs (Haugen and Klungseth 2017). This approach ensured full control over solutions and guaranteed alignment with internal processes. However, as digital solutions have become increasingly complex, the resources and expertise required to develop and maintain them have outstripped the capacity of many public administrations. This shift marks a turning point for the public sector, which must now depend on external expertise to meet the demands of modern smart city infrastructures, for example, in the area of the Internet of Things or artificial intelligence (Almarri and Boussabaine 2023).

While some nations, such as Estonia, have successfully circumvented these challenges, their success is rooted in specific historical and structural contexts (Veljković et al. 2014). Estonia's digital infrastructure was conceived alongside its governmental structures with a "*blue ocean*" approach, allowing for integrated and future-proof solutions from the outset. Conversely, in historically developed and federally organized states like Germany, the digital infrastructure has grown piece by piece over decades (Kattel et al. 2019, 2022; Mazzucato and Kattel 2020). This results in a patchwork of systems that can be difficult to modernize cohesively. For such countries, procuring smart city solutions requires careful navigation of legacy systems, diverse regional requirements, and the federal structure of governance. This complexity amplifies the need for solutions that fit not only different technical infrastructures but also align with broader visions and financial limitations.

The procurement process itself often presents significant hurdles. Long procurement cycles—spanning two years or more—are commonly defended on the grounds of strict legal frameworks and limited market options (Harland et al. 2012). These delays, however, are not an inevitable consequence of public sector processes. Innovative initiatives like "*Code for America*"[1] in the United States or the German hackathon "*Wir vs. Virus*"[2] have proven that procurement can be accelerated without compromising quality or compliance (Harrell 2013; Schrock 2016a, b). By leveraging agile methodologies, open innovation platforms, and closer collaboration between public and private actors, these initiatives illustrate how procurement cycles can be shortened to meet the urgency of technological adoption (Gerrard 2001).

The complexity of procurement is further compounded by a lack of knowledge and exposure to best practices (Harland et al. 2012). Decision-makers often struggle to identify models or case studies that have successfully addressed similar challenges. This underscores the importance of sharing knowledge, fostering collaboration, and documenting successful implementations to inspire confidence and innovation in procurement strategies (Schrock 2016a, b). Lighthouse projects and case studies serve as critical tools for this purpose, providing a tangible demonstration of what is possible when procurement processes are optimized.

Moreover, there is a growing recognition of the need to rethink procurement frameworks at an institutional level. This involves revisiting traditional practices, exploring flexible procurement instruments, and enabling greater experimentation with public–private partnerships (Gerrard 2001). Modernizing procurement practices also calls for a cultural shift within administrations, emphasizing agility, iterative development, and risk tolerance (Klassen et al. 2023). These shifts are essential to empower public administrations to respond effectively to the rapidly evolving demands of smart city development.

The intersection of legal, technical, and organizational considerations complicates procurement for smart cities. Legal frameworks, often designed for traditional infrastructure projects, are not always well-suited to the dynamic and iterative nature of digital solutions. This mismatch necessitates a more adaptive legal and regulatory environment that allows for innovation while maintaining transparency and accountability (Hoekstra et al. 2023).

Engaging GovTech solutions can assist in overcoming the aforementioned challenges. Encouraging the growth of GovTech ecosystems and nurturing relationships

[1] https://codeforamerica.org/.

[2] https://wirvsvirus.org/.

with innovative companies can help expand the range of available options, ensuring a competitive and diverse marketplace (Dener et al. 2021; Niehaves and Klassen 2024). This proactive approach to market development can reduce dependency on a small number of vendors and enhance digital sovereignty by fostering homegrown technological capabilities.

This chapter aims to provide practical insights and recommendations for the procurement of innovative solutions that support the development of smart cities and collaboration between the public sector and GovTechs. By examining successful practices, exploring innovative procurement models and highlighting case studies, we aim to equip public administrations with the knowledge needed to effectively procure innovative solutions to address the current challenges of smart cities.

3 The Case for GovTech

GovTech companies present a compelling alternative to traditional providers like large multinational corporations (big tech) and established mid-sized companies (Bharosa 2022; Commission of the European Union 2024). In the following, we will highlight the most important advantages of GovTechs for smart cities and provide an overview of use cases.

(1) Price and Performance: GovTechs often combine affordability with high technological capabilities, frequently derived directly from recent advancements in academic and industrial research. Particularly young start-ups are modest when it comes to pricing, as they are usually heavily reliant on acquiring new customers first. This makes them an appealing option for public administrations seeking to modernize their infrastructure while optimizing costs.
(2) Customization and flexibility: In addition to cost-effectiveness, engaging GovTechs offers public administrations a greater degree of influence over solution configurations, ensuring that technologies are tailored to specific local needs. As an early customer, the administration can increase its influence on providers through continuous feedback and participate in product design.
(3) Citizen-centricity: GovTechs tend to prioritize the needs and experiences of citizens, designing solutions that directly address urban challenges and enhance quality of life. Quite often, the companies emerge from the challenges that their founders themselves face in real life.
(4) Digital Sovereignty: By reducing reliance on foreign multinational providers, GovTechs can empower governments to retain control over their digital infrastructure and data. On the one hand, this is due to the diversification of the market. On the other hand, many GovTechs are creating or building on open-source solutions, thus contributing transparency, interoperability and accessibility. The local storage of data also increases data security.

Table 1 Smart city use-cases from Germany

Technology	Context	Example
Platform for managing data-heavy administrative processes	The platform supports public authorities in digitizing processes, including data integration, workflows, case management, and reporting	polyteia
(Urban) Digital Twinning	Digital twin technologies enable cities to simulate the state of infrastructure using geodata	urbanistic.
Inclusivity and inclusive language	Solutions promote inclusivity by turning bureaucratic language into easy, barrier free speech	SUMM AI
Infrastructure mapping	Detailed mapping tools provide an overview of critical urban infrastructure, such as utilities, roads, and public facilities	vialytics
IoT (e.g. Air quality monitoring)	Internet of things (IoT) solutions allow cities to monitor air quality in real time, using networks of sensors deployed across urban areas	BREEZE TECHNOLOGIES

...and more: The versatility of GovTech solutions ensures that their applications are limited only by the imagination of their developers and the needs of public administrations. From waste management optimization to advanced water monitoring systems, GovTechs are constantly expanding the scope of their solutions offered for the public market (Selke et al. 2024)

(5) Strengthening the economy: Finally, commissioning GovTechs can help to strengthen local technology companies and regional start-up scenes. The public sector is considered to be the strongest potential customer for technologies and generates considerable revenue for its contractors. Commissioning local companies can help them grow as well as encouraging the creation of new start-ups.

In Germany, GovTechs have pioneered a diverse array of applications, addressing critical needs in smart cities (Kuziemski et al. 2022; Selke et al. 2024). The use cases listed below demonstrate the breadth of GovTech's contributions and their potential to transform public services, see Table 1.

4 Smart City Procurement Through the Levels of Public Administration

Public procurement operates in a fundamentally different manner from private sector procurement, primarily due to the emphasis on transparency, accountability, and compliance with legal frameworks. While private companies prioritize efficiency and cost-effectiveness, public administrations must balance these goals with the need to ensure fairness, equal access, and adherence to regulatory requirements. These differences often introduce additional complexity and length to public sector procurement processes, particularly when compared to the agility of private sector counterparts.

It is also important to distinguish between procurement and tendering, as the two terms are often used interchangeably despite their distinct meanings. Procurement refers to the broader process of acquiring goods, services, or works and includes planning, market research, and contract management. Tendering, on the other hand, is a specific phase within procurement, encompassing the preparation and issuance of requests for proposals, evaluation of bids, and awarding of contracts. Understanding this distinction is critical to navigating the intricacies of public sector procurement and identifying opportunities for optimization.

There are several methods of public procurement, ranging from open competitive bidding to negotiated or direct awards. Each method has specific advantages and limitations, depending on the complexity, value, and urgency of the solution being procured. For instance, competitive bidding ensures maximum transparency and fairness, but it is often time-intensive and rigid. Conversely, negotiated procedures or direct awards allow for greater flexibility and speed but may raise concerns about fairness and accountability if not appropriately justified.

The procurement of smart city solutions, particularly those developed by GovTechs, often exposes the limitations of traditional public sector procurement processes. These solutions are frequently iterative, rapidly evolving, and require close collaboration between public administrations and providers. Traditional procurement frameworks, which can span multiple years from initiation to contract award, are ill-suited to the fast-paced and dynamic nature of these technologies. Lengthy processes risk rendering solutions obsolete or misaligned with the rapidly changing needs of urban environments by the time they are implemented.

To address these challenges, alternative procurement approaches tailored to smart city solutions and GovTechs should be considered. Agile procurement practices, pre-commercial procurement models, and innovation partnerships are examples of methodologies that enable administrations to acquire cutting-edge solutions more efficiently while maintaining transparency and accountability.

The following sections will illustrate practical examples and case studies of smart city procurement across different levels of public administration. From municipal initiatives to federal and national programs, these examples will demonstrate the value added by optimized procurement practices.

5 Examples from Germany

5.1 Communal

In North Rhine-Westphalia, the "*GovTech Tage*"[3] serve as an example as public private partnership launched event-based platform for fostering collaboration between GovTechs and municipal administrations. These events, held in various cities across the region, aim to explore how innovative solutions can shape the future of local governance.

The GovTech Tage facilitate direct engagement between public sector representatives and private sector innovators. Through workshops, panel discussions, and networking sessions, participants collaboratively address pressing challenges in municipal administration. This interaction enables municipalities to gain insights into emerging technologies and their practical applications, while startups and tech companies receive valuable feedback on the specific needs and constraints of public administration.

A key feature of these events is the emphasis on co-creation. By involving diverse stakeholders in the solution development process, the GovTech Tage ensure that resulting partnerships are user-centric and effectively address the challenges faced by local governments. This participatory approach not only enhances the relevance of the solutions but also fosters a sense of ownership and commitment among all parties involved.

As a result, several municipalities have initiated pilot projects with startups they connected with during the events, leading to the implementation of solutions that improve service delivery, enhance citizen engagement, and increase operational efficiency. For instance, some cities have adopted digital platforms for citizen feedback, while others have implemented smart infrastructure management systems, all stemming from collaborations initiated at the GovTech Tage.

Lessons Learned: The GovTech Tage in North-Rhine-Westphalia demonstrate the effectiveness of creating interactive platforms that bring together public administrations and technology innovators. Such events facilitate mutual understanding, foster collaborative solution development, and accelerate the adoption of GovTech innovations at the communal level. By prioritizing co-creation and direct engagement, municipalities can more effectively leverage technological advancements to meet the evolving needs of their communities.

[3] https://open.nrw/govtech-tag-2024.

5.2 Federal State

GovTechHH,[4] launched in 2022 by the City of Hamburg, provides a notable example of innovative procurement at the federal state level. This initiative is organized as a venture clienting unit, a model designed to foster collaboration with startups and emerging technology providers. The venture clienting approach positions GovTechHH as an "*early adopter*" of cutting-edge solutions, enabling Hamburg to test and refine technologies before broader implementation. This model ensures that GovTech solutions are rapidly integrated into public administration processes, enhancing Hamburg's ability to respond to evolving digital needs.

A pivotal development in Hamburg's approach to smart city procurement was the introduction of an experimental clause in 2024. This regulatory instrument enables contracts valued at up to EUR 100,000 to bypass traditional procurement procedures. Instead, the City of Hamburg can directly invite a single company to submit a tender. However, several conditions must be met: the city's GovTechHH unit must be involved, and the contract must focus on the adoption of innovative technologies to modernize local government processes. This clause accelerates the procurement of novel solutions and allows for faster testing and integration of technologies, empowering Hamburg to maintain its leadership in smart city innovation.

Lessons Learned: Hamburg's experience highlights the importance of regulatory flexibility and targeted initiatives in fostering innovation. By creating structures like GovTechHH and leveraging experimental clauses, public administrations can expedite procurement and rapidly integrate emerging technologies into their workflows.

5.3 Inter-Federal State

At the inter-federal state level, Dataport,[5] Germany's largest public IT provider, exemplifies a collaborative approach to smart city procurement. Through its dedicated GovTech program, Dataport leverages economies of scale to deliver innovative solutions efficiently and cost-effectively to multiple federal states. This model not only reduces redundancy in procurement efforts but also ensures that public administrations across different regions have access to cutting-edge technologies.

Dataport's role goes beyond procurement. It conducts extensive market research and solution presentations for its clients, pre-vetting GovTech providers and offering a curated selection of innovative options. This preparatory work reduces administrative burdens on individual public sector clients, enabling them to focus on strategic implementation rather than provider selection.

---

[4] https://startupcity.hamburg/de/trends-themen/govtechh.

[5] https://www.dataport.de/services-produkte/consulting/kooperation-mit-start-ups-foerdern/.

Additionally, Dataport offers hosting services for GovTech solutions on its secure servers, ensuring compliance with stringent data protection standards while facilitating faster rollout and scaling of solutions. By acting as both a central procurement platform and an IT infrastructure provider, Dataport accelerates digital transformation while fostering collaboration among its clients.

Dataport's innovative approach also extends to the development of spin-offs from its own IT divisions and those of its clients. These spin-offs contribute to the broader GovTech ecosystem by creating tailored solutions that address specific administrative needs while reinforcing Germany's position as a leader in public sector digitalization.

Lessons Learned: Dataport demonstrates the value of centralized platforms and economies of scale in inter-federal procurement. By consolidating efforts and fostering collaboration, inter-federal entities can streamline procurement, drive innovation, and strengthen the public sector's capacity to adopt advanced technologies.

5.4 National

At the national level, the GovTech Campus[6] represents an initiative aimed at creating a cohesive ecosystem for solution providers and public sector users. Unlike many other GovTech programs, the GovTech Campus adopts a broader definition of GovTech, including not only startups and SMEs but also established tech companies and public–private collaborations.

The Campus is designed to serve as a platform for exchange and co-creation, bringing together public administrations from municipal to federal levels with a diverse array of technology providers. Its mission extends beyond procurement, emphasizing the creation and scaling of innovative solutions that address complex administrative challenges.

Initially headquartered in Berlin, the GovTech Campus has begun expanding to other cities, including Heilbronn, where it focuses on artificial intelligence applications, and Hamburg, which emphasizes urban digitalization. This geographic expansion reflects the Campus's ambition to foster innovation across Germany while leveraging the unique strengths of each location.

The GovTech Campus also emphasizes fostering best practices at a national level. By promoting direct procurement and unrestricted awarding models, it accelerates the adoption of GovTech solutions and streamlines collaboration between public and private stakeholders. Through its activities, the Campus creates an integrated ecosystem where ideas are shared, solutions are co-developed, and digital transformation is realized at scale.

[6] https://govtechcampus.de/.

Lessons Learned: The GovTech Campus illustrates the power of national-level initiatives in unifying fragmented efforts and driving innovation. By serving as a platform for collaboration and scaling solutions across administrative levels, national initiatives can build cohesive ecosystems that support long-term digital transformation.

6 Best Practice Recommendations

Based on the examples and insights presented in this paper, the following recommendations are designed to provide practitioners with actionable strategies for successfully leveraging GovTech solutions in smart city development (Niehaves and Klassen 2024). These best practices target public administrators, policymakers, and other stakeholders aiming to streamline procurement processes and foster innovation, see Table 2.

6.1 Active Procurement and Market Research

One of the foundational steps in procuring smart city solutions is engaging in proactive procurement and market research. Unlike traditional reactive procurement approaches, which often wait for solutions to be presented through formal bidding processes, active procurement involves exploring market trends, identifying emerging GovTech providers, and understanding the evolving landscape of public sector technology.

Table 2 Summary of best practices

Best practice	Outcomes
Active procurement and market research	• Understanding of market trends • Early Adoption of innovation
Building local ecosystems	• Reliable local ecosystems • Strengthened local tech economy
Scaling ahead of local needs	• Longevity of purchased solutions • Increased scalability
Co-develop solutions	• Tailored solutions • Increased understanding of technology and capabilities
Adjusting procurement instruments	• Fast procurement • Strengthened collaboration • Maintained transparency

Practitioners should establish dedicated teams or units responsible for market intelligence, attending innovation events like *GovTech Tage*, and engaging with startups, SMEs, and research institutions. This ensures early identification of promising technologies that align with local needs, reducing the time and effort required during formal procurement. Proactive market research also fosters competitive ecosystems, preventing over-reliance on a few large providers and enhancing digital sovereignty by diversifying solution sources.

6.2 Building Local Ecosystems

Fostering local GovTech ecosystems is essential for ensuring sustainable innovation. Municipalities and regions can act as incubators for startups and small tech companies by providing platforms for collaboration, mentorship, and pilot projects. Initiatives like the *GovTech Tage* in North Rhine-Westphalia demonstrate the value of co-creation spaces where public administrations work directly with GovTech innovators.

Building local ecosystems also has long-term economic benefits. By nurturing homegrown GovTech industries, public administrations can stimulate job creation, retain talent, and enhance regional competitiveness. To achieve this, municipalities should prioritize partnerships with local providers, invest in public–private innovation hubs, and create incentives for startups to focus on public sector challenges.

6.3 Scaling Ahead of Local Needs

Effective GovTech strategies involve anticipating future demands and scaling solutions proactively rather than reactively. Scaling ahead of local needs ensures that technology implementations remain relevant and capable of accommodating growth or changing circumstances.

Practitioners can leverage the inter-federal model exemplified by Dataport, which uses economies of scale to deliver cost-effective solutions to multiple clients. By collaborating across regions, pooling resources, and aligning procurement goals, public administrations can develop scalable platforms that serve a broader range of stakeholders. Additionally, creating modular solutions that can be expanded or adapted over time ensures flexibility and longevity.

6.4 Co-developing Solutions

Co-development between public administrations and GovTech providers is a powerful approach for creating tailored solutions that address specific administrative challenges. Unlike off-the-shelf products, co-developed solutions are designed collaboratively, incorporating the expertise of both public sector stakeholders and technology innovators.

This process requires close engagement with end-users, such as municipal staff or citizens, to ensure that solutions are user-centric and practical. Agile methodologies, iterative feedback loops, and pilot testing are essential components of successful co-development. Events like the GovTech Campus and Smart Urban Labs demonstrate the potential of co-creation platforms in fostering innovation and delivering impactful outcomes.

6.5 Adjusting Procurement Instruments

Traditional procurement frameworks are often rigid and slow, making them ill-suited for GovTech solutions. To address this, practitioners should consider modernizing procurement instruments to align with the iterative and dynamic nature of GovTech. Key adjustments include:

Increased Allocation Values: Introducing higher thresholds for direct awards or simplified procurement processes can accelerate decision-making and reduce administrative overhead for smaller, high-impact projects.

Coopetitions: Combining elements of competition and collaboration, coopetitions allow multiple providers to work together while vying to deliver the best solution. This approach encourages innovation and fosters partnerships between vendors.

Co-development Contracts: Procurement instruments should enable shared risk and shared reward models, incentivizing providers to deliver tailored solutions that meet the unique needs of public administrations.

Reusability: Emphasizing open-source standards and modular design ensures that solutions can be reused across different administrations or adapted for new purposes, maximizing return on investment.

By adopting these innovative procurement instruments, public administrations can streamline processes, foster collaboration, and ensure transparency and accountability.

7 Perspectives: What a Coherent Smart City GovTech Vision Could Accomplish

A coherent Smart City GovTech vision has the potential to transform urban governance, positioning cities as hubs of innovation, sustainability, and citizen empowerment (Nam and Pardo 2011). Such a vision aligns public administration processes with technologies, ensuring cities can adapt while meeting the evolving needs of their communities (Harrell 2013).

This vision prioritizes integrated smart city ecosystems where public services—such as mobility, energy, waste management, and citizen engagement—operate seamlessly through interoperable platforms. By fostering collaboration between public administrations, GovTech providers, and local communities, cities can implement solutions that are practical, scalable, and designed with citizens at the core (Bharosa 2022).

Investing in local GovTech ecosystems is essential to this vision. Supporting regional startups and SMEs not only enhances digital sovereignty but also stimulates local economies. These partnerships can encourage open-source development, ensuring solutions remain adaptable and reusable across various contexts, potentially reducing long-term costs and dependencies (Niehaves and Klassen 2024).

Data plays a critical role in realizing smart city goals. A GovTech-driven approach ensures that cities can harness real-time data to enhance urban planning, optimize resource use, and predict challenges before they escalate. For example, predictive analytics can enable better traffic flow management, targeted energy-saving measures, and tailored public services.

Procurement must also evolve to align with this vision. Flexible procurement instruments—such as agile contracts and pre-commercial models—are needed to support iterative development, co-creation, and rapid scaling. This adaptability ensures that technologies remain relevant and aligned with both immediate and long-term urban strategies.

Finally, a national framework for smart city GovTech, connecting municipal, regional, and federal efforts, can amplify the impact of individual initiatives. By fostering knowledge sharing, promoting standardized practices, and leveraging shared infrastructure, such a framework ensures that all cities, regardless of size, can access and benefit from transformative technologies.

A coherent Smart City GovTech vision will not only enhance operational efficiency and sustainability but also redefine urban living by creating cities that are inclusive, resilient, and prepared for the future.

References

Almarri, K., & Boussabaine, H. (2023). Critical success factors for public–private partnerships in smart city infrastructure projects. *Construction Innovation.*

Bharosa, N. (2022). The rise of GovTech: Trojan horse or blessing in disguise? A research agenda. *Government Information Quarterly, 39*(3), 101692.

Collington, R. (2022). Disrupting the welfare state? Digitalisation and the retrenchment of public sector capacity. *New Political Economy, 27*(2), 312–328.

Commission of the European Union. (2024). *Regulation (EU) 2024/903 of the European Parliament and of the Council of 13 March 2024 laying down measures for a high level of public sector interoperability across the Union (Interoperable Europe Act).* Bruxelles: Commission of the European Union.

Dai, Y., Hasanefendic, S., & Bossink, B. (2023). A systematic literature review of the smart city transformation process: The role and interaction of stakeholders and technology. *Sustainable Cities and Society*, 105112.

Dener, C., Nii-Aponsah, H., Ghunney, L. E., & Johns, K. D. (2021). *GovTech maturity index: The state of public sector digital transformation.* World Bank Publications.

Gerrard, M. (2001). Public-private partnerships. *Finance and Development, 38*(3), 48–51.

Harland, C., Telgen, J., Knight, L., & Callender, G. (2012). Challenges facing public procurement. In *Public Procurement* (pp. 379–386). Routledge.

Harrell, C. (2013). The Beginning of a Beautiful Friendship: Data and Design in Innovative Citizen Experiences. *Beyond Transparency: Open Data and the Future of Civic Innovation*, 151–163.

Haugen, T. B., & Klungseth, N. J. (2017). In-house or outsourcing FM services in the public sector: a review of 25 years research and development. *Journal of Facilities Management, 15*(3), 262–284.

Herdiansyah, H. (2023). Smart city based on community empowerment, social capital, and public trust in urban areas. *Global Journal of Environmental Science and Management, 9*(1), 113–128.

Hoekstra, M., Van Veenstra, A. F., & Bharosa, N. (2023). Success Factors and Barriers of GovTech Ecosystems: A case study of GovTech ecosystems in the Netherlands and Lithuania. ACM International Conference Proceeding Series.

Jarvers, S., & Schachtner, C. (2024). Linking smart city projects to the internal e-government projects of municipalities in Germany. *Central and Eastern European e| Dem and e| Gov Days 2024*, 61.

Kattel, R., Drechsler, W., & Karo, E. (2019). Innovation bureaucracies: How agile stability creates the entrepreneurial state.

Kattel, R., Drechsler, W., & Karo, E. (2022). *How to make an entrepreneurial state: Why innovation needs bureaucracy.* Yale University Press.

Klassen, G., Palombo, R., Bauer, L. T., & Niehaves, B. (2023). Overcoming Inefficiency in Public Procurement: An OpenData Approach.

Klassen, G., Bauer, L. T., Fritzsche, R., Kordyaka, B., Weber, S., & Niehaves, B. (2024). Turning Tenders into Tinder: How AI and Open Data can spark Bidding Matches.

Kogan, N., & Lee, K. J. (2014). Exploratory research on the success factors and challenges of Smart City projects. *Asia Pacific Journal of Information Systems, 24*(2), 141–189.

Krause, T. A., Schachtner, C., & Thapa, B. E. (2023). *Handbuch Digitalisierung der Verwaltung.* utb.

Kuziemski, M., Mergel, I., Ulrich, P., & Martinez, A. (2022). *GovTech practices in the EU: A glimpse into the European GovTech ecosystem, its governance, and best practices.*

Mazzucato, M., & Kattel, R. (2020). COVID-19 and public-sector capacity. *Oxford review of economic policy, 36*(Supplement_1), S256-S269.

Mergel, I., Ulrich, P., Kuziemski, M., & Martinez, A. (2022). *Scoping GovTech dynamics in the EU.*

Nam, T., & Pardo, T. A. (2011). Conceptualizing smart city with dimensions of technology, people, and institutions. Proceedings of the 12th annual international digital government research conference: digital government innovation in challenging times.

Neri, A. (2022). *The public sector must accelerate digital transformation—or risk losing sovereignty and trust.* World Economic Forum. https://www.weforum.org/agenda/2022/05/the-public-sector-must-accelerate-digital-transformation-or-risk-losing-sovereignty-and-trust/

Niehaves, B., & Klassen, G. (2024). *GovTech: influencing factors, common requirements and recommendations-Supporting the development of cross-border, interoperable GovTech practices in the European landscape* (9268218887). J. R. C. o. t. E. Commission.

Schrock, A. R. (2016a). Case Study: Code for America—Scaling Civic Engagement through Open Data and Software Design.

Schrock, A. R. (2016b). Civic hacking as data activism and advocacy: A history from publicity to open government data. *New media & society*, *18*(4), 581–599.

Selke, D., Kilian, M., & De Jonge, R. (2024). *GovTech in Europe—The GovTech Opportunity*. Sopra Steria & GovMind. Retrieved 23.04 from https://www.venture-insights.de/european-govtech-publication

Veljković, N., Bogdanović-Dinić, S., & Stoimenov, L. (2014). Benchmarking open government: An open data perspective. *Government information quarterly*, *31*(2), 278–290.

Luxembourg: Free Transit Fails to Dethrone Cars

Falk Fernbach

1 Introduction

Is free public transport the key to addressing Luxembourg City's unique mobility challenges?

This question lies at the heart of an innovative approach that Luxembourg has embraced to manage its complex and evolving transportation needs. Luxembourg, a small landlocked country in Western Europe, is bordered by Belgium, Germany, and France. Its capital, Luxembourg City, is a key EU hub hosting major European institutions. Despite its modest size of just 2586 km^2, Luxembourg plays a significant role in European politics and finance, distinguished by its multilingual population and high standard of living.

Luxembourg City experiences a significant daily influx of around 230,000 cross-border workers from neighbouring countries; France, Germany, and Belgium. This unique demographic dynamic creates specific mobility demands that are not commonly seen in other European cities. Luxembourg City covers a relatively small area, much of which is constrained by its topography with steep valleys and plateaus. This geographical setup makes expansive road networks impractical and emphasizes the need for efficient public transport and soft mobility solutions like cycling and walking paths, which are less space-consuming and can be seamlessly integrated into the existing urban fabric. As the capital city of one of the smallest yet wealthiest countries in Europe, and as a key EU hub hosting several European institutions, Luxembourg City's mobility needs are disproportionately high relative to its size. Luxembourg has a strong commitment to environmental sustainability, which aligns with the European Union's goals for carbon reduction and sustainable urban development. Despite its international significance, Luxembourg City maintains a focus on the quality of life for its residents. The existing mobility plan supports this by

F. Fernbach (✉)
LuxMobility, Luxembourg, Luxembourg
e-mail: f.fernbach@luxmobility.eu

C. Schachtner (ed.), *The European Smart City Movement*,
https://doi.org/10.1007/978-3-031-96700-9_19

reducing traffic congestion, promoting quieter, less polluted urban environments, and facilitating easier and safer access to green spaces and recreational areas.

Luxembourg is well-positioned to invest in innovative transport technologies and infrastructure, such as electric buses or advanced tram systems, due to its economic strength. Luxembourg made headlines worldwide by becoming the first country to offer free public transport across the entire nation starting March 1, 2020. This bold move aimed to address various challenges, including the reduction of traffic congestion, environmental sustainability, and social equity. The initiative was part of a broader strategy to tackle climate change by reducing car use, a significant contributor to greenhouse gas emissions. Luxembourg's government was motivated by the need to accommodate more commuters without relying on private vehicles, aiming to cut the number of people commuting by car by nearly half from its 2017 levels.

Luxembourg's free public transport initiative is not a standalone solution, but part of a larger, comprehensive strategy encapsulated in the National Mobility Plan 2035 (PNM 2035). This plan outlines a detailed strategy to handle 40% more trips than in 2017. Key aspects include transitioning from a reactive to a proactive approach in anticipating future mobility demand, determining the most appropriate transport means based on the number of people and destinations, and engaging all stakeholders in mobility planning. This plan is a testament to Luxembourg's commitment to sustainable and accessible transportation for all.

2 The PNM 2035-National Mobility Plan

The National Mobility Plan 2035 emphasizes the need for a cohesive vision to guide the development of transportation infrastructure and services. It advocates for strategic planning to ensure that mobility solutions are not only sustainable and efficient but also aligned with broader societal and environmental objectives. This approach is crucial for orchestrating future developments in a way that meets the increasing and evolving mobility demands of Luxembourg's population.

The PNM 2035 aims to enable a 40% increase in trips compared to 2017 by implementing Luxembourg's strategy for sustainable mobility, Modu 2.0. This strategy includes transitioning to a proactive approach in anticipating future demand, strengthening the most appropriate transport modes for given contexts, and involving key stakeholders, namely, the state, municipalities, employers, and citizens, in the mobility ecosystem. The plan identifies key challenges such as managing traffic flows in relation to Luxembourg City, prioritizing urban transport modes within metropolitan areas, and differentiating mobility approaches in rural areas. Opportunities include leveraging the success of high-quality mobility offers, the popularity of cycling, public transport capacity, carpooling on cross-border motorways, and effective parking regulation and management.

PNM 2035 sets specific modal share targets across various trip contexts (principal centers like Luxembourg City, secondary centers, and rural regions) to guide the country towards achieving its mobility objectives by 2035. This includes significant investments in public transport, cycling infrastructure, and road network adjustments to make the transportation system more multimodal and efficient. The plan also considers the role of technological advancements in shaping future mobility solutions, emphasizing the need for realistic opportunity assessments. It outlines the importance of continuously updating mobility plans to adapt to societal changes and infrastructure development timelines.

The PNM 2035 does not have statutory force, meaning specific projects will require legislative action. The plan calls for political consensus and timely decision-making to ensure the proposed infrastructure and policy measures are implemented effectively within the necessary timeframes. It advocates for a coherent multimodal concept that will be refined in the coming years, suggesting that successive governments should adjust the national mobility strategy every five years based on updated data and projections. This strategic document underscores Luxembourg's commitment to enhancing its mobility infrastructure and services, focusing on sustainability, efficiency, and preparedness for future growth. It highlights the complexity of planning for national mobility and the importance of adopting an integrated, forward-looking approach to transportation policy and infrastructure development.

2.1 The Strengths of the PNM 2035 and Free Public Transport

The PNM 2035 for Luxembourg embodies "smart" planning in several key aspects, focusing on sustainability, innovation, and efficiency to address the future mobility needs of the country. Below are some key elements of the strategic approach:

Anticipating Future Demand: The plan is proactive, aiming to anticipate and accommodate a 40% increase in mobility demand by 2035. This forward-looking approach ensures that infrastructure and services evolve in sync with population and economic growth, preventing future congestion and mobility challenges.

Multimodal Network Approach: The PNM 2035 promotes intermodality as a core principle, facilitating seamless transitions between different modes of transport. This integration ensures a more efficient and user-friendly mobility system, encouraging the use of public transport and reducing reliance on private vehicles.

Emphasis on Cycling and Public Transport: The plan identifies significant opportunities in cycling and public transport, planning to expand infrastructure and services in these areas. By doing so, it addresses not only the sustainability aspect but also health and accessibility, making cycling and public transport more attractive options for daily commutes.

Strategic Infrastructure Development: Infrastructure developments are planned strategically to enhance network efficiency, such as the expansion of tram networks, improvement of rail services, and the creation of multimodal hubs. These initiatives are designed to improve connectivity and reduce travel times across the country.

Leveraging Technological Innovations: The PNM 2035 is open to incorporating technological advancements that promise to enhance mobility. This includes digital innovations that could improve traffic management, public transport scheduling, and real-time information systems for users. The smart incorporation of technology aligns with the goal of creating a flexible and responsive mobility ecosystem.

Sustainable and Inclusive Approach: The plan's emphasis on sustainable transport modes and its goal to reduce car dependency reflect a commitment to environmental stewardship. By providing high-quality transport options that are accessible to all segments of the population, the plan supports social inclusion and equity.

Continuous Planning and Adaptation: Recognizing the dynamic nature of mobility needs, the plan advocates for continuous updates and adaptations. This iterative planning process ensures that Luxembourg's mobility strategy remains relevant and responsive to emerging trends, challenges, and opportunities. By integrating these smart elements, PNM 2035 aims to create a sustainable, efficient, and user-friendly mobility system that supports Luxembourg's growth and enhances the quality of life for its residents and visitors.

2.2 The Challenges of the PNM 2035 and Free Public Transport

The PNM 2035 for Luxembourg is ambitious and forward-thinking, but like any comprehensive strategic document, it may have potential weaknesses or areas of concern that need careful consideration:

Implementation and Funding Challenges: The plan involves a wide array of infrastructure projects, technological advancements, and policy changes, all of which require substantial investment. Securing consistent and adequate funding over the plan's timeframe might be challenging, especially in the face of economic fluctuations or changing political priorities.

Risk of Technological Overestimation: While the plan smartly incorporates the potential of technological innovations to improve mobility, there's a risk of overestimating the speed at which these technologies can be adopted or their effectiveness. This includes advancements in digital infrastructure, autonomous vehicles, and smart traffic management systems.

Intermodality and Coordination Complexities: The success of the multimodal approach heavily depends on seamless coordination between different modes of transport, agencies, and levels of government. Achieving this level of integration and ensuring that all parts of the mobility system are well-aligned can be complex and challenging.

Environmental and Urban Space Constraints: Expanding public transport and cycling infrastructure in a way that is environmentally sustainable and minimizes disruption to existing urban spaces requires careful planning. There's a potential weakness in balancing development with the preservation of natural and built environments, especially in denser urban areas.

Changing Mobility Patterns: The plan is based on projections of mobility demand and patterns that could change unpredictably due to factors like demographic shifts, changes in work habits (e.g., increase in remote working), or evolving consumer preferences. These changes could render some of the plan's assumptions and solutions less relevant over time.

Dependency on Cross-Border Cooperation: Given Luxembourg's unique geographic and economic context, the plan's success is partly dependent on cooperation with neighboring countries. Achieving alignment on cross-border transport initiatives can be challenging due to differing priorities, regulations, and funding mechanisms among these countries.

Public Acceptance and Behavioral Change: A critical aspect of the plan involves encouraging a shift from private car use to public and active transport modes. Overcoming entrenched behaviors and preferences for car use requires not only infrastructure and services but also effective public engagement and incentives, which can be difficult to implement successfully.

Risk of Underutilization: There is a risk that after the implementation of new infrastructure or services, they might be underutilized if they do not meet the actual needs or preferences of the population. Continuous monitoring and flexibility in adapting to real usage patterns will be crucial.

While the National Mobility Plan 2035 sets a visionary path for Luxembourg's future mobility, addressing these potential weaknesses will be key to its success. Ongoing evaluation, adaptability, and engagement with stakeholders will be essential in navigating these challenges.

3 Luxembourg City's Mobility Plan

Luxembourg City's Mobility Plan (VDL 2024) outlines a comprehensive mobility strategy for Luxembourg City, focusing on enhancing public transportation, cycling, pedestrian infrastructure, and managing car traffic and parking. It provides a comprehensive model for urban mobility that prioritizes sustainable transport, public engagement, and smart city principles. The plan tackles future transport challenges stemming from population growth, high commuter inflow, and geographical constraints. It focuses on expanding public transportation, enhancing soft mobility options like cycling and walking, reducing car dependency, and improving intermodal connectivity. The approach integrates transport planning with broader urban development, ensuring adaptability and alignment with environmental and social goals.

The Public Transportation Optimization aspect highlights its strengths, such as efficient and frequent services with a good spread of direct connections, clean, safe, and user-friendly systems. However, weaknesses include a lack of redundancy routes for the tram, insufficient direct cross-town connections, and at-capacity lines during peak times. Future plans aim to enhance network capacity and coverage, including a new high-capacity north–south public transport axis, minimizing transfers, and avoiding parallel routes between different services.

Regarding Cycling and Pedestrian Infrastructure, strengths include progressive improvements in the network, high-quality public spaces, and barrier-free access. Weaknesses involve incomplete cycling paths, insufficient public bike parking, and older pedestrian pathways not meeting current width standards. Future plans focus on expanding and optimizing the bike and pedestrian network to increase its safety and coverage, separating traffic streams, and improving connectivity with neighboring areas.

For the Road Network and Parking, strengths are noted in the well-structured city road network and a good supply of parking in lots and Park & Ride (P&R) facilities. Weaknesses include inner-city traffic congestion and limited potential for network expansion due to geographical and infrastructural constraints. Future plans include integrating traffic calming measures, revising parking management (tariff structures, parking durations), and enhancing peripheral connections to reduce inner-city congestion.

Cities should develop integrated strategies that encompass all modes of transport, public transit, personal vehicles, and non-motorized traffic, to create a seamless and efficient urban mobility landscape. Engaging the public in the planning process helps in understanding their needs and getting buy-in for new initiatives. Luxembourg City's approach includes public events and solicitations for feedback, crucial for successful policy implementation. Infrastructure must adapt to evolving urban demands. Taken together, the plan strives to weave every mode of transport into a coherent living system that keeps the city moving while safeguarding the environment and quality of life, though its success will ultimately depend on overcoming structural bottlenecks, behavioural inertia, and the city's limited spatial flexibility.

4 Analysis of Impacts and Challenges

Paradoxically, the same wealth that enables Luxembourg to finance fare-free transit also sustains one of Europe's highest rates of car ownership, revealing a behavioural inertia that infrastructure alone cannot overcome. Unless this cognitive dissonance is addressed, by making driving less convenient and alternative modes culturally aspirational, even the most generous mobility budgets risk merely gilding the gridlock. Congestion levels in May 2022 were observed to be like, or even higher than, those before the policy's implementation. This phenomenon highlights a crucial aspect: free public transport alone cannot entice people away from their cars if the public transport system faces issues like overcrowding or delays. Furthermore, the most enthusiastic adopters of free travel policies tend to be individuals who would otherwise walk or cycle, indicating that the policy may not be reaching its intended audience of car users.

According to projections (European Commission 2022), the number of cars on Luxembourg's roads is expected to rise by 6% by 2035 due to population growth, underscoring the challenges of relying solely on free public transport to reduce car dependency.

In Luxembourg, the transition to free public transport, which commenced on February 29, 2020, was funded through general taxation. This approach was adopted to democratize access to mobility, with the overarching goals of reducing the country's high car density, mitigating traffic congestion, and addressing environmental concerns due to vehicle emissions. Before the introduction of free public transport, ticket sales generated €41 million annually, a small fraction compared to the over €500 million required to run the country's public transport system (Symons 2023). The shortfall is essentially covered by taxpayers, aligning with a broader principle of social equity: those who pay more taxes contribute more to the funding of public transport, whereas individuals with lower or no income, who pay little to no taxes, benefit from the system without direct costs.

The implementation of free public transport in Luxembourg was part of a larger strategic vision to transform the country's mobility landscape. The initiative was not just about eliminating fares but was accompanied by significant investments in improving and expanding transport infrastructure. Luxembourg has been investing approximately €500 per citizen per year on modernizing its railway network, emphasizing the need for a quality, reliable, and integrated transport system that can effectively serve as a viable alternative to private car use (Brown 2023). The country's Deputy Prime Minister, François Bausch, highlighted the multifaceted approach required to address mobility challenges, underscoring the necessity of a multimodal transport system where public transport plays a central role alongside cars in a balanced mobility ecosystem. Luxembourg's free public transport is recognized not just for its immediate practical benefits, such as ease of travel and environmental positivity, but also as a "good form of freedom" and even a "fundamental right" for residents. This sentiment reflects the transformative impact of the initiative on societal norms regarding mobility and access to public services.

Despite these achievements, challenges remain, including the persistence of high car ownership rates and the cultural dominance of car use (Luxembourg Times 2023). In fact, Luxembourg faced significant traffic congestion challenges due to high car ownership, robust population growth, and numerous cross-border commuters. A study's findings (Bigi et al. 2023) underscore the complexity of altering transportation behaviors and the challenge of reducing reliance on cars in densely populated areas. While free public transport policies can boost public transport usage, especially among specific population segments, their effectiveness in significantly reducing traffic congestion may be limited without additional measures to discourage car use or enhance public transport efficiency and connectivity.

5 The Interplay Between Mobility Plans, Strategic Frameworks, and Human Behaviour

When mobility strategies treat people as creative partners rather than predictable variables, ideas leap off the drawing board and take root in daily life. Recognising that the most important transport corridor runs through our minds unlocks genuine, measurable behavioural change. The relationship between mobility plans, strategic frameworks, and human behaviour is both intricate and dynamic. Urban mobility strategies are not just about the physical infrastructure or transportation systems; they are deeply connected to how individuals and communities behave, make decisions, and ultimately, how they live their lives. Understanding this interrelation is crucial for the success of any mobility plan, particularly in a city as unique as Luxembourg.

Mobility Plans and Behavioural Influence: Mobility plans are designed with the intention of guiding and influencing human behaviour. By creating efficient public transport systems, expanding cycling and walking paths, and managing traffic flows, these plans aim to encourage people to make choices that are more sustainable and beneficial for the community. For instance, the introduction of free public transport in Luxembourg was not merely a financial decision but a strategic move to influence travel behaviour by making public transport more attractive than private car use. These strategies often rely on the assumption that if the right infrastructure is provided, people will naturally adjust their behaviors to align with the new opportunities. This can include shifting from car travel to public transport, increasing the use of bicycles, or opting for walking over driving for short distances. The idea is that convenience, cost savings, and environmental benefits will motivate people to adopt new habits.

Conversely, human behaviour plays a significant role in shaping the effectiveness and direction of mobility strategies. People's travel habits, preferences, and resistance to change can all impact the success of a mobility plan. For example, even with free public transport, many residents in Luxembourg continue to rely on cars due to ingrained habits, cultural norms, or the perceived convenience of personal vehicles. This persistence of car use despite new alternatives highlights the complexity of altering established behaviors. Mobility strategies must therefore be adaptive, recognizing that infrastructure alone may not be sufficient to change behaviors. Engaging communities through education, incentives, and even disincentives for less sustainable practices becomes essential. This also involves understanding the psychological and cultural factors that drive transportation choices, such as the status associated with car ownership, the perceived reliability of public transport, or the individual's comfort with using new modes of transport.

While mobility plans and strategies attempt to influence behaviour, it is essential to recognize that these behaviors are ultimately rooted in cultural norms and values. Culture dictates what is considered acceptable, desirable, and practical in terms of how people move around their cities. For instance, in car-centric cultures, driving is often seen as a symbol of freedom and independence, making it difficult to shift towards public transport or other forms of mobility, regardless of the infrastructure provided. In Luxembourg, the cultural attachment to cars persists despite the availability of free public transport. This attachment is not merely about transportation; it reflects deeper cultural values related to personal space, time management, and social status. Therefore, altering transportation behavior in such a context requires more than just infrastructure changes, it demands a cultural shift.

Strategies that successfully integrate cultural understanding into mobility planning are more likely to achieve their goals. This includes not only providing alternatives but also reshaping the narrative around these alternatives. For example, promoting cycling as a healthy and environmentally responsible choice, or framing public transport as a communal and socially equitable option, can help shift cultural perceptions and, in turn, behaviors. Ultimately, while mobility plans and strategies can provide the framework and incentives for behavioural change, they have limitations. Human behaviour is complex and influenced by a myriad of factors beyond just mobility infrastructure, such as social norms, personal values, and cultural context. This means that even the most well-designed mobility plans may not fully achieve their intended behavioural changes if they do not address the underlying cultural factors.

For Luxembourg, this underscores the importance of a holistic approach to mobility planning, one that considers not only the physical and logistical aspects of transportation but also the cultural dimensions that drive human behaviour. It suggests that for sustainable mobility strategies to be truly effective, they must be accompanied by efforts to cultivate a cultural shift towards more sustainable, community-oriented values.

6 Conclusion

At the heart of Luxembourg's mobility strategy lies not yet a transformation, but a challenge: to move beyond infrastructure planning and confront the cultural and behavioural patterns that still shape everyday mobility choices. While the country has made bold headlines with the introduction of fare-free public transport, it still faces one of the highest car ownership rates in Europe, a reality that underlines how deeply the private car is embedded in daily routines, convenience, and identity. Luxembourg's unique geographic and socio-economic context, a wealthy state with a small territory, a highly mobile population, and over 230,000 daily cross-border commuter, creates a complex backdrop. The physical constraints of the capital, combined with political fragmentation and entrenched habits, make it difficult to realise the full potential of policies that look progressive on paper. The move to make public transport free was courageous and symbolically powerful. It positions mobility as a social right and aligns with broader sustainability goals. But it is not transformative on its own. Studies and congestion data suggest that fare-free transit alone does not substantially reduce car traffic, especially when the alternatives are not always perceived as faster, more convenient, or more prestigious. The deeper issue is behavioural inertia. If car travel remains the most flexible and socially reinforced option, many people will not switch, regardless of price or sustainability arguments. Infrastructure improvements are necessary, but not sufficient. A true shift will require changing the way people relate to mobility: how they perceive public transport, how they experience walking or cycling, and how urban space is structured to support these choices. That's where Luxembourg still has work to do. The strategy must become more than a set of infrastructure projects or service upgrades. It needs to actively shape norms, expectations, and daily habits, through urban design, smart disincentives, better intermodal connections, and a strong narrative that makes sustainable choices not just possible, but aspirational. Mobility planning must evolve into cultural policy. Citizens should be seen not just as passengers, but as participants in a broader societal shift. This includes transparent engagement, education, and the political courage to make driving less convenient in favour of collective benefit. In the end, the question is not whether Luxembourg can afford a mobility transition, as it clearly can. The real question is whether it will go beyond symbols and infrastructure to deliver the deeper change required. A smart city is not only measured by the efficiency of its systems, but by its ability to reflect the values it wants to live by, fairness, sustainability, and shared responsibility. The path forward will depend on whether the country is willing to treat mobility not just as movement, but as a mirror of what kind of society it aims to become.

References

Bigi, F., Schwemmle N. and Viti, F. (2023). Evaluating the impact of free public transport using agent-based modeling: the case-study of Luxembourg. https://transp-or.epfl.ch/heart/2023/abstracts/hEART_2023_paper_4604.pdf

Brown, H. (2023). Europe's richest country made public transport free. Could other countries do the same? Euronews. https://www.euronews.com/green/2023/06/07/europes-richest-country-made-public-transport-free-could-other-countries-do-the-same

European Commission. (2022). Luxembourg's experience with free public transport. Urban Mobility Observatory. https://urban-mobility-observatory.transport.ec.europa.eu/news-events/news/luxembourgs-experience-free-public-transport-2022-07-19_en

Luxembourg Government. (2022). PNM 2035. https://gouvernement.lu/en/dossiers/2022/pnm2035.html

Luxembourg Times. (2023). New car registrations rise sharply in Luxembourg. https://www.luxtimes.lu/businessandfinance/new-car-registrations-rise-sharply-in-luxembourg/7164284.html

Symons, A. (2023). The world's richest country made public transport free. Here's what happened next. Euronews. https://www.euronews.com/green/2023/03/22/the-worlds-richest-country-made-public-transport-free-heres-what-happened-next

VDL. (2024). Plan de mobilité Ville de Luxembourg [Mobility plan for the City of Luxembourg]. https://www.vdl.lu/fr/media/360856

North Macedonia—Smart Solutions of Transport and Logistics

Christian Schachtner

1 Introduction

This article was written on the basis of a conversation with project managers of the initiative Nalas-Network of associations of local authorities of South-East Europe. The Republic of North Macedonia is facing complex challenges in the development of its transport infrastructure, which include both national and international dimensions.

As a strategically important hub in the Balkan transport network, the country positions itself at the intersection of two pan-European transport corridors: Corridor 8 (East–West) and Corridor 10 (North–South), which enable central transport links through the country (Ministry of Transport and Communications, 2023).

North Macedonia's current transport infrastructure has significant development potential. According to Eurostat data, road network density in 2019 is only 13 m per square kilometre for motorways—4 m less than the EU average. The road network includes local roads with a total length of approx. 19,000 km, municipal and regional road sections with a total length of 906 km, and motorways and expressways with a total length of 228 km. The government is pursuing an ambitious strategy to improve transport infrastructure, with the aim of optimising connectivity to European transport networks (World Bank 2021).

The National Transport Strategy 2018–2030 defines clear objectives for the further development of the transport sector. Key aspects of this strategy include full integration into the European TEN-T network, harmonisation with EU directives and regulations, the development of a sustainable transport sector and the implementation of Intelligent Transport Systems (ITS).

The two most important transport corridors are as follows: Corridor VIII (East–West) stretches from Varna (Bulgaria) to Durrës (Albania) with a length of 253 km

C. Schachtner (✉)
RheinMain University of Applied Sciences, Wiesbaden, Germany
e-mail: Christian.schachtner@hs-rm.de

C. Schachtner (ed.), *The European Smart City Movement*,
https://doi.org/10.1007/978-3-031-96700-9_20

in North Macedonia and a daily traffic frequency of about 1,500 to 2,500 vehicles. Corridor X (north–south) connects Athens via Skopje to Belgrade, Zagreb and Munich and has a length of 170.4 km in North Macedonia.

The implementation of the transport strategy requires significant investments and structural reforms. Important fields of action include the improvement of the road infrastructure, the modernisation of railway connections, the expansion of international air connections and the strengthening of transport links with neighbouring countries. Recent policy developments under the new VMRO-DPMNE-led government show an increased focus on transport projects, with the aim of increasing national mobility and connectivity capabilities (China-CEE 2024).

2 Smart Mobility Planning

The development of intelligent transport systems in North Macedonia requires a strategic and multidimensional approach. The National Transport Strategy 2018–2030 defines clear goals for the integration of innovative transport technologies and harmonisation with European standards (Kopevski et al. 2023).

The implementation is based on a holistic approach that combines technological innovation, infrastructural improvements and regional networking. North Macedonia positions itself as a key hub in the pan-European transport corridors 8 and 10, which provides an optimal starting point for technological innovation (Dimitrieski et al. 2017). The IoT network infrastructure forms the backbone of the development, using various communication protocols such as Wi-Fi, LoRaWAN, and 5G (Petrovski et al. 2019).

The implementation steps include several key aspects. Initially, the focus is on traffic management, whereby the use of IoT technologies enables precise control and optimization of traffic flows. Smart sensor systems can provide real-time information about traffic volume, parking availability, and route efficiency (Jovanovik et al. 2018).

Another core area is infrastructural networking. The rehabilitation of road corridors, especially those between Skopje and neighboring capitals, is crucial in this regard (World Bank Report 2020). In parallel, the development of a national e-mobility strategy requires special attention. Recent studies show that the existing power grid can support up to 40% electric vehicles by 2040 (EBRD Infrastructure Assessment 2022).

The technological implementation is based on close cooperation between public institutions and private sector actors. Key implementation steps include the installation of IoT-enabled traffic management systems, the development of a central data platform for traffic information, the introduction of real-time traffic monitoring, and the integration of smart mobility solutions such as car sharing and intelligent parking guidance systems.

Various challenges must be overcome: limited financial resources, the need for technological training and the complex integration of existing infrastructures. Strategic partnerships between government, technology companies, and international organizations such as the EBRD are crucial to overcome these obstacles (Nikolov and Trajkovic 2021).

3 Integrated Mobility Management

Adaptive traffic control systems offer enormous potential for optimizing traffic flow in urban areas. These technologies enable dynamic adjustment of traffic signals in real time, based on current traffic conditions. North Macedonia is aiming for full harmonization with EU transport directives and is making targeted investments in smart mobility technologies to improve the efficiency and sustainability of the transport system (Kopevski et al. 2023).

The main benefits include improved traffic efficiency by reducing congestion and waiting times, reducing emissions through reduced fuel consumption, as well as increased road safety through adaptive signal control (FasterCapital 2023).

The project aims to develop an integrated mobility system that seamlessly connects different modes of transport. The introduction of a national multimodal ticket collection system is a key aspect of the strategy. This system is intended to provide users with a uniform identification and payment option across different modes of transport (Kopevski et al. 2023).

A major focus is on improving the transport infrastructure. Concrete measures include the optimisation of safety at level crossings, the installation of 42 modernised crossing points and the implementation of real-time traffic information systems. The introduction of AI-driven traffic light systems demonstrates the potential of technological innovations (Hermann 2023).

Comparable systems in other European cities have already shown impressive results: a reduction in travel time by up to 20%, a reduction in stops by 15% and improved integration of different road users such as public transport, cyclists and pedestrians (Hermann 2023).

4 Best Practice of Urban Mobility Challenges: Smart Transport in Skopje

The city of Skopje was faced with the challenge of modernizing its public transport system while increasing efficiency and reducing environmental impact. As one of the most polluted cities in Europe, Skopje developed a holistic approach to transforming the urban mobility system (Makedonski Telekom 2016).

The Smart City project involved the implementation of a smart bus system with several core components such as automatic vehicle tracking system, electronic ticketing, real-time passenger information and route optimization. The partnership between Makedonski Telekom, the city transport company and the municipality enabled the development of an innovative transport solution (European Bank for Reconstruction and Development, 2021). The system provides real-time tracking of buses and communicates timetables via info displays, SMS and internet platforms, significantly reducing passenger waiting times.

The implementation yielded remarkable results: over 2.5 million transactions in just three months, significant fuel savings, reduction of traffic congestion and reduction of air pollution. The project included the introduction of a Bus Rapid Transit (BRT) system with priority bus lanes, energy-efficient and hybrid buses, cashless ticketing and dynamic service customization. The innovative solution has been internationally recognized and won the WITSA Global ICT Excellence Award in the Sustainable Growth Solution category in 2016 (European Bank for Reconstruction and Development, 2021).

Makedonski Telekom and the City of Skopje plan to scale the project to other urban areas, including bike sharing systems, smart parking solutions and electronic ticketing systems for urban facilities. This forward-thinking strategy underlines the city's commitment to sustainable and smart mobility solutions.

5 Conclusion

The transformation of the transport system in North Macedonia into a smart, sustainable mobility network requires a holistic, iterative approach. The City of Skopje has developed a comprehensive approach to modernizing its public transport system to increase efficiency and reduce environmental impact. The Smart City project included the introduction of an intelligent bus system with automatic vehicle tracking, electronic ticketing and real-time information for passengers. The implementation resulted in remarkable results, including over 2.5 million transactions in just three months, significant fuel savings, and a reduction in traffic congestion and air pollution. The project has been internationally recognized and won the WITSA Global ICT Excellence Award in the Sustainable Growth Solution category in 2016. With plans to expand the project to other urban areas, including bike sharing systems and smart parking solutions, Skopje underlines its commitment to sustainable and smart mobility solutions. The implementation of smart transport is not only a technological challenge, but a comprehensive social and infrastructural transformation.

References

China-CEE. (2024). Policy Briefing: From Stagnant to Strategic Mobility of Macedonia.

Dimitrieski, V., et al. (2017). Smart Cities Transport: Challenges of the Macedonian Capital City. ResearchGate.

EBRD Infrastructure Assessment. (2022). Mobility Transformation in Southeast Europe.

European Bank for Reconstruction and Development. (2021). Skopje to introduce first Bus Rapid Transit system in Western Balkans. Department of Urban Infrastructure.

FasterCapital. (2023). Adaptive Traffic Control: RTTC System Analysis. Technical Research Report on Traffic Optimization.
Hermann, W. (2023). AI-controlled traffic lights in Ellwangen: Innovative traffic control. Ministry of Transport Baden-Württemberg.
Jovanovik, M., et al. (2018). IoT-basierte Verkehrsmanagementsysteme. International Journal of Smart Urban Planning.
Kopevski, M., Hristovski, J., & Lazarevski, J. (2023). National Transport Strategy Presentation. Ministry of Transport and Communications, North Macedonia.
Makedonski Telekom. (2016). Smart City Award for Intelligent Bus System in Skopje. Technologieabteilung.
Ministry of Transport and Communications of North Macedonia. (2017). Annual Report on Urban Mobility Development. Skopje.
Ministry of Transport and Communications. (2023). Presentation of the National ITS Strategy for the Republic of North Macedonia.
Nikolov, R., & Trajkovic, L. (2021). Strategic Technology Transformation in Südosteuropa. Urban Technologies Review.
Petrovski, A., et al. (2019). Kommunikationsprotokolle in intelligenten Verkehrssystemen. Telecommunications Journal.
Popovski, K. & Nikolovski, M. (2018). Intelligent Transport Systems in South-East Europe: A Comparative Analysis. Verkehrstechnologisches Journal, 45(2), 112-129.
WITSA Global Awards Committee. (2016). Innovative ICT Solutions in Urban Mobility. Award documentation.
World Bank Report. (2020). Infrastructure Development in North Macedonia.
World Bank. (2021). Press release: Key transport infrastructure in North Macedonia receives support.

Turkey: A Needs Based Top-Down Approach for Tailor Made Smart City Roadmap

Mustafa Eruyar, Fatih Gundogan, and Fatih Kafali

1 Introduction

Globalization has been upending the status quo for centuries and it has accelerated in recent decades because of trade liberalization and emerging market growth. These trends disrupt existing business models, hence economies, by arise of new competitors, restructuring supply chains management systems. Complexity will be increased by the next waves and it will require flexible business models to respond to global shifts (EY 2016).

"Smart" takes an asset, infrastructure, or even transaction, ensures it is up to date and analysed for being autonomous and effective. Smart is the level of knowledge and decision-maker in the era of internet of things. These smartness revolution could fundamentally change how the world works since as the aging world becomes more resource-constrained, the need for investing in smart solutions increases (EY 2016).

New generation of citizens and workers are in search for experiences that are connected to a greater purpose in their lives (Leberecht and Vanderbeck 2014). Thus, cities should be able to modify themselves and develop new features to be able to support its residents on the road to more a purposeful and meaningful life. Regarding this trend, it is important for the sake of a successful Istanbul Smart City Transformation process to have a humanistic reason for being to be able to inspire all of the process owners, stakeholders and citizens (Collins and Porras 1996).

For a purposeful smart city transformation having a vision plays an important role to define the desired state. According to BSI Group, a smart city requires a joint vision, a set of goals and an overall strategy which all stakeholders agree on (BSI 2015). Having a vision is also critical for strategy and roadmap implementation,

M. Eruyar (✉) · F. Kafali
Hinda Smart City Technologies, Istanbul, Türkiye
e-mail: mustafa.eruyar@hinda.city

F. Gundogan
Streadlab, Istanbul, Türkiye

C. Schachtner (ed.), *The European Smart City Movement*,
https://doi.org/10.1007/978-3-031-96700-9_21

respectively. Since smart city strategies and roadmap will shape the "future Istanbul", it is important that they are aligned with each other and serve a shared goal.

2 Literature Review

Success in a smart city transformation requires a broad and flexible common vision that will encourage different stakeholder groups, engage all city elements, and bridge the differences in communication and outcomes. In line with this very reason, smart cities around the world have been developing "smart city visions" and strategic plans to have a set of guiding principles.

The main objective of strategy is to bring the mission of an organization to the same direction with its vision. Each city has its own way of creating strategy related to its identity and values.

Below are the vision statements and main strategies of leading smart cities (Table 1).

3 Methodologies

Istanbul smart city is evaluated under eleven different conceptual areas namely, mobility, environment, energy, governance, economy, living, people and security as domains and information communication technologies, human resources, finance as enablers.

In this study, determination of the manifest was the first stage. After the determination of the manifest of Istanbul smart city, the strategy for achieving the vision was formulated and finally a roadmap that takes into account the time aspect of the works that are planned was generated.

3.1 Concept of Istanbul Smart City Manifest

5 concepts, namely purpose, values, mission, vision, and strategy are deeply interconnected with each other, sometimes defined together in groups (purpose + values under the name of purpose, mission + vision under the name of mission, etc.) in the literature, leading to a number of results:

- They are easily and constantly being confused with each other. Therefore it is critical to agree on the same scope that each concept covers. Mutual agreement between stakeholders of this study is more important than coming up with the right definitions, as long as the definitions are mutually exhaustive and collectively exhaustive (MECE).

Table 1 Vision and main strategy statement examples of smart cities

City	Vision	Main Strategy
New York	"We envision a city with a dynamic, thriving economy that is also a responsible steward of the environment and is resilient against shocks both natural and manmade" (NYC Mayor's Office of Tech Innovation 2015)	While improving communication for everyone, planning and scaling smart technologies, growing the innovation economy, and making sure that city development is human-centric (NYC Mayor's Office of Tech Innovation 2015)
Berlin	"Berlin's smart vision of the future—a super smart city with green roofs, an ecentralized energy supply, multimedia health management and electric transport" (Berlin Partner für Wirtschaft und Technologie GmbH 2017)	Supporting the communication of stakeholders by using the ICT infrastructure in economy, increasing the synergy of the city and creating plans to compete with other leading cities on environmental issues in international arena (Berlin Partner für Wirtschaft und Technologie GmbH 2017)
Dubai	"Our vision is to make Dubai the happiest city on Earth" (Smart Dubai Office 2017)	By using ICT, digitizing state services available to all stakeholders in the areas of transportation, communications, infrastructure, energy, economy and urban planning, and eliminating physical state service centres (UAE Government 2017)
London	"A modern management approach that is based on innovation, productivity, transparency and common action approach, which is held in the foreground and is processed with new technologies to solve problems in London" (ARUP 2016)	Supporting technological developments through collaborations and investing in education, training and the city's digital infrastructure. Encouraging open data and further sharing of ideas and views (ARUP 2016)
Copenhagen	"Being a carbon-free, blue and green capital; clean, healthy and the most ideal city for cyclists and make the city a more livable place for everyone" (Copenhagen Solutions Lab 2016)	To bring innovative solutions to the city's water, heating, transportation, waste management and alternative energy resources, and to pursue their zero emission objective (Copenhagen Solutions Lab 2016)
Singapore	"Through harnessing the power of networks, data and info-communication technologies, we seek to improve living, create economic opportunity and build a closer community" (Smart Nation and Digital Government Office 2017)	Leading the economy and connecting people at a high speed with its sophisticated and innovative structure while supporting Singapore's Smart Nation goal by building a world leader ICT infrastructure and educating citizens about this technology (Smart Nation and Digital Government Office 2017)

Fig. 1 Template of Istanbul smart city manifest

- They can be used together under the same name. For example, there are a number of mission statements that actually consists of mission, purpose, and values altogether (David et al. 2014). Some researchers even suggest to include vision in mission statements (Van Zyl 2011).
- It is not easy to work on them isolated from each other, as vision statement discussions easily evolve to purpose statement, purpose statement discussions to mission statement or values, and so on.

As a result; the concepts in the literature are selected to provide;

- Compliance with MECE rules,
- A modular structure that can easily be handled throughout the project.

Finally, to make sense in the minds of the people; "Manifest of Smart Istanbul" merging all of these five statements together is developed, as the final output of the study in a format as Fig. 1.

3.2 *Smart City Manifest Survey*

After the formation of the above template, a stakeholder survey was designed in order to provide input from the stakeholders of Istanbul Smart City. A wide literature review was conducted in order to determine the set of questions that will be asked to the stakeholders in respect to purpose, vision, mission, strategy, and values. In the end, a questionnaire which included five main questions and five back-up questions for each field was prepared. The questions and their rationale will be detailed under the following headings.

3.2.1 Vision

The literature review suggests that people's vision can be determined by asking them about the changes they would like to see in their companies, schools, organizations etc. or the results of the impact that they want to see their organizations make on the external world (Nagy and Fawcett 2017). The city, or specifically Istanbul, was determined to be the equivalent term for the environment that will be impacted by the smart city organization in this scenario and it was decided to ask the participants about the city they dream of. Moreover, because the vision is defined as almost a visual concept (Collins and Porras 1996), the word "illustrate" was intentionally injected into the question instead of "dream of", which is a more common and straightforward use in Turkish.

Ultimately, the question on vision was set as "If we asked you to illustrate the Istanbul of your dreams freely without any limitations, what would you say?".

3.2.2 Strategy

The participants were asked "Please indicate, in order of preference, which three areas out of eight should be focused in order to make your dream Istanbul come true." Asking the question, the participants were to be informed about eleven different conceptual areas.

3.2.3 Purpose

Upon examination, it was deemed appropriate to use the question "If you woke up tomorrow morning with enough money in the bank to retire, would you nevertheless keep working here? Why?" as suggested by Collins and Porras in the process of assessing the participants' purpose (Collins and Porras 1996). This question would allow the participants to state their true motivations by removing monetary concerns out of the equation.

3.2.4 Values

Values must be non-negotiable and timeless. The concept of children were therefore incorporated into the question, as it is personal and also involves a long-term perspective. Ultimately, the question on values was set as "What are two values or qualities that you would never sacrifice and would pass on to your children?".

3.2.5 Mission

The interpretation of mission as the "the business you are in" was determined as the suitable approach in gathering participant ideas on mission. Consequently, the fifth question was set as "Assume that a company named Istanbul Smart City was established. What business would this company be in?".

Telephone interviews were conducted with 203 C-level stakeholders based on a prewritten script and responses were recorded as text in an interview template and conclusions were drawn to be incorporated as inputs for the following sections.

3.3 Development of Istanbul Smart City Manifest

This section details the workshops that were organized in order to develop and finalize the manifest of Istanbul Smart City. Workshops were heavily inspired by the concept of hackathons and the Purpose City workshop where numerous stakeholders come together to produce an output (Leberecht and Vanderbeck 2014). Lateral and vertical thinking methodology for the proceedings of the workshop was developed in order to maximize the creativity of the participants and the variety of the outputs. The Istanbul Smart City technical manifest was obtained at the end of the workshops. The creation of the final version of the Smart City Manifest has benefited from the perspectives of two independent authors in order to increase emotion and feeling of manifest.

3.4 Citizen Needs Analysis

The purpose of this study is to take the views of people who live in Istanbul into account while establishing the smart city's strategic roadmap. The results of this study will provide crucial inputs to the following phases of the project, mainly the strategy determination. Therefore, using in-depth interview techniques, interviews were conducted with people who would represent the city and the results of those interviews were organized to be analyzed. For this purpose; profiles were created, questions and key words were prepared, face-to-face interviews were held, collected data was categorized and prepared for analysis, and data analysis activities were carried out.

In the beginning of the study, the important characteristics of the profiles that can affect or shape the study were determined. These features are listed as follows: Gender, occupation, location, marital status, means of transportation, health status, indigenous or visitor status, socioeconomic status, household size, accommodation and lifestyle.

A detailed list of all the possible profiles were created by using every possible combination of the specified properties, resulting in 2.7 million profiles. Then, the profiles that were impossible in nature were eliminated. For example, a primary school student can't be married at the same time, so there should be no match between these two properties. Amongst the remaining profiles, using the grouping, screening and selection methods, 150 profiles with the highest representation ability were determined and 3 people were identified for each of these profiles, creating a list consisting of 450 people. After detecting overlaps between the profiles and realizing that the view of some profiles could be obtained from the other profiles, 25 profiles were removed and finally 375 people's needs are analyzed.

Each interview was conducted face-to-face and lasted approximately 20 to 40 min. The gathered information was noted on paper by hand. The raw data that was collected in free text format during the interviews was examined and categorized based on root causes. The answers that were given by the participants were matched to these categories as needs and problems and the data was made ready for analysis. In terms of software, Microsoft Excel and IBM SPSS Modeler were used during these categorization and analysis phases.

The social media analysis is carried out to support other surveys and face-to-face studies conducted within the scope of the project. It has been determined that the most appropriate social media platform for this study is Twitter because it includes relatively less visual items and more written messages which will provide an easier access to information (Akshay Java 2007). Accordingly, to be able to retract viable information from the database, keywords have been determined and given to a third party firm to compile tweets related to our research from Twitter. 700 keywords have been determined to extract the data related to domains from the database. Keywords were scanned in the database using conditional searches, logical queries and parsing techniques. The tweets were categorized into subject areas by means of special codes and made ready for analysis.

3.5 Development of Istanbul Smart City Domain and Enabler Strategies, Strategic Objectives, Programs and Projects

For each enabler and domain, an individual strategy was set. In order for the strategies to be effective, it is necessary to make a good analysis of the current situation, identify the strengths and weaknesses, and consider important opportunities and threats (Allison and Kaye 2015). In this direction, the domain main strategies were based on findings that were revealed in the current state analysis. For this purpose, a SWOT analysis was conducted and needs in the relevant areas were determined and corresponding strategies were developed. Additionally, these individual strategies benefited from the literature review, survey and needs analysis studies.

In light of the individual strategies, strategic objectives were defined for each domain and enabler. Strategic objectives are used both to guide and track the programs and projects that will meet the needs of the relevant domain (Nagy and Fawcett 2017). Strategic objectives were established based on Istanbul's current state and the opinions of the related subject matter experts. Strategic objectives, like the strategies, have been formed in line with the main strategy of Istanbul Smart City. These strategic objectives will be detailed in the following sub-sections.

Later, strategic targets were detailed for each strategic objective to lead the related programs and projects. The targets will answer the question of "Where do we want to be when the programs and projects are completed?". A well-established target will be determined in line with the mission statement and objectives, and will express the desired results clearly. This information was taken into account when strategies, strategic objectives and targets were determined for the Istanbul Smart City project.

A gap analysis was carried out on the basis of the determined strategic objectives. Gap analysis aimed to reveal the differences between the current state and the target state. As a result of this gap analysis, programs were identified for strategic objectives. Programs include thematically relevant projects to be carried out in direction with the strategic objectives. The gap analysis and identified programs will be detailed in the following sections.

The final component of the strategy determination phase is the projects. Under the programs described, projects that are compatible with the strategies and can make Istanbul a smart city were prepared. The projects have been developed by the relevant subject matter experts. The activities, risks, benefits, and responsible parties for each project were defined in the relevant project cards.

Stakeholder involvement is thought to play a critical role when the smart city strategy is taken into consideration, and stakeholders were included in the key points while this work was being conducted. Stakeholder evaluation meetings and stakeholder visits were carried out in this direction.

3.6 Forming the Istanbul Smart City Roadmap

To be able to prioritize the Smart City projects, "prioritization matrix" tool was utilized which is "a simple tool that provides a way to sort a diverse set of items into an order of importance (Gosenheimer 2012)". The developed prioritization methodology for Istanbul Smart City Project involves the following set of criteria under two main areas: impact (addressing the current problems, responding the needs and problems of the citizens, supporting published strategies, relation with the Istanbul Smart City Index indicators, public relations effect) and feasibility (duration of project, financial resource requirement, number of stakeholders, recent maturity of project). Each criteria has its own weight for calculating the average score.

After evaluating each project in terms of above criteria; all of the average scores was normalized on a scale from 1 to 5 for standardization. Projects were placed on the matrix according to their average impact and average feasibility scores and prioritized based on their region on the matrix. The roadmap, which is the road to the smart city vision; is divided into three main time periods: short term (one year), medium term (five years) and long term (ten years). Finally, dependency relationships

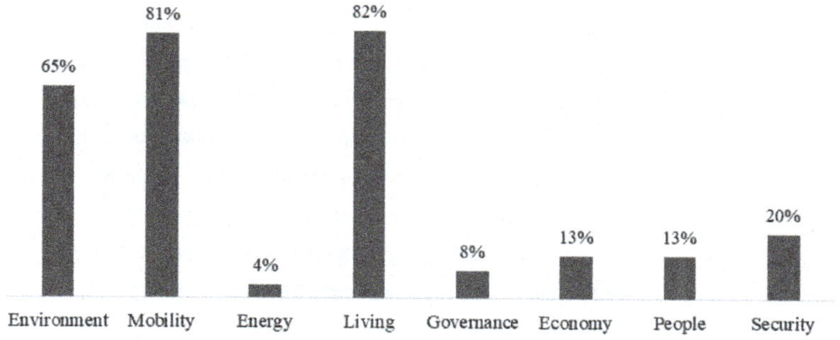

Fig. 2 Domain mention frequency in vision and strategy surveys

(projects acting as pre-requisites to one another) between projects in different areas are identified based on time intervals.

4 Results

4.1 Smart City Manifest Survey

It is highly likely that the participants of the survey focused on the latest issues of Istanbul when articulating their opinions regarding purpose, vision, and strategy. Therefore, it is suggested to incorporate the results assuming that they are "the statements/opinions derived from the current biggest issues of Istanbul" while setting the vision, strategy, and purpose. Vision and strategy survey results indicate that stakeholders envision an Istanbul with improvements in domains of Environment, Mobility, and Living and in pursuing such vision they believe in the adoption of an approach which places enabler of Human Resources at the center. In purpose survey "Benefiting other people" and in values survey "Being honest" should be referred in the purpose and values statements to enable all the participants embrace it in the widest possible way.

Mission surveys show that eliminating human deficiencies and increasing quality of the city services must be focus of mission statement (Fig. 2).

4.2 Citizen Needs Analysis

Citizen needs interviews and social media analysis have high importance to the strategic roadmap as it mainly uncovered problems and needs relating to Istanbul and as it provided input to the design and development stages of various products and services.

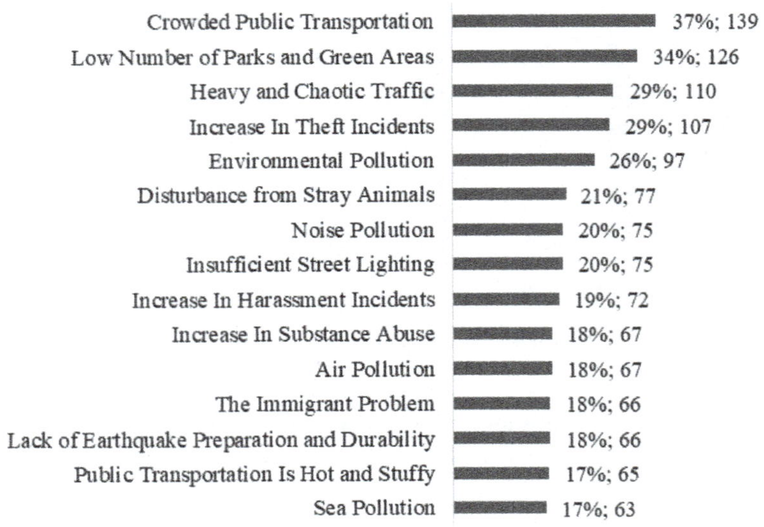

Fig. 3 Top 15 citizen problems

It can be observed from the analysis results that repetition quantities for problems and needs differed amongst domains and also that the number of unique needs and problems were also different amongst domains. As each domain occupies a different portion of people's lives, it is natural that different domains would produce differing results.

Result of the analysis of the data that was gathered through the conducted studies are shown in Figs. 3 and 4. It becomes clear that smart systems should be implemented based on the needs and the problems of the citizens.

4.3 Development of Istanbul Smart City Domain and Enabler Strategies, Strategic Objectives, Programs and Projects

The statement in the Istanbul Smart City Manifest "We will achieve this goal by producing with stakeholders and citizens, using technologically innovative methods and focusing on productivity" has been determined as the main strategy. Domain/ enabler strategies, strategic objectives, programs, and projects have been set based on that statement. In this section, strategies based on the eight smart city domains and the three smart city enablers are explained.

Mobility

Despite the large investments, transportation systems cannot be used efficiently because of the unproductive and uncoordinated planning of the transportation systems. Therefore, mobility main strategy is that dynamic and intelligent systems

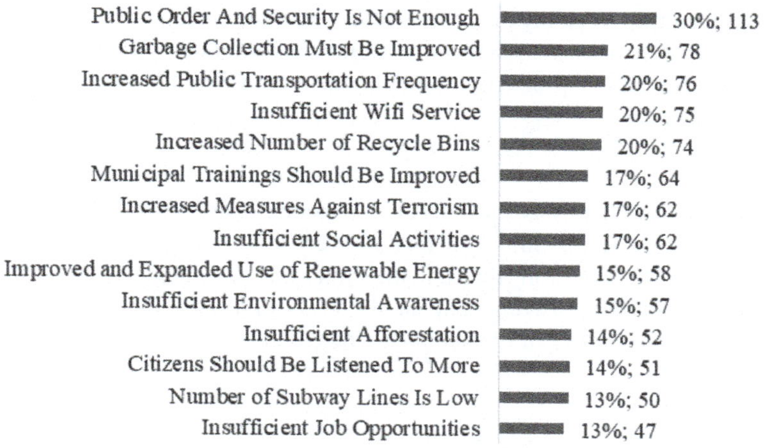

Fig. 4 Top 15 citizen needs

are required to determine the needs and discomforts of the citizens in the field of Mobility and use the available resources in the most efficient way to provide more comfortable, environmentally-friendly and safe transportation.

Environment

In accordance with the conducted analysis, it is realized that environmental and technological investments for managing natural resources, and citizens' environmental awareness are not at an adequate level. Therefore, environment main strategy is that to realize a complete return on environmental and technological investments, citizens awareness must be increased.

Energy

Although there are some energy programs, the use and coordination of new technologies are not sufficient in these programs. Therefore, energy main strategy is that smart grids and renewable energy production must be supported, and emissions of greenhouse gases must be reduced and energy efficiency must be increased.

Governance

The lack of coordination and data sharing makes it difficult to develop creative and innovative services. For this reason, governance main strategy is that the management of data collected from the city and the inclusion of the urban ecosystem into the decision-making process should be ensured to maximize the effectiveness of these processes.

Economy

Creating a smart economy, which is based on data and knowledge and which stands out with open data and public–private sector partnerships, is a development area for

Istanbul. Therefore economy main strategy is that, if the economic data is converted into information and transferred to the economy; opportunities for entrepreneurship, incentives, cooperation, productivity, employment growth and investment attraction should emerge.

Living

In light of conducted analysis, the imbalance in income distribution and the increasing population density results in the unequal delivery and difficult coordination of the services provided. Therefore living main strategy is that, it is suggested that the services provided in the Living area should be coordinated with the technological support so that the quality of life should be increased.

People

Provision of education to all citizens with innovative practices is anticipated to maximize productivity and social co-operation. People main strategy is the need for greater equal opportunity in terms of access to quality and contemporary educational opportunities, and the participation of all citizens in urban life.

Security

Although the use of intelligence, experience, human resources and technology has already been important in Istanbul and Turkey for many years due to various security problems originating from its geopolitical position, communication, coordination and use of smart solutions are not in sufficient levels in the whole city. Therefore, security main strategy is that smart solutions in the field of Security aim to quickly measure and evaluate possible security risks and to make communication and coordination more effective.

Information and Communication Technologies (ICT)

Low interoperability and integration, along with the lack of strategy, planning, institutional architecture, and investments are detected as a result of the analysis. Therefore, it can be stated that it is necessary to invest in the ICT area according to a pre-determined strategy and to adopt a holistic approach which enables integrated solutions to work together.

Human Resources (HR)

Considering findings of lack of understanding about objectives and strategies amongst stakeholders and citizens, low level of motivation of the staff in the organizations working for smart city, lack of competent and focused human resource in smart city initiatives, in the Human Resources area, there is a need for effective execution of the projects, increase in coordination between the units and sectors, increase in competent workforce and in competencies of the existing human capital.

Finance

The share of publicly funded investments in Istanbul is still high and there are many examples of creative companies. It is essential to overcome the difficulties of funding

Table 2 Number of strategic objectives, programs and projects for each domain/enabler of smart city

Domain/Enabler	Number of strategic objectives	Number of programs	Number of projects
Mobility	3	9	20
Environment	2	6	14
Energy	2	5	8
Governance	2	2	6
Economy	2	4	8
Living	2	4	5
People	2	5	11
Security	5	9	18
ICT	2	3	6
HR	1	2	4
Finance	1	1	1
Total	24	50	101

projects with high social but relatively low financial benefits. For this reason, it is recommended to focus on EU funds with creative PPP and special financing solutions besides the public financing for smart city investments in Istanbul.

In order to realize these domain/enabler strategies; number of developed strategic objectives, programs and projects are given in Table 2.

4.4 Forming the Istanbul Smart City Roadmap

101 smart city projects were placed on the below matrix according to their average impact and feasibility scores and prioritized based on their region on the matrix. Region 1 and 2 are short term projects, region 3 is midterm projects and region 4 and 5 are long term projects (Fig. 5).

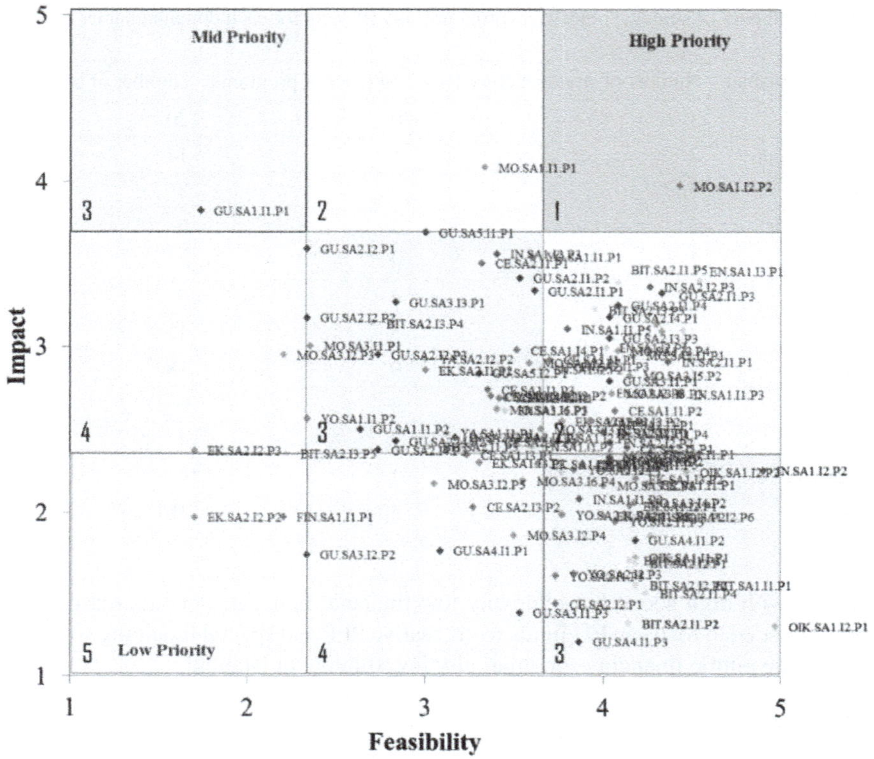

Fig. 5 Impact-feasibility matrix of smart city roadmap projects

5 Conclusion

During all of above studies, project team worked closely with around 200 smart city stakeholders in Istanbul. Stakeholder engagement is important for the quality of the project but each one potentially has the ability to slow down the project. In this project, stakeholders were not the driver but extremely useful sponsors and agents of smart city transformation. This study is the first comprehensive study for smart city roadmap in Turkey. Besides drawing a roadmap for Istanbul, awareness of stakeholders towards smart cities have been increased.

This study has a top-down approach following manifest, strategy, strategic objectives, programs and projects, respectively. Although the manifest and strategy is created by public authority, strategic objectives, programs and projects are determined with a gap analysis between the strategy and needs/problems of citizens, which are extracted from citizen needs analysis. Al in all, a needs based top-down approach for tailor made smart city roadmap for Istanbul is designed.

In the end, Istanbul has successfully surpassed the stages of literature review, current state analysis, and determination of the vision, strategy, and roadmap towards becoming a smart city. In this process, it has carried out the necessary activities to organize and strategically manage the already existing smart city practices and improved the smart city consciousness to a higher level. Istanbul is ready to walk confidently towards the manifest that is set out to become a smart city.

References

Akshay Java, X. S. (2007). Why We Twitter: Understanding Microblogging. In Proceedings of the 9th WebKDD and 1st SNA-KDD 2007 workshop on Web mining and social network analysis, 56–65.

Allison, M., & Kaye, J. (2015). Strategic Planning for Nonprofit Organizations: A Practical Guide for Dynamic Times. New Jersey: John Wiley & Sons.

ARUP (2016). Smart City Opportunities for London. London: Greater London Authority (GLA).

Berlin Partner für Wirtschaft und Technologie GmbH. (2017, September 13). Retrieved from http://reason-why.berlin/super-smart-city-berlin

BSI Standards Publication, Department for Business Innovation & Skills. (2015, February). Smart cities overview.

Collins, J. C., & Porras, J. I. (1996, September 1). Building your company's vision. Harvard Business Review, pp. 65–77.

Copenhagen Solutions Lab. (2016). Smart City Copenhagen. Copenhagen: City of Copenhagen.

David, M., David, F., & David, F. (2014). Mission Statement Theory and Practice: A Content Analysis and New Direction. International Journal of Business, Marketing, and Decision Sciences, 7(1), 95–110.

EY (2016). The Upside of Disruption: Megatrends Shaping 2016 and Beyond.

Gosenheimer, C. (2012). Project Prioritization: A structured Approach to Working on What Matters Most. Madison: University of Wisconsin-Madison Office of Quality Improvement.

Leberecht, T., & Vanderbeck, D. (2014). The Purpose City. New Cities Foundation, NBBJ, Imperative.

Nagy, J., & Fawcett, S. B. (2017). An Overview of Strategic Planning or "VMOSA"(Vision, Mission, Objectives, Strategies, and Action Plans).

NYC Mayor's Office of Tech Innovation. (2015). Building a Smart Equitable City. New York.

Smart Dubai Office. (2017, September 13). Retrieved from https://smartdubai.ae/

Smart Nation and Digital Government Office (2017). Smart Nation: The Way Forward.

UAE Government (2017, September 13). Retrieved from https://government.ae/en/about-the-uae/the-uae-government/smart-uae/smart-dubai

Van Zyl, J. (2011). Built to Thrive: Using Innovation to Make Your Mark in a Connected World: Using Innovation to Make Your Mark in a Connected World.

Finland—Smart Urban Air Mobility

Christian Schachtner

1 Introduction

This article was written in consultation with the contributing sub-project leader Vadim Kramar, Senior Researcher at VTT Technical Research Centre of Finland, after the author had insight into the role of a digital manager of a replicator city.

The idea of using urban airspace for passenger transport is not new, but it is gaining traction again due to technological advances and the pursuit of sustainable mobility solutions. Urban Air Mobility (UAM) promises to complement existing transport systems and contribute to the decarbonisation of the transport sector (Straubinger 2024).

UAM offers the potential to relieve urban traffic and reduce travel times. However, studies show that UAM's market share in urban passenger transport is expected to remain limited. Estimates put it at around 1%, due to higher operating costs due to infrastructure investments and personnel costs. Despite this low market penetration, the introduction of UAM could lead to a significant number of daily flights, as eVTOL (electric Vertical Take-Off and Landing) technology only allows for smaller payloads of a maximum of 9 passengers in urban areas. A major challenge for the operational implementation of UAM is the required ground infrastructure. The overall throughput of a UAM transport system is severely limited by the number and size of vertiports. This requires careful planning and integration into existing urban infrastructures. The environmental sustainability of UAM is a critical aspect that needs to be looked at closely. Compared to other electrically powered modes of transport, UAM has a lower environmental sustainability. This is mainly due to the fact that vertical take-off and landing is the most energy-intensive way of flying, resulting in higher electrical energy consumption and larger battery masses (Bauhaus Luftfahrt 2024).

C. Schachtner (✉)
RheinMain University of Applied Sciences, Wiesbaden, Germany
e-mail: christian.schachtner@hs-rm.de

C. Schachtner (ed.), *The European Smart City Movement*,
https://doi.org/10.1007/978-3-031-96700-9_22

Concrete figures illustrate this difference: A multicopter requires about 0.408 kWh/km of electricity in flight plus an additional 6.08 kWh per flight for take-off and landing. In comparison, an electric car consumes only 0.2 kWh/km. Assuming average emissions of 0.06 kg CO2 per kWh of electricity for electrically powered vehicles, it can be seen that while UAM can reduce CO2 emissions when it replaces trips with gasoline-powered cars, it is less energy-efficient compared to electric cars on the same route and leads to higher CO_2 emissions. The introduction of UAM also poses social challenges. Studies show that the social impact can vary greatly depending on the income group. For cost reasons, parts of society will not be able to use the new transport service at all or only rarely. There is a risk that negative externalities such as noise and visual impairment will have a disproportionate impact on these population groups (Straubinger 2024).

To fully exploit the potential of UAM and at the same time overcome the challenges, intensive research and development efforts are required. The Clean Sky RTI initiative, a European public–private partnership for clean aviation, has been working on improvements in lightweight construction and propulsion technology since 2008. These efforts have reduced global CO_2 emissions per passenger-kilometer by 53% from 1990 to 2019. For the future of UAM, further innovation in areas such as electric propulsion, alternative fuels and new energy sources is essential. In addition, new approaches to autonomy, certification, infrastructure and operation must be researched (Austrian Federal Ministry for Climate Action, Environment, Energy, Mobility, Innovation and Technology 2021).

2 Civil Acceptance

Public acceptance and engagement play a crucial role in the success of energy transition projects. The published results on the topic of "Public and stakeholder acceptance" provide important insights into the public's perception and attitude towards this new technology (European Commission 2023).

The subchapter "D4.6 Report on evolution of public and stakeholder acceptance and citizen engagement strategies" will focus on strategies for citizen participation. The analysis shows that the acceptance of renewable energies is generally high among the population, but that local resistance can occur in concrete projects (Müller 2020). It follows that the acceptance of energy transition projects can vary greatly from region to region. Factors such as the local economic structure, cultural peculiarities and previous experience with similar projects play an important role (Müller and Weber 2022). This underlines the need to adapt participation strategies to local conditions and to design them flexibly. However, this also illustrates the need to enter into dialogue with citizens at an early stage and to take their concerns seriously. Studies show that transparent communication and participation opportunities can significantly increase acceptance (Schmidt and Weber 2023).

Traditional formats such as information events are increasingly being supplemented by innovative digital approaches. Online platforms and social media enable low-threshold access to information and participation opportunities (Bauer et al. 2022). The analysis of social media data is proving to be a valuable tool for recording sentiments and opinions in the population at an early stage. By using sentiment analysis and machine learning algorithms, we can identify trends and potential areas of conflict before they escalate (Wagner and Hoffmann 2023). This makes it possible to proactively address concerns and adapt communication strategies.

However, personal contacts and local networks continue to play an important role, especially in building trust (Hoffmann 2021). The role of multipliers and opinion leaders must be taken into account here. Local personalities who are highly respected in the community can have a significant impact on the acceptance of energy transition projects (Klein et al. 2023). Involving these key people in the communication and participation process often proves to be a promising strategy.

The acceptance of general energy transition projects depends heavily on the perceived distributive justice. The studies show that models that enable financial participation by the local population meet with significantly higher approval (Klein and Groß 2024). Citizen energy cooperatives and municipal participation models are proving to be promising approaches to increasing acceptance and generating local value creation at the same time. The development of public opinion on energy transition projects is dynamic and influenced by various factors. Current events such as energy crises or extreme weather events can change perception in the short term. In the long term, however, we observe a trend towards greater support for the energy transition, especially among younger generations (Becker et al. 2025).

An interesting aspect of the project documentation is the increasing importance of digital technologies for citizen participation. Virtual reality and augmented reality applications, for example, make it possible to visualize planned projects vividly and make potential effects tangible (Schmidt et al. 2024). These innovative approaches can help to convey information in a more understandable way and improve the basis for discussion.

An interesting aspect of the study is the different perception of drones depending on the intended use. While the use in emergency services is largely welcomed, respondents are more skeptical about commercial applications (European Commission 2023). The project documentation shows that the general attitude towards drones in the rescue service is predominantly positive. Many respondents see the potential for faster emergency care and more efficient rescue operations (Smith and Johnson 2024). However, there are also safety, privacy, and noise concerns that need to be addressed in order to achieve broad societal acceptance (Brown et al. 2023).

Another important aspect is the consideration of different stakeholder groups. In addition to the general public, interest groups, local companies and political decision-makers also play an important role. This includes local authorities, residents, emergency responders and aviation authorities (Wilson 2024). From this, it can be deduced that tailor-made communication strategies are needed for different target groups in order to be able to address their specific needs and concerns (Wagner and Schmidt 2023). Based on the study results, several recommendations are made for the successful implementation of drones in the emergency services:

1. Transparent communication: It is important to fully inform the public about the benefits and possible risks of drone operations (European Commission 2023).
2. Security measures: Developing and communicating robust security protocols can reduce public concerns (Taylor and Lee 2025).
3. Noise reduction: Technical innovations to minimize noise pollution should be driven forward to increase acceptance in densely populated areas (Brown et al. 2023).
4. Data protection: Clear privacy policies need to be developed and communicated to address concerns about possible surveillance (Smith and Johnson 2024).
5. Stakeholder involvement: Continuous involvement of all relevant stakeholders in the decision-making process is critical to long-term success (Wilson 2024).

3 Business Model "Air Mobility"

The subchapter "D5.2 EMS UAM business models for companies and public sector" deals with the development of innovative business models for Urban Air Mobility (UAM) in the context of companies and the public sector. UAM represents a promising approach to extending urban mobility into the third dimension and holds significant potential for the transformation of urban transport (Roland Berger 2024).

As part of our project, we analyzed various business models that could be relevant for UAM applications. A particularly interesting model is the "freemium" concept, where customers can first use a free basic version of a product or service before opting for a paid premium version (WEKA E-Learning Magazine 2024). In the UAM context, this model could be adapted for booking platforms or information services, for example.

Another promising business model is the subscription model, in which customers make regular payments for continuous access to a service (WEKA E-Learning Magazine, n.d.). In the UAM sector, this could be relevant for regular commuter flights or logistics services. For the development of successful UAM business models, it is crucial to precisely define the specific value propositions and customer segments. The Business Model Canvas can serve as a useful tool here (Schachtner 2024). In the UAM context, value propositions such as fast, congestion-free connections or innovative logistics solutions could be in the foreground.

When analyzing potential customer segments for UAM services, both the private and public sectors must be considered. In the private sector, business travellers, commuters in metropolitan areas or logistics companies in particular could benefit from UAM solutions. In the public sector, we see potential for applications in the field of emergency services, urban planning or public transport. A critical aspect in the development of UAM business models is integration into existing transport systems and infrastructures. Innovative solutions must be found here to enable seamless transitions between different modes of transport. This requires close collaboration between UAM providers, urban planners and transport authorities.

In addition, regulatory frameworks play a crucial role in the feasibility of UAM business models. The development of suitable regulations for the operation of air taxis and drones in urban areas is a key challenge that must be tackled in close coordination between industry and authorities.

Another important aspect is acceptance among the population. UAM providers need to develop strategies to address potential noise, security, or privacy concerns. Approaches such as community engagement or transparent communication could play an important role here. For companies looking to enter the UAM market, it is recommended to carefully analyze the profit and revenue levers. Both direct revenue streams from UAM services and indirect effects such as efficiency gains or new business opportunities should be taken into account (Roland Berger 2024).

In the public sector, we see great potential for UAM applications in the field of disaster control and medical care. Here, drones and autonomous flight systems could make an important contribution to improving public safety and health care.

4 Classification of Results

The project documentation shows that the acceptance of drones in the rescue service is basically there, but challenges still remain. The identified concerns provide important starting points for future research and development (European Commission 2023). It is recommended to conduct pilot projects to gain practical experience and evaluate the real impact on emergency care. Future studies should focus on the long-term effects of drone operations on the quality of emergency care. The development of best practice models for integrating drones into existing rescue systems is also seen as an important next step (Taylor and Lee 2025).

5 Conclusion

In conclusion, UAM has the potential to become a valuable building block in a diversified and sustainable transport system of the future. The AiRMOUR project provides important insights into the public and social acceptance of drones in the rescue service. The results form a solid basis for the further development and implementation of this promising technology in urban airspace. However, further research and technological developments are needed to address the operational and environmental challenges and achieve broad societal acceptance. It can be said that the development of acceptance and citizen participation in energy transition projects is a complex and dynamic process. The results show that a combination of transparent communication, innovative participation formats and fair participation opportunities is the key to success. Although Urban Air Mobility is not a revolutionary innovation for urban passenger transport, it offers potential for sustainable business models for OEMs and operators. The challenges in terms of energy efficiency, infrastructure and social acceptance are considerable, but not insurmountable. For a successful integration of UAM into future transport systems, it is crucial that municipalities and decision-makers carefully weigh up the benefits and costs. The introduction of UAM should only be supported if the mobility concept is in line with overarching climate policy goals. The continuous adaptation and further development of these strategies will remain a key challenge for the successful implementation of the energy transition in the future.

References

Austrian Federal Ministry for Climate Action, Environment, Energy, Mobility, Innovation and Technology. (2021). RTI Strategy Aviation 2040plus. https://open4aviation.at/resources/pdf/FTI-Strategie-Luftfahrt-2040plus.pdf

Bauer, L., Müller, K., & Schmidt, T. (2022). Digital Participation in Energy Transition Projects: Opportunities and Challenges. *Journal of Environmental Policy and Sustainability, 15*(3), 178–195.

Bauhaus Luftfahrt. (2024). What can Urban Air Mobility do and what can't it do? Our summary whitepaper. https://www.bauhaus-luftfahrt.net/de/forschungsbereiche/urban-regional-air-mobility/was-kann-urban-air-mobility-und-was-nicht-unser-zusammenfassendes-whitepaper

Becker, A., Klein, M., & Wagner, S. (2025). Intergenerational acceptance of the energy transition: A long-term study. *Journal of Sustainable Development, 28*(1), 45–62.

Brown, A., Davis, C., & Evans, E. (2023). Drone acceptance in urban environments: Challenges and opportunities. *Journal of Urban Technology*, 15(3), 245–260.

European Commission. (2023). AiRMOUR Project: Public and stakeholder acceptance interim report. https://ec.europa.eu/research/participants/documents/downloadPublic?documentIds=080166e5xxxxx&appId=PPGMS

Hoffmann, R. (2021). Building trust through personal networks in the energy transition. *Sozialwissenschaftliche Rundschau, 42*(2), 87–103.

Klein, F., & Groß, M. (2024). Financial citizen participation as the key to the acceptance of energy projects. *Economy and Environment, 19*(4), 312–328.

Klein, F., Wagner, T., & Müller, S. (2023). The influence of local opinion leaders on the acceptance of energy transition projects. *Journal of Environmental Psychology, 17*(2), 89–105.

Müller, K., & Weber, L. (2022). Regional differences in the acceptance of energy transition projects. *Spatial Research and Spatial Planning, 80*(3), 267–284.

Müller, T. (2020). Local resistance in the implementation of energy transition projects. *Journal of Energy Policy, 12*(1), 34–52.

Roland Berger. (2024). Global Topic Urban Air Mobility. https://www.rolandberger.com/de/Insights/Global-Topics/Urban-Air-Mobility/

Schachtner, C. (2024). On the Way to a Data-Based Governance Model for Swiss Public Institutions – Focus Fields for Empowerment in Innovation and Identification. *Swiss Yearbook of Administrative Sciences*, 15(1), 25–38. https://doi.org/10.5334/ssas.214.

Schmidt, A., & Weber, M. (2023). Transparency and participation as success factors for the acceptance of energy projects. *Energiewirtschaftliche Tagesfragen, 73*(5), 28–35.

Schmidt, T., Bauer, L., & Hoffmann, R. (2024). Virtual Reality in Citizen Participation: New Ways of Project Visualization. *Digital Transformation, 8*(2), 156–173.

Smith, J., & Johnson, L. (2024). Public perception of drones in emergency services: A comprehensive analysis. *International Journal of Emergency Management*, 12(2), 178–195.

Straubinger, A. (2024). Urban Air Mobility: More of a castle in the air than a high-flyer! ZEW policy brief, 15 https://www.econstor.eu/bitstream/10419/303499/1/1904287522.pdf

Taylor, R., & Lee, S. (2025). Integrating unmanned aerial vehicles in urban airspace: Safety protocols and public trust. *Aviation Safety Journal*, 18(4), 412–428.

Wagner, S., & Hoffmann, T. (2023). Use of AI-powered sentiment analysis to capture public opinion in the energy transition. *Big Data and Society, 6*(1), 78–95.

Wagner, T., & Schmidt, A. (2023). Target group-specific communication strategies for energy transition projects. *Communication and Sustainability, 14*(3), 201–218.

WEKA E-Learning Magazine. (2024). 10 examples of successful business models. https://magazin.weka-elearning.de/new-work/10-beispiele-fuer-erfolgreiche-business-modelle

Wilson, M. (2024). Stakeholder engagement in drone implementation projects: Best practices and lessons learned. *Journal of Technology Management*, 9(1), 67–82.

Czech Republic: Sustainable Urban Planning: Approaches to Neighbourhood-Level Sustainability Design in Prague

Licia Felicioni, Barbora Hejtmánková, Michal Kuzmič, Pavla Pelčíková, Nohemi Ramirez Aranda, Kristýna Schulzová, Mahulena Svobodová, Tomáš Vácha, Martin Volf, and Klára Zámečníková

1 Introduction

Currently, cities show a rapid urbanisation process that puts pressure on the infrastructure and necessary resources to supply a growing population (Ortiz-Fernández et al. 2023). In fact, the demand for new housing units in Prague and other Czech cities is steadily increasing. Interest in new apartments in Prague in 2024 increased by 80% compared to the end of 2023 (Newstream.cz 2024). Therefore, planning suitable urban spaces and neighbourhoods becomes vital and urgent, also considering the essential functions of living, working, commerce, healthcare, education and entertainment by decentralizing urban functions and services (Khavarian-Garmsir et al.

L. Felicioni (✉) · B. Hejtmánková · M. Kuzmič · K. Schulzová · T. Vácha · M. Volf · K. Zámečníková
Czech Technical University in Prague, University Centre for Energy Efficient Buildings, Buštěhrad 273 43, Czech Republic
e-mail: Licia.Felicioni@cvut.cz

B. Hejtmánková
e-mail: barbora.hejtmankova@cvut.cz

M. Kuzmič
e-mail: michal.kuzmic@cvut.cz

K. Schulzová
e-mail: kristyna.schulzova@cvut.cz

T. Vácha
e-mail: tomas.vacha@cvut.cz

M. Volf
e-mail: martin.volf@cvut.cz

K. Zámečníková
e-mail: klara.dvorakova@cvut.cz

C. Schachtner (ed.), *The European Smart City Movement*,
https://doi.org/10.1007/978-3-031-96700-9_23

2023). This kind of neighbourhood can be dense and socially connected and functionally mixed neighbourhoods based on the human scale design to encourage walking and cycling (Balletto et al. 2021; Noworól et al. 2022).

Planners must take into account not only future residents, but also the current urban population and potential housing developers, whose investments are crucial for the municipality. As a result, tensions may arise between the interests of citizens and developers. Planners play a key role in balancing these competing interests (Wenander 2024). In this general context, urban planners face a clear demand for projects of higher environmental performance from both the public and private sectors (Lotteau et al. 2015).

In 2022, the City of Prague became an important member of the New European Bauhaus (NEB). This initiative, coined by European Commission President Ursula von der Leyen, highlights the importance of core values driving the green transition: sustainability, inclusivity, and aesthetics (European Union 2021).

In order to evaluate urban development in terms of sustainability and NEB values, indicators can serve as metrics to simplify complex concepts, offering valuable information about various phenomena (Khodakarami et al. 2023). This is demonstrated by rating systems that use a criteria-based structure with indicators to assess the quality of neighbourhoods (Zoghi et al. 2024).

This chapter seeks to enhance the understanding of how planners in Prague navigate and balance various interests when developing visions and goals for new sustainability-focused neighbourhoods and what this means for achieving sustainability objectives that the city has set.

The following research questions guide the investigation:

1. How do planners in Prague manage different forms of participation when formulating visions and goals?
2. How are the interests of housing developers and residents expressed, and how do planners address any potential conflicts between these interests during the formulation of visions and goals? What urban planning tools are used to address participation and housing developers' interest?
3. How do the tools contribute to NEB values (environmental, social, and aesthetic)?

To address these questions, a few different tools utilized by urban planners in Prague have been presented to determine whether they can be effectively used in

N. Ramirez Aranda
Faculty of Architecture and the Built Environment, Department of Urbanism, Environmental Technology and Design Section, Delft University of Technology (TU Delft), Delft 2628, Netherlands
e-mail: n.e.ramirezaranda@tudelft.nl

P. Pelčíková · M. Svobodová
Prague Institute of Planning and Development, Prague 12800, Czech Republic
e-mail: pelcikova@ipr.praha.eu

M. Svobodová
e-mail: svobodova.m@ipr.praha.eu

practice for designing sustainable, livable and attractive urban spaces. Some of these tools have been tested within the framework of European projects or at the request of the municipality of Prague, which wanted to explore different ways of sustainable urban planning.

The chapter finally contributes with additional knowledge of the early planning phase, including the formulation of visions and goals of sustainability-profiled neighbourhoods.

2 Urban Design Ecosystem in Prague

In the context of urban planning in Prague, multiple actors play a critical role in ensuring the effective implementation of development projects. The Prague Institute of Planning and Development (IPR) is responsible for overseeing urban planning, architectural oversight, strategic development, and land use management. Its mission is to guide the city's sustainable growth, making Prague a livable, inclusive, and efficient metropolis. IPR acts as a key advisor to the City of Prague, its districts, and stakeholders such as developers and investors. It provides expertise in areas such as urban planning, public spaces, strategic planning, and sustainable development, and it is equipped with tools such as Geographic Information Systems (GIS), 3D models, and simulations. Additionally, IPR operates the Centre of Architecture and Municipal Planning (CAMP), which serves as a platform to engage the public, professionals, and policymakers in discussions about Prague's urban future. A significant part of IPR's work addresses the historical division created by the city's main rail lines, which, while connecting Prague to the rest of Europe, disrupted neighbourhood connectivity.

The Prague Development Company (Pražská developerská společnost, PDS) is another municipal entity focused on managing and developing publicly owned land and real estate in alignment with the city's strategic objectives. PDS works closely with IPR to promote affordable housing and support sustainable integrated urban development. PDS also facilitates public–private partnerships to synchronize private investments with Prague's development strategy, contributing to large-scale urban projects.

Operátor ICT (OICT), a municipal joint-stock company wholly owned by the city, leads the city's Smart City initiatives. It manages ICT projects for Prague and its districts, offering innovation management services and implementing smart urban solutions. OICT oversees platforms like the urban data hub Golemio (OICT, n.d.), the mobility app Lítačka (OICT, n.d.), and the Prague citizen portal. The Golemio team includes IT experts, programmers, data analysts, and consultants to provide complex in-house services for the city of Prague. Golemio provides public dashboards that allow citizens and other stakeholders to access and use the available data. Golemio also creates custom dashboards and reports for municipal departments and organizations. It focuses on a wide range of use cases including mobility, environmental monitoring, energy, tourism, and more. The team is capable of flexible response to emergency situations, for example, Golemio helped to monitor and manage the COVID pandemic in Prague. The platform is committed to open-source and modular design. This enables flexible integration of available tools, support for third party developers, and mutual exchange with other cities.

Since 2017, OICT has been publishing the annual Smart Prague Index yearbook (Smart Prague 2023). Using specific measurable indicators and yearly evaluations, it analyzes Prague's development, tracks trends, summarizes the capital's achievements, and serves as a valuable foundation for its further development.

The University Centre for Energy Efficient Buildings of the Czech Technical University in Prague (CTU UCEEB) is another key actor, collaborating with the aforementioned stakeholders mostly conducting research. It contributes to the development of specific criteria and indicators for sustainable urban development, such as assessment methods for projects Letňany neighbourhood, Dolní Počernice neighbourhood or for the Prague 8 City District. The CTU UCEEB also plays a role in participatory planning activities, working with stakeholders and citizens to raise awareness about creating livable and attractive neighbourhoods. Noteworthy examples include the "Water in the City" methodology (Sýkorová et al. 2021), which provides innovative approaches to rainwater management, integrating nature-based solutions within urban infrastructure to enhance resilience and sustainability. Additionally, CTU UCEEB has produced a "Methodology for Planning Public Spaces in Small Municipalities," a practical guide that assists smaller communities in creating high-quality public spaces (Hainc et al. 2021; Malá et al. 2022).

3 Assessment Tools for Urban Areas

With the rapid pace of urbanization and the growing social and political drive to minimize the environmental impact of cities, urban areas face significant challenges, including environmental concerns, social sustainability, and critical economic issues. To address these challenges, a wide range of urban assessment tools have been developed globally since the early 2000s. These tools have sought to conceptualize urban sustainability in diverse and sometimes contrasting ways, providing different approaches to understanding this topic (Lucchi and Buda 2022).

These frameworks assess sustainability issues and solutions on various scales, from neighbourhoods to entire cities and rural areas. By doing so, they guide decision-makers and policymakers in determining which actions to pursue or avoid promoting greater societal sustainability (Tam et al. 2018).

Indeed, these tools serve to achieve and maintain balanced, environmentally sound, socially equitable, and economically sustainable development of urban neighbourhoods. By achieving specific criteria through certification systems, projects, and areas are certified at different levels. These certificates are awarded on the basis of meeting set standards in areas such as ecology, energy, social inclusion, and others. Certificates provide a visual representation of the level of sustainability of a neighbourhood. For municipalities and developers, certification is a tool to visualize, maximize, and highlight the sustainability aspects of an area. Investors may view the certificate as assurance that an area is achieving specific sustainability standards and may find it helpful in making future investment decisions.

The primary goal of certification is to provide a clear and consistent assessment of sustainability, allowing for comparisons between different neighbourhoods. This approach helps identify and compare development projects labeled as 'sustainable,' offering better guidance and information about the quality of life in these areas.

In response to market trends, the building and construction industry has introduced various Neighbourhood Sustainability Assessment (NSA) tools. These include the Building Research Establishment Environmental Assessment Method for Communities (BREEAM-C) (BRE 2012), Leadership in Energy and Environmental Design for Neighbourhood Development (LEED-ND) (USGBC 2018), the Comprehensive Assessment System for Building Environmental Efficiency for Urban Development (CASBEE-UD) (CASBEE), and Green Star Communities (Green Building Council of Australia 2016).

In Europe, LEED certification is used primarily in western and southern countries. Apart from the UK, the BREEAM methodology is mainly adopted by northern and eastern European countries. DGNB (2020) is predominantly used in Germany and its neighboring regions. However, many countries have also developed their own tools tailored to their specific regulations and standards.

In the Czech Republic, neighbourhood certification is not yet widely practiced. However, among the first certified areas are:

- Urbanity Campus Tachov, BREEAM Communities, "Excellent"—awarded in 2022, it is the first certified industrial campus in the Czech Republic and the first certified industrial campus with a predominantly manufacturing function in the world (Urbanity, 2022).
- Nová Zbrojovka, Brno-Židenice: 22 hectares of unused brownfield land, the project includes 2500 apartments, 200,000 m^2 of office space and 100,000 m^2 of commercial and retail space and services—the first urban district in the Czech Republic to achieve a BREEAM Communities rating. The project is expected to be completed in 2035 and has so far successfully achieved provisional certification at the design stage.

In 2023, a comparative analysis of four certification systems—BREEAM-C, DGNB District, LEED-ND, and SNTool (International Initiative for a Sustainable Built Environment 2020)—was conducted by CTU UCEEB for the Institute of Planning and Development (IPR). The analysis provided essential information about each system, such as the availability of the methodology, its origin, the number of rating categories and criteria, and details of the assessment process—some information is presented in Table 1. It also examined the applicability of each system within the Czech context. A key component of the study involved applying two of these methodologies to a concrete case: the draft spatial study for the Letňany-Kbely large development area, located in the northern part of Prague. This neighbourhood, designed by IPR with the involvement of citizens (during 2023, a series of participatory events took place in Letňany, Kbely and Prosek for collecting feedbacks and comments), aims to follow sustainable development principles as closely as possible. Therefore, the analysis allowed an evaluation of how well the district design aligns with sustainability principles.

Finally, a preliminary assessment of the Letňany-Kbely development area in Prague was conducted using the LEED ND (Leadership in Energy and Environmental Design for Neighbourhood Development) and DGNB (Deutsche Gesellschaft für Nachhaltiges Bauen) District rating systems. This assessment aimed to identify how well the project currently meets the sustainability targets of each rating system and where improvements could be made.

Under the LEED ND assessment, the site initially scored 29 points, which fell below the certification threshold. However, with additional documentation and design modifications—such as improvements to district heating, rainwater management, and the inclusion of social housing—it could reach 84 points, qualifying for the highest platinum certification.

Similarly, the DGNB District assessment showed that the site, in its current state, would not achieve certification. However, with thorough documentation and design adjustments, the project could potentially score 75%, earn a Gold certificate, and possibly achieve the Platinum level with 80%. The DGNB evaluation places a strong emphasis on detailed analyses, adherence to local regulations, and input from experts.

Both assessments underscore the site's potential for high sustainability ratings, given the right modifications and expert guidance. They also highlight the critical importance of integrating sustainability criteria early in the planning process to ensure the success of the project in meeting international certification standards.

This work served as a starting point to highlight the need for a rating system specifically tailored to the Czech context. Such a system would be highly beneficial for urban planning projects to ensure alignment with the objectives and strategies of the respective district and broader development plans. In response, IPR began developing a rating system for the city of Prague that focuses on neighbourhood assessments (as discussed in Sect. 4), while some of the individual administrative districts of the city began working on their own systems, as detailed in Sect. 5.

Furthermore, it is essential to highlight that the Resilient Built Environment tool for Central Europe (RBE-CE tool) is currently under development as part of the SuPeRBE Interreg project (Supporting Cross-scale Planning and Policy Readiness for a Resilient Built Environment) [2024–2026] (Interreg Central Europe 2024).). This multi-scale assessment tool will evaluate the resilience levels of buildings, neighbourhoods, and communities, identifying key parameters to effectively reduce vulnerability to specific hazards. For the city of Prague, the main hazards are outlined in the Prague Climate Plan 2030 (Prague City hall—Environment Protection Department 2021), which serves as Sustainable Energy and Climate Action Plan (SECAP) (Bertoldi 2018).

During the project, the tool will be tested in the Prague 5 municipality districts and subsequently utilized by stakeholders to evaluate the resilience of their projects, including neighbourhoods. The findings will inform specific adaptation measures,

Table 1 Description of neighbourhood assessment tools

Rating system	Description
BREEAM [UK] Building Research Establishment Environmental Assessment Method (BRE 2012)	BREEAM evaluates various sustainability criteria, including energy efficiency, water management, material selection, indoor environmental quality, and ecological impact. The evaluation process is divided into three steps: establishing development principles, determining the layout, and designing details In the Czech Republic, BREEAM is already used to build ratings, making it familiar and potentially more acceptable to local stakeholders
LEED [USA] Leadership in Energy and Environmental Design (USGBC 2018)	LEED has two versions for urban areas: LEED ND (Neighbourhood Development) and LEED for Cities and Communities. The evaluation process involves registering the project and meeting criteria across categories such as Smart Location and Linkage, Neighbourhood Pattern and Design, Green Infrastructure and Buildings, Innovation and Design Process, and Regional Priority Credits. LEED is user-friendly, with a clear point structure, and does not require an authorized person for evaluation. However, it has higher certification costs and is based on non-European standards, which may require adaptation for use in the Czech Republic
DGNB [DE] Deutsche Gesellschaft für Nachhaltiges Bauen (Green Building Council of Australia 2016)	DGNB emphasizes European standards, and its evaluation process consists of three phases: pre-certification at the planning level, partial certification during construction, and final certification upon completion. The categories of DGNB include Environmental Quality, Economic Quality, Sociocultural and Functional Quality, Technical Quality, and Process Quality. The system is noted for its comprehensive approach and the ability to create national modifications However, it requires extensive documentation, which can be resource intensive. Given similar geographic and legislative conditions, DGNB is suitable for use in the Czech Republic
SNTool Sustainable neighbourhood tool (International Initiative for a Sustainable Built Environment 2020)	SNTool is a customizable framework to assess neighbourhood sustainability. It requires adaptation to local conditions, including setting performance benchmarks and weighting criteria. SNTool's categories include Context and Vulnerabilities, Built Urban Systems, Economics, Energy, Non-renewable Resources, Environment, and Social Aspects The tool is highly adaptable and can be tailored to specific regional conditions. However, it is still in development and technically complex. The potential to create a Czech version exists, but it requires significant customization and resources

facilitating data-driven decision making as needed. This approach will ensure that the neighbourhoods are not only sustainable, but also capable of meeting the resilience targets to a significant extent.

4 Prague Sustainable Neighbourhoods Tool

Recognizing the need for a comprehensive sustainability assessment tool tailored to neighbourhood-specific and locally relevant criteria, IPR is leading the development of the Prague Sustainable Neighbourhood Tool. This initiative aims to create a framework that reflects the unique environmental, social, and economic conditions of Prague, ensuring that new developments and existing communities align with high standards of sustainability.

Indeed, the main goal of the project is to navigate new construction in Prague to meet comprehensive standards of sustainable development, i.e. mitigate climate change or adapt to it, preserve the environment, save non-renewable resources and materials, be economically and socially sustainable, and therefore contribute to increasing the quality of life in Prague.

The focus of the Tool is currently on new construction, with the perspective of focusing on existing neighbourhoods in the next phase. The reason is that within new construction, more sustainability requests can be made and the focus on existing neighbourhoods will necessitate other measures.

The first objective of the project is to define what a sustainable neighbourhood is, to declare the vision and requirements of Prague within a defined set of topics.

The target scale is a neighbourhood, at the minimum that means a block of buildings including public space. The main focus is not individual buildings, as there are many tools concerned with the sustainability of individual buildings.

The Tool is aimed at various target users, ranging from the public actors of the city, such as IPR and other municipal organizations and departments, city districts, private and public developers. One of the key goals is to achieve coordination within the Prague city administration, in order to share a common vision on what is a sustainable neighbourhood and how to achieve it.

In terms of enforceability, the Tool will be a soft tool, not a regulation or a certification tool. It should instead serve as motivation and also as a signpost, pointing to existing guidelines and regulations that lead to sustainable neighbourhoods.

Most likely, there will be several types of output (offline and online)—tailored for different stakeholders and containing more of vision or of practical information.

The Tool should be regularly updated to stay aligned with the latest advancements in the field.

Collaboration with diverse stakeholders is essential to the development of the tool, which is expected to be completed by mid-2025. As part of the NEB-STAR European project, the Tool is being benchmarked through a twinning exercise with practices from Utrecht (NL) and Stavanger (NO) (NEB-STAR project 2023). This process involves a series of workshops held in person with stakeholders.

While focusing on global topics, the Tool takes into account local conditions, environmental, economic, social, and legal. In particular, the Tool explores the following topics:

- Structure: natural environment, land use, built environment, public space, urban climate, use;
- Infrastructure: mobility, energy and materials, water, smart infrastructure, waste;
- Society: housing, community, health, wellbeing;
- Process and Governance: planning, participation, governance, economics, monitoring.

By focusing on neighbourhood-level planning and considering the diverse needs of various stakeholders, the Tool will serve as a practical guide to promote resilient and liveable communities throughout the city.

5 Design of a Rating System for a Sustainable District

Drawing on the knowledge of existing rating systems, a customized rating system for districts can be developed. This approach was exemplified by the creation of the "Sustainable Prague 8" manual (Udržitelná Praha 8, 2022), a comprehensive evaluation system developed by CTU UCEEB in collaboration with the Prague 8 Municipality. The manual outlines an assessment framework for construction projects in the Prague 8 municipality, ensuring alignment with the district's strategic goals and the Adaptation Strategy of Prague (Prague City hall—Environment Protection Department 2021).

This system facilitates a transparent evaluation of construction projects, assessing their compliance with the Prague District Development Plan and district-specific objectives. Its development was preceded by extensive research into global evaluation tools like LEED, BREEAM, DGNB, and SBTool, alongside documents related to Prague's spatial development. The objective of this rating system is to provide a clear assessment of the architectural, urban planning, technical, and economic quality of a project, while also accounting for the adaptation to climate change and encouraging a participatory approach.

The evaluation framework is structured around four key areas: public spaces, architectural and technical design, participation, and climate change. Public spaces are evaluated based on criteria such as permeability, accessibility, civic responsibility, traffic impact, and sustainable maintenance. Architectural and technical design assessments focus on environmental value, architectural merit, and cultural and social qualities. The participation dimension emphasizes public participation in planning, integrated design, and community engagement. Climate change criteria include adaptation measures such as stormwater management and green infrastructure, alongside mitigation measures such as energy efficiency, use of renewable energy, and support for the circular economy. Table 2 presents five criteria as an example.

Table 2 Example of five criteria of the evaluation system

Criterion	Description
Permeability through the area and satisfactory routing of paths	*Indicator:* Ensures that the project does not create dead ends and promotes easy pedestrian movement *Evaluation:* Projects are evaluated to determine whether they allow for direct and convenient passage through the area, preventing enclosed or impenetrable spaces
Impact of traffic on public spaces	*Indicator:* Assesses the integration of parking solutions and support for alternative transport modes *Evaluation:* Projects are evaluated based on the percentage of parking spaces available within a certain distance, the use of permeable materials for parking areas and the provision of facilities for cyclists
Public participation	*Indicator:* Encourages active public participation in the planning process *Evaluation:* The projects are evaluated on the method and extent of public participation, including workshops, roundtables, and the incorporation of public feedback into the design
Adaptation to climate change	*Indicator:* Focuses on stormwater management, green infrastructure use, and water conservation *Evaluation:* The projects are evaluated for the implementation of nature-friendly solutions for stormwater management, the use of treated rainwater or wastewater, and the amount of greenery within the project
Promoting the circular economy	*Indicator:* Encourage the use of recycled materials and designs that facilitate future adaptability *Evaluation:* The projects are evaluated in terms of the use of recycled materials, the preparation of material passports, and the inclusion of future expansion or changes in the project design

The evaluation process involves a detailed questionnaire tailored to each project, which scores various criteria on a scale of 0–10, with the final score serving as the basis for discussions with investors.

This system rewards projects that go beyond legal requirements, contributing positively to the community and the environment, ultimately promoting sustainable development in Prague 8. The overarching goal is to create an evaluation system that is easily understood by the applicants, ensuring that their projects align with the strategic objectives of the municipality and contribute to the sustainable development of Prague 8.

The evaluation system adopted by the Prague 8 Municipality provides a transparent and structured approach to assess the sustainability and community impact of construction projects. By focusing on key areas such as public spaces, architectural and technical design, participation, and climate change, the system ensures that projects not only comply with legal requirements, but also contribute positively to the environment and the community. This framework rewards projects that go beyond the minimum standards, promoting higher quality and more sustainable development. The system is adaptable to different sizes and types of projects, allowing for tailored evaluations that meet the specific needs of the city.

Ultimately, this evaluation system supports the municipality's goal of creating a more sustainable, liveable, and resilient urban environment for current and future Prague 8 residents.

6 Barcode Methodology

The Barcode methodology was developed by the municipality of Utrecht (NL), a rapidly growing city with limited expansion opportunities due to the presence of highways, waterways, and heritage sites that restrict its borders (NEB-STAR project 2023). Faced with a significant housing shortage, the city of Utrecht needed to devise a strategy for urban growth within the confines of its existing city limits. In response, the Barcode methodology was introduced as a solution to balance growth while managing the city's spatial limitations and ensuring space for more than just housing.

The Barcode methodology serves as a tool for guiding urban development within existing cities by helping planners determine how to utilize limited space best—each city should develop its own because it strictly relies on the context. It is a theoretical framework that demonstrates how the provision of approximately 10,000 new housing units requires a corresponding set of services, such as green spaces, schools and playgrounds, work areas, infrastructure, and renewable energy production. The core concept of the Barcode methodology is to integrate these amenities into urban expansion in a way that enhances the city. Typically, these services are quantified in terms of square meters and land use (Fig. 1).

The development of the Barcode begins by determining the parameters for each service based on national norms and local legislation. Essentially, the Barcode is a visual representation that reflects, according to these standards, the amenities required to support 10,000 new dwellings. It incorporates established policies, along with

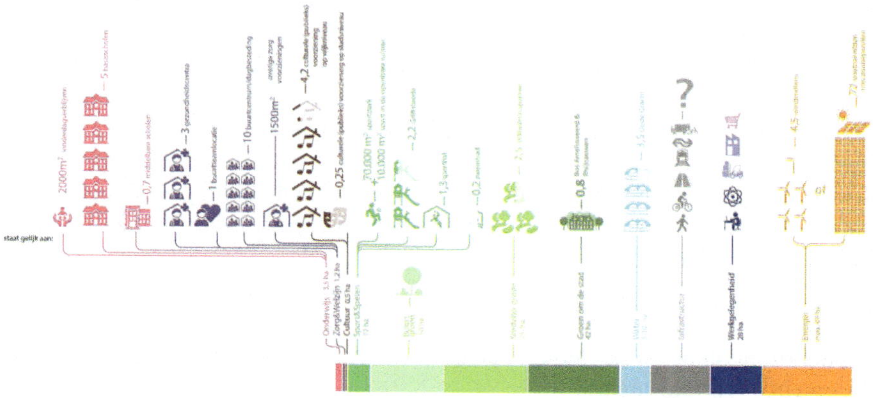

Fig. 1 Example of a Barcode diagram. *Source* (Utrecht Municipality 2021a, b)

emerging trends and developments, to ensure a holistic approach to urban planning. Each element is interdependent, with an increase in one often necessitating an increase in another.

While the Barcode is primarily used as a communication tool among planners to facilitate discussions on spatial planning, it is not intended as a prescriptive blueprint for urban design. Not every project is suited to the Barcode approach, and its application must be carefully considered. The Barcode tool can operate at two levels:

- City-scale Barcode—the standard model.
- Neighbourhood-scale Barcode—adapted to the principles of a 10- or 15-min city, ensuring essential services and amenities are within close proximity in every neighbourhood.

The Barcode can be applied to different urban functions:

- Preservation—incorporating new elements such as energy systems while maintaining similarities with the current urban layout.
- Densification—increasing housing density within the same land area, replacing single-family homes with higher-density housing.
- Transformation—converting brownfields into residential areas, shifting land use while introducing more greenery and services for residents.
- Expansion—converting agricultural land into urban spaces for development.

The Barcode also allows for creative problem-solving, such as combining functions to optimize land use and reduce the space required.

Notably, within the framework of the NEB-STAR European project (New European Bauhaus Stavanger, 2022–2025) (NEB-STAR 2022), the Barcode methodology has been implemented in a case study focused on the Letňany neighbourhood in the northeastern part of Prague. A second application of the Barcode framework occurred in the Netherlands in 2023. Students from the Geodesign Course on Circular Economy in Urban Regions (GDCEUR), part of the Urbanism Master track at TU Delft (NL), applied the framework to a case study in the Merwede neighbourhood of Utrecht (Fig. 2). Their task involved inventorying and analyzing the area using the Barcode methodology. In addition, they were required to propose strategies to facilitate the accommodation of 6000 new high-density dwellings in the coming years (Fig. 3). Similarly, the city of Stavanger, located in the southwestern corner of Norway, served as another test site for the Barcode methodology. In collaboration with various stakeholders, a series of workshops was held in April 2024 to assess critical areas in Pedersgata and Ryfylkegata (see Fig. 4).

The Barcode is expressed quantitatively, using metrics such as square meters and land use rather than qualitative data. It can also be used to visualize the existing city, helping discussions on key issues such as density, green space allocation, and the placement of parking areas.

While the Barcode is a powerful tool, its success depends heavily on the expertise of the planners using it. It requires well-trained urban planners who can interpret and apply it effectively. In Prague, municipalities typically do not develop plans

Fig. 2 Project impression of the future Merwede area. *Source* Municipality of Utrecht. *Source* (Utrecht municipality 2021a, b)

Fig. 3 Biophilic design proposed interventions in section, interior, and exterior views by students from Group 2 of the GDCEUR 2023. *Source* (Ramirez Aranda 2023)

independently; they often collaborate with public and private partners or rely on private developers to create the plans.

Fig. 4 NEB-STAR Pedersgata and Ryfylkegata Barcode workshop's result. *Source* (NEB-STAR Project 2024)

7 Interconnecting Urban Design with Prague's Digital Twin

Prague's digital twin, managed by IPR, plays a pivotal role in advancing sustainable urban planning. Using open data, a semantic and photorealistic 3D city model in CityGML format, and future real-time simulation capabilities, it supports data-driven decision-making, efficient resource management, and enhanced resilience through climate mitigation modelling.

At the moment, Prague does not have a true digital twin with a bidirectional automatic data flow, but rather an advanced digital model (Fig. 5). While both tools are designed to represent the city reality, the twin aims are to allow direct control of city processes through a digital interface. The intermediate step in-between those concepts is a digital shadow, which allows real-time monitoring without enabling direct control.

Currently, OICT manages the dynamic data within the Golemio platform. Together with IPR, they are working towards integrating the Geoportal city model with real-time data to create a fully functional digital twin of Prague, featuring bidirectional automatic data flow.

At its core, the digital twin is built on a semantic 3D city model in CityGML format, ensuring compatibility with a variety of systems. Unlike standard 3D models, this semantic version is enriched with detailed information about urban elements, such as buildings, infrastructure, and public spaces. Additionally, the Prague photorealistic model offers highly detailed visual representations of the city's landscape.

A defining feature of Prague's digital twin is its commitment to open data and formats, allowing for diverse applications. By providing accessible and interoperable data, IPR enables developers, researchers, and policymakers to use the digital

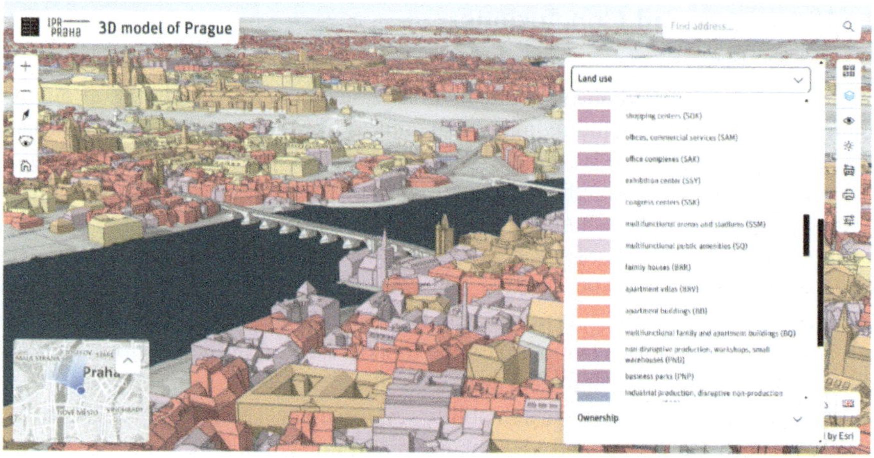

Fig. 5 Screenshot of the semantic 3D model of Prague from the geoportal platform (IPR Praha, n.d.)

model for everything from environmental modelling to infrastructure management, fostering innovation, and expanding its utility.

The use of the CityGML standard facilitates seamless data exchange and integration with other geospatial (GIS) and Building Information Modelling (BIM) systems.

Looking to the future, Prague aims to enhance its digital twin by incorporating real-time data. This will transform it from a static or semi-static model into a dynamic tool that reflects current conditions (aka a digital shadow). Real-time data on population movement, transportation, mobility, microclimates, and energy consumption will enable planners to perform up-to-date analyses of environmental and infrastructure parameters. This enhancement will further improve urban planning decisions aligned with sustainability goals.

In addition to supporting sustainable development, Prague's digital model also aids in resilience planning by allowing the simulation of climate mitigation and adaptation strategies. Planners can test and prioritize solutions before implementation, optimizing resource allocation, and increasing resilience against climate change.

Detailed mapping of public assets, including streetlights, benches, waste bins, traffic signs, and advertising structures, facilitates efficient asset management and maintenance planning. This helps to optimize resource use and extend the lifespan of the infrastructure. Furthermore, the digital model promotes public engagement by enabling citizens to visualize and interact with proposed developments through applications such as "Prague Tomorrow 3D", enhancing transparency and encouraging civic participation.

8 Discussion and Conclusions

Effective sustainable urban planning in Prague necessitates collaboration among multiple stakeholders, including municipal entities like the Prague Institute of Planning and Development (IPR), Prague Development Company (PDS), and Operátor ICT (OICT), who work together to align urban development projects with the city's strategic objectives and sustainability goals.

The use of advanced tools such as Geographic Information Systems (GIS), 3D models, and the Prague digital twin enhances data-driven decision making and resource management, supporting the planning and development of sustainable neighbourhoods—Table 3 summarized the tools presented in the previous sections along with their strengths and weaknesses.

Various Neighbourhood Sustainability Assessment (NSA) tools, including BREEAM, LEED, DGNB, and SNTool, provide frameworks for evaluating sustainability criteria such as energy efficiency, water management, and social inclusion, visually representing the sustainability level of neighbourhoods, guiding investment decisions, and highlighting commitments to sustainable development. These tools facilitate the visualization of an area's value, which is essential for both public and private stakeholders. However, most of the tools discussed in this chapter have limitations that must be acknowledged to achieve a more accurate overview. One limitation of these tools is that they are primarily, if not entirely, quantitative in nature. The problem with this approach is that, by reducing value to quantitative terms—often monetary—it compromises the broader significance of a place, overlooking potentially rich qualitative dimensions. This reliance on one-dimensional metrics can be misleading when informing public policies, as it implicitly assumes that certifications will automatically translate to increased wealth, education, environmental sustainability, health, and happiness. However, this perspective often frames these attributes as mere by-products of economic progress, rather than directly addressing the comprehensive needs of human beings.

To overcome this limitation, it would be beneficial to complement these tools with more qualitative approaches. Such a strategy would require active engagement of municipal authorities and public stakeholders in citizen science initiatives that contribute to the urban planning process.

Customized rating systems, such as the 'Sustainable Prague 8' system, ensure that urban projects meet local sustainability standards by evaluating criteria such as public spaces, architectural design, participation, and adaptation to climate change.

In recent years, significant advancements have enabled a deeper understanding of qualitative aspects such as emotions using social media data and artificial intelligence. Through advanced natural language processing applications, data streams from sources like Golemio and zmente.to could offer a more detailed and spatial perspective on what people value and need in their city. This approach helps validate subjective measures and bridges environmental and psychological factors.

Participatory planning is crucial to creating attractive and livable neighbourhoods that balance the interests of residents, developers, and municipality. While the current

Table 3 Overview of urban planning tools for the city of Prague

Tool	Description
International NATs (Neighbourhood assessment tools)	*Strengths:* Structured, clear frameworks that enable municipalities and developers to evaluate and highlight sustainability achievements in neighbourhoods, promote comparability and establish clear sustainability benchmarks in various aspects, making these tools valuable for informed decision-making *Weaknesses:* Many of these tools require significant adaptation for non-native regions (e.g., LEED in Europe), which can increase costs and complexity, and some systems demand extensive documentation and specialized knowledge, which can be resource-intensive and limit accessibility for broader applications
Prague sustainable neighbourhoods tool	*Strengths:* It covers global and local dimensions, addressing diverse topics while aiming to unify Prague's stakeholders—city actors, developers, and the public—under a shared vision *Weaknesses:* As a nonbinding tool, its influence may be limited compared to regulatory frameworks, potentially reducing its impact. The primary focus on new construction could also delay sustainability improvements in existing neighbourhoods, which may require different and more complex measures
Customized assessment tool (e.g., Sustainable Prague 8)	*Strengths:* A comprehensive and locally tailored framework for evaluating construction projects that addresses specific district goals while incorporating global standards *Weaknesses:* As a non-regulatory tool, its impact relies on voluntary adherence, potentially limiting its influence if developers do not prioritize sustainability measures beyond legal requirements. However, developers are incentivized to meet the assessment requirements due to the prospect of facilitating the building permit process or securing approval for the necessary modifications to the city master plan
Barcode methodology	*Strengths:* A flexible and context-specific tool for urban planning that balances housing needs with essential services, ensuring a holistic approach to city growth. Its visual and quantitative format helps planners and stakeholders understand spatial needs and optimize land use. Applicable on both city-wide and neighbourhood scales, it supports diverse urban projects, from preservation to expansion, while aligning with 10- or 15-min city principles *Weaknesses:* Its effectiveness relies heavily on skilled urban planners who can adapt it to local needs. Its quantitative focus may overlook qualitative factors, and its success depends on collaboration among public and private partners, which can be challenging in municipalities like Prague that rely on external planning expertise

(continued)

Table 3 (continued)

Tool	Description
Digital twin of the city	*Strengths:* This model promotes innovation and public participation by allowing accessible visualizations and simulations of urban changes *Weaknesses:* Prague's model lacks real-time interactivity, functioning more as an advanced digital model than as a fully interactive digital twin. Full realization will require further integration of real-time data, which involves technical and resource challenges

focus is on new construction, there are plans to extend sustainability efforts to existing neighbourhoods, which require different measures and adaptations.

The ongoing development of the Prague digital twin and the integration of real-time data will further enhance urban planning capabilities, supporting efficient resource management, climate mitigation, and public participation, contributing to the city's long-term sustainability goals.

Acknowledgements This research has been supported by the New European Bauhaus STAvangeR project (NEB-STAR) [2022–2025], under Grant Agreement No. 101079952, NEB-STAR// Beautiful. Sustainable. Together. (nebstar.eu).

References

Balletto G, Ladu M, Milesi A, Borruso G (2021) A Methodological approach on disused public properties in the 15-minute city perspective. Sustainability 13:593. https://doi.org/10.3390/SU13020593

Bertoldi P. (ed) (2018) Guidebook "How to develop a Sustainable Energy and Climate Action Plan (SECAP)." Luxembourg: Publications Office of the European Union.

BRE (2012) BREEAM Communities—Technical Manual 2012.

CASBEE CASBEE—Tools for urban scale. https://www.ibecs.or.jp/CASBEE/english/toolsE_urban.htm. Accessed 30 Oct 2024

DGNB (2020) DGNB Criteria set—Districts version.

European Union (2021) New European Bauhaus. https://new-european-bauhaus.europa.eu/index_en. Accessed 1 Oct 2024

Green Building Council of Australia (2016) Green Star—Communities. https://new.gbca.org.au/green-star/rating-system/communities/. Accessed 30 Oct 2024

Hainc J, Kandusová V, Sýkorová M, et al (2021) Methodology for planning public spaces of small municipalities

International Initiative for a Sustainable Built Environment (2020) SNTool. https://www.iisbe.org/taxonomy/term/344. Accessed 30 Oct 2024

Interreg Central Europe (2024) Supporting Cross-scale Planning and Policy readiness for a Resilient Built Environment (SuPeRBE). https://www.interreg-central.eu/projects/superbe/. Accessed 30 Oct 2024

Khavarian-Garmsir AR, Sharifi A, Sadeghi A (2023) The 15-minute city: Urban planning and design efforts toward creating sustainable neighbourhoods. Cities 132:104101. https://doi.org/10.1016/J.CITIES.2022.104101

Khodakarami L, Pourmanafi S, Mokhtari Z, et al (2023) Urban sustainability assessment at the neighbourhood scale: Integrating spatial modellings and multi-criteria decision making approaches. Sustain Cities Soc 97:104725. https://doi.org/10.1016/J.SCS.2023.104725

Lotteau M, Yepez-Salmon G, Salmon N (2015) Environmental assessment of sustainable neighborhood projects through NEST, a decision support tool for early stage urban planning. Procedia Eng 115:69–76. https://doi.org/10.1016/J.PROENG.2015.07.356

Lucchi E, Buda A (2022) Urban green rating systems: Insights for balancing sustainable principles and heritage conservation for neighbourhood and cities renovation planning. Renewable and Sustainable Energy Reviews 161:112324. https://doi.org/10.1016/J.RSER.2022.112324

Malá M, Sýkorová M, Felicioni L (2022) How municipalities should approach the transformation of public spaces. In: Acta Polytechnica CTU Proceedings. Czech Technical University in Prague, pp 635–641

NEB-STAR project (2023) What is Barcode? https://nebstar.eu/news/what-is-barcode/. Accessed 30 Oct 2024

NEB-STAR project (2024) D1.5 NEB-STAR twinning reports (initial version)—Deliverable 1.5. https://wp.innocode-cdn.com/nebstar.eu/uploads/2024/03/D1.5-NEB-STAR-Twinning-report-initial-version-FINAL-1.pdf. Accessed 15 Oct 2024

NEB-STAR (2022) NEB-STAR // Beautiful. Sustainable. Together. https://nebstar.eu/. Accessed 30 Oct 2024

Newstream.cz (2024) Byt na investici. Nejvíc se vyplácí koupě těch nejmenších v centru Prahy. https://www.newstream.cz/reality/byt-na-investici-nejvic-se-vyplaci-koupe-tech-nejmensich-v-centru-prahy. Accessed 30 Oct 2024

Noworól A, Kopyciński P, Hałat P et al (2022) The 15-minute city—the geographical proximity of services in Krakow. Sustainability 14:7103. https://doi.org/10.3390/SU14127103

OICT (n.d.) Lítačka portal | Lítačka. https://pidlitacka.cz/en. Accessed 30 Oct 2024b

OICT (n.d.) Golemio. https://golemio.cz/english. Accessed 30 Oct 2024

Ortiz-Fernández J, Astudillo-Cordero S, Quesada-Molina F (2023) Spatial neighbourhood sustainability assessment for urban planning, Cuenca, Ecuador. Environmental and Sustainability Indicators 20:100307. https://doi.org/10.1016/J.INDIC.2023.100307

Prague City hall—Environment Protection Department (2021) Prague Climate Plan 2030. Prague

Smart Prague (2023) Smart Prague Index. https://smartprague.eu/en-smart-prague-index. Accessed 30 Oct 2024

IPR Praha (n.d.) Geoportál Praha. https://geoportalpraha.cz/mapove-aplikace. Accessed 30 Oct 2024

Ramirez Aranda NE (2023) Geodesign for a Circular Economy in Urban Regions: Report on Master Students' Output. Delft University of Technology, Faculteit Bouwkunde

Sýkorová M, Tománek P, Šuš1íková L, et al (2021) VODA VE MĚSTĚ Metodika pro hospodaření s dešťovou vodou ve vazbě na zelenou infrastrukturu

Tam VWY, Karimipour H, Le KN, Wang J (2018) Green neighbourhood: Review on the international assessment systems. Renewable and Sustainable Energy Reviews 82:689–699. https://doi.org/10.1016/J.RSER.2017.09.083

Udržitelná Praha 8 (2022) Udržitelná Praha 8: postup a struktura hodnoticího systému. Nákladem městské části Praha 8 vydala Česká zemědělská univerzita v Praze. ISBN: 8021332328

USGBC (2018) LEED v4 for Neighbourhood Development

Utrecht Municipality (2021) Spatial Strategy Utrecht 2040. https://omgevingsvisie.utrecht.nl/dekoers/ruimtelijke-strategie-utrecht-2040. Accessed 21 Oct 2024

Utrecht Municipality (2021) Merwedekanaalzone. https://www.utrecht.nl/wonen-en-leven/bouwprojecten-en-stedelijke-ontwikkeling/bouwprojecten/merwedekanaalzone. Accessed 21 Oct 2024

Wenander H (2024) Whose visions and goals for sustainability? Planning, participation and depoliticisation in new urban neighbourhoods in Sweden. Cities 152:105202. https://doi.org/10.1016/J.CITIES.2024.105202

Zoghi M, Ferrari S, Paganin G, Dall'O' G (2024) A novel framework to measure and promote smartness in neighbourhoods. Sustain Cities Soc 101:105206. https://doi.org/10.1016/J.SCS.2024.105206

Netherlands3D—Digital Open Source-Twining in Co-creation

Christian Schachtner

1 Introduction: Project Goals and Vision

The Netherlands3D project was presented to the author on the basis of an organized project excursion of the German Federal Institute for Research on Building, Urban Affairs and Spatial Development (BBSR) as part of a EuroCities tandem visit.

One of the main goals of Netherlands3D is to create a uniform and consistent 3D representation of the entire Dutch territory. This is achieved by integrating various basic datasets, including the 3D BAG (Basisregistratie Adresseen Gebouwen), the 3D Basisvorziening and tree data from the BGT (Basisregistratie Grootschalige Topografie) (van der Meulen et al., 2020). This approach ensures coherent visualization that can serve as the basis for a wide range of projects and applications.

Another key objective of the project is to promote data integration and interoperability. Netherlands3D allows users to combine datasets from different sources and formats and present them in a unified 3D context (Stoter et al., 2021).

This ability to aggregate and visualize data provides significant value to decision-makers, urban planners, and other stakeholders who rely on comprehensive spatial information. The platform stands out for its flexibility and adaptability. Users can decide for themselves which data should be visualized and which functionalities they need. This user-centered approach differentiates Netherlands3D from other visualization tools and allows for a tailored experience for each project (de Kleijn et al., 2019).

An important goal of the project is also to promote cooperation and knowledge sharing. The platform actively invites people to participate in its further development, whether by developing new functionalities or integrating additional data sources (Zlatanova et al., 2018).

C. Schachtner (✉)
RheinMain University of Applied Sciences, Wiesbaden, Germany
e-mail: Christian.schachtner@hs-rm.de

C. Schachtner (ed.), *The European Smart City Movement*,
https://doi.org/10.1007/978-3-031-96700-9_24

This open and collaborative approach fosters innovation and ensures that the platform is continuously improved and adapted to the needs of users. Netherlands3D strives to be a powerful tool for supporting decision-making processes. The ability to visualize complex spatial data enables decision-makers to make informed decisions in areas such as urban planning, environmental management, and infrastructure development (Biljecki et al., 2022).

Another goal of the project is to improve the accessibility of 3D geodata. By providing a user-friendly platform that does not require specialized technical knowledge, Netherlands3D democratizes access to complex spatial information. This can lead to broader participation in spatial decision-making processes and promote public understanding of spatial relationships (Çöltekin et al., 2020).

The vision of Netherlands3D goes beyond pure data visualization. The project aims to play a central role in the digital transformation of Dutch spatial planning and management. By providing a unified platform for the visualization and analysis of spatial data, Netherlands3D can serve as a catalyst for innovative solutions in areas such as smart cities, climate adaptation, and sustainable urban development (Verbree et al., 2022).

2 Project Partners and the Sharing of Responsibilities

The Netherlands3D project is characterized by its unique ability to link different data sets from different sources and in different file formats and visualize them in a coherent 3D environment. This functionality allows users to present and analyze complex spatial information in an intuitive and engaging way (van der Meulen et al., 2021).

A central aspect of the project is the cooperation of various partners, who contribute their respective skills and resources. The main players include public institutions, research institutes and private companies in the field of geoinformatics and 3D modelling (Bakker & de Vries, 2022). The responsibilities are divided as follows:

1. Data Provision: Various government institutions, including the Kadaster (Dutch Land Registry) and the Centraal Bureau voor de Statistiek (CBS), provide basic data sets. These include, among others, the 3D BAG (basic registry addresses en Gebouwen) and the BGT (basic registry Grootschalige Topografie) with tree data[1]. Providing uniform and consistent data for the whole of the Netherlands is a core task of these partners (Stoter et al., 2020).
2. Technical Development: The development of the platform is driven by a team of software engineers and GIS specialists. This group is responsible for integrating different data formats, implementing visualization tools, and ensuring the usability of the platform (van Oosterom et al., 2019).

3. Scientific Support: Research institutions such as TU Delft and the University of Twente contribute to the continuous improvement and expansion of the platform with their expertise in the field of 3D GIS and urban modelling. They also conduct studies on the usability and added value of 3D visualizations in different application areas (Ledoux et al., 2021).
4. Application Development: Private companies from the GIS and IT sectors are responsible for developing specific applications and extensions. They work closely with end-users to create tailor-made solutions for different industries and use cases (de Kleijn et al., 2022).
5. Coordination and Project Management: The overall coordination of the project is in the hands of a central project office. This is responsible for coordination between the partners, adherence to schedules and quality assurance. It also organises regular meetings and workshops to promote knowledge sharing and collaboration between partners (van der Spek & van Nes, 2023).

A special feature of Netherlands3D is the openness to contributions and extensions by the user community. The project actively invites people to participate in the further development of the platform, whether by integrating new datasets or developing additional functionalities. This participatory approach not only fosters innovation, but also ensures that the platform meets the real needs of users (Brink et al., 2021).

The flexibility of Netherlands3D is also evident in the ability to create customized 3D environments for specific projects or organizations. This requires close collaboration between the technical developers and the respective end-users to integrate the desired functionalities and datasets (Jansen et al., 2020).

3 Project Successes and Future Developments

The Netherlands3D project has established itself as an innovative platform for the three-dimensional visualization of geodata in the Netherlands. As project manager, I would like to explain the successes to date and future development potential of this groundbreaking project. A key success of Netherlands3D lies in its ability to integrate different data sets from different sources and formats and present them in a coherent 3D environment. This functionality enables users to intuitively grasp and analyze complex spatial relationships. The platform supports common geospatial data formats such as GeoJSON and WFS, which underlines the interoperability and flexibility of the system (van der Meulen, 2024).

Another milestone is the nationwide availability of uniform and consistent 3D basic data. By integrating datasets such as the 3D BAG (Basisregistratie Adresseen Gebouwen), the 3D Basisvoorziening and tree data from the BGT (Basisregistratie Grootschalige Topografie), Netherlands3D provides a solid foundation for a wide range of applications (Bakker et al., 2023). This standardisation not only facilitates use, but also promotes the comparability of analyses and projects at national level.

The platform is particularly characterized by its user orientation. Unlike many other geodata viewers, users can decide for themselves which data should be visualized and which functionalities should be used. This flexibility makes it possible to optimally adapt Netherlands3D to the specific needs of different projects and organizations (de Vries, 2024).

An important aspect for the success of the project is the openness of the system to extensions and adjustments. Netherlands3D offers the possibility to create custom 3D environments, for example with customized starting locations or preset functionalities and datasets. This adaptability makes the platform attractive to a wide range of users, from urban planners to environmental researchers (Jansen & Pieterse, 2025).

For the future development of Netherlands3D, we see great potential in the increased integration of simulation and analysis tools. The platform can serve as the basis for computational models and simulation software or visualize their results. To take full advantage of these possibilities, we actively invite you to co-develop interfaces between Netherlands3D and various analysis tools (van den Berg, 2024).

Another focus of future development is the improvement of data quality and timeliness. Although uniform data is already available nationwide, we are continuously working to increase the accuracy and level of detail of the 3D models. This includes the integration of new data sources and the refinement of presentation methods (Smits & van der Linden, 2025).

Promoting collaboration between different stakeholders is another key goal for the future of Netherlands3D. By creating an open platform for the exchange of 3D geospatial data and analytics, we aim to intensify interdisciplinary collaboration in areas such as urban planning, environmental protection, and infrastructure development (de Groot et al., 2024).

Future development will focus on further expanding functionalities, improving data quality, and fostering collaboration between different users and disciplines. With these efforts, we are confident that Netherlands3D will continue to play a key role in spatial data visualization and analysis in the Netherlands in the future.

In conclusion, Netherlands3D has already achieved considerable success in the visualization and integration of 3D geospatial data. The platform has established itself as a valuable tool for various application areas.

4 Conclusion

Netherlands3D aims to create a powerful, flexible and collaborative platform for 3D visualization of spatial data. By providing a unified and consistent 3D image of the whole of the Netherlands, promoting data integration and interoperability, and supporting decision-making processes, the project aims to make a significant contribution to spatial planning and management in the Netherlands. The clear division of responsibilities and the open, collaborative structure of the project ensure continuous development and improvement of the platform. The open and user-centered approach of the project promises to make Netherlands3D an important tool for a wide range of stakeholders and to change the way we visualize and use spatial information in the long term. This makes Netherlands3D a pioneering project in the field of 3D geospatial data visualization that can serve as a model for similar initiatives in other countries.

References

Bakker, A., Jansen, P., & de Vries, M. (2023). Standardization of 3D geospatial data: The Netherlands3D approach. Journal of Spatial Information Science, 18(2), 45–62.

Bakker, L., & de Vries, M. (2022). Collaborative 3D city modeling: Insights from the Netherlands3D project. ISPRS International Journal of Geo-Information, 11(3), 156.

Biljecki, F., Lim, J., Crawford, J., Moraru, D., Tauscher, H., Konde, A., ... & Frantz, D. (2022). Extending 3D city models for urban digital twins: A systematic review of open-source tools and data. Computers, Environment and Urban Systems, 96, 101825.

Brink, L., Stoter, J., & Zlatanova, S. (2021). Establishing a national 3D geo-data asset in The Netherlands: Challenges and opportunities. Transactions in GIS, 25(2), 613–633.

Çöltekin, A., Lochhead, I., Madden, M., Christophe, S., Devaux, A., Pettit, C., ... & Hedley, N. (2020). Extended reality in spatial sciences: A review of research challenges and future directions. ISPRS International Journal of Geo-Information, 9(7), 439.

de Groot, R., van der Meer, K., & Bakker, L. (2024). Fostering interdisciplinary collaboration through 3D geospatial platforms. Urban Planning and Development, 42(3), 210–225.

de Kleijn, M., van Manen, N., Kolen, J. C. A., & Scholten, H. J. (2019). Towards a user-centric SDI framework for historical and heritage European landscape research. International Journal of Spatial Data Infrastructures Research, 9, 1–35.

de Kleijn, M., van Manen, N., Kolen, J., & Scholten, H. (2022). Enhancing urban planning using 3D geospatial technologies: A case study with Netherlands3D. Journal of Cultural Heritage, 53, 222–234.

de Vries, J. (2024). User-centric design in geospatial visualization: Lessons from Netherlands3D. International Journal of Cartography, 60(1), 78–95.

Jansen, E., Stoter, J., Ledoux, H., & Commandeur, T. (2020). Exploring the potential of 3D city models in the Netherlands: The what, why and how. International Journal of 3-D Information Modeling, 9(1), 1–24.

Jansen, M., & Pieterse, S. (2025). Customizable 3D environments for urban planning: A case study of Netherlands3D. Cities, 108, 103352.

Ledoux, H., Arroyo Ohori, K., Kumar, K., Dukai, B., Labetski, A., & Vitalis, S. (2021). CityJSON: A compact and easy-to-use encoding of the CityGML data model. Open Geospatial Data, Software and Standards, 6(1), 1–16.

Smits, B., & van der Linden, G. (2025). Enhancing 3D geospatial data quality: Challenges and opportunities. ISPRS International Journal of Geo-Information, 14(5), 235.

Stoter, J., Commandeur, T., Ledoux, H., Pijpker, S., Klooster, R., & Peters, R. (2021). 3D BGT: Where point clouds meet topology. International Archives of the Photogrammetry, Remote Sensing & Spatial Information Sciences, 46.

Stoter, J., Peters, R., Commandeur, T., Dukai, B., Kumar, K., & Ledoux, H. (2020). Automated reconstruction of 3D input data for noise simulation. Computers, Environment and Urban Systems, 80, 101424.

van den Berg, R. (2024). Integrating simulation tools with 3D visualization platforms: The Netherlands3D experience. Computers, Environment and Urban Systems, 95, 101825.

van der Meulen, F. (2024). Interoperability in 3D geospatial data: A review of current standards and practices. Transactions in GIS, 28(2), 324–341.

van der Meulen, M., Tuithof, N., & van Oosterom, P. (2021). Visualizing large-scale 3D geospatial data in web browsers. ISPRS International Journal of Geo-Information, 10(2), 62.
van der Meulen, M., Tuithof, N., van Oosterom, P., & Stoter, J. (2020). 3D Cadastral visualization in the Netherlands. Land Use Policy, 98, 104418.
van der Spek, S., & van Nes, A. (2023). Spatial analysis and 3D modeling for urban design: Lessons from the Netherlands3D project. Journal of Urban Design, 28(2), 239–258.
van Oosterom, P., Martinez-Rubi, O., Ivanova, M., Horhammer, M., Geringer, D., Ravada, S., ... & Gonçalves, R. (2019). Massive point cloud data management: Design, implementation and execution of a point cloud benchmark. Computers & Graphics, 78, 99–125.
Verbree, E., Zlatanova, S., van Oosterom, P., & Quak, W. (2022). The role of 3D in urban digital twins. Computers, Environment and Urban Systems, 96, 101831.
Zlatanova, S., Yan, J., Wang, Y., Diakité, A., Isikdag, U., Sithole, G., & Barton, J. (2018). Spaces in spatial science and urban applications—state of the art review. ISPRS International Journal of Geo-Information, 7(12), 456.

Smart, but Cybersecurity—Vulnerable Cities

Lasse Berntzen

1 Introduction

This chapter discusses safe and secure smart cities in the context of the changed cyber threat landscape caused by new threat actors and new threat vectors. A smart city is smart due to its use of technology to improve city operations. A city is a complex organization with numerous systems operating in parallel and communicating with each other. Such complex organizations are vulnerable to cyberattacks. Cyberattacks may target the information systems used for running the city, e.g., tax collection, payrolls, and accounting. However, what is more worrying is attacks against critical infrastructure.

2 Smart Cities

Most definitions of smart cities involve using information and communication technology (ICT) to improve the quality of life for their citizens through better services and operations and by reducing their environmental footprint. Citizen participation is often mentioned as an important characteristic of smart cities.

One often-cited definition proposed by Dameri (2013) is: "*A smart city is a well-defined geographical area, in which high technologies such as ICT, logistic, energy production, and so on, cooperate to create benefits for citizens in terms of well-being, inclusion and participation, environmental quality, intelligent development; it is governed by a well-defined pool of subjects, able to state the rules and policy for the city government and development.*"

L. Berntzen (✉)
School of Business, University of South-Eastern Norway, Horten, Norway
e-mail: lasse.berntzen@usn.no

C. Schachtner (ed.), *The European Smart City Movement*,
https://doi.org/10.1007/978-3-031-96700-9_25

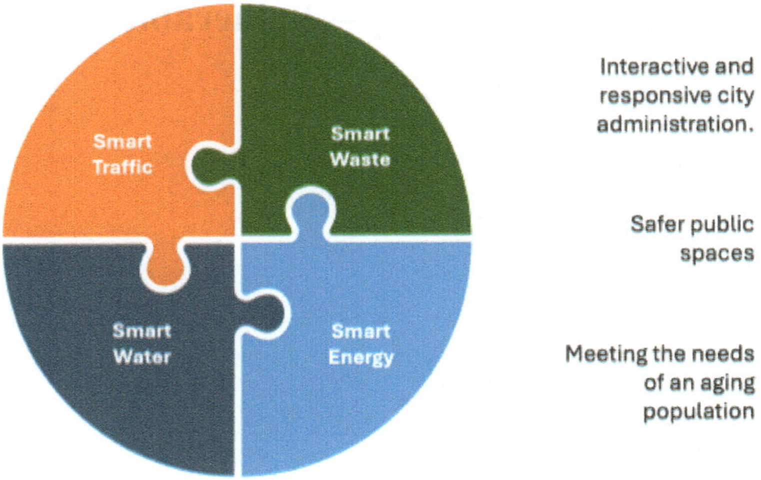

Fig. 1 Smart city characteristics (European Commission n.d.)

According to the European Commission (n.d.), a smart city is: "*a place where traditional networks and services are made more efficient with the use of digital solutions for the benefit of its inhabitants and businesses. A smart city goes beyond the use of digital technologies for better resource use and less emissions. It means smarter urban transport networks, upgraded water supply and waste disposal facilities and more efficient ways to light and heat buildings. It also means a more interactive and responsive city administration, safer public spaces and meeting the needs of an aging population.*"

This more recent definition focuses more on critical infrastructure such as transport networks, water supply, waste management, and energy. In the context of cybersecurity, critical infrastructure is vulnerable, and therefore, this chapter finds this definition highly relevant (Fig. 1).

However, citizen participation also relies on technology, and there is a danger of cyberattacks against technology that is intended to facilitate participation. Fake news is spread through social media, and voting platforms may be manipulated. Citizens, particularly elders, are targets for cyberattacks and disinformation or fake news (Wu et al. 2019).

3 IoT in Smart Cities

Smart cities are primarily driven by technology. In most cases, smart cities have several application areas where digitalization is used to improve the efficiency of city operations. The introduction of the Internet of Things, where new types of devices are connected to the Internet, is rapidly expanding. IoT Analytics estimates

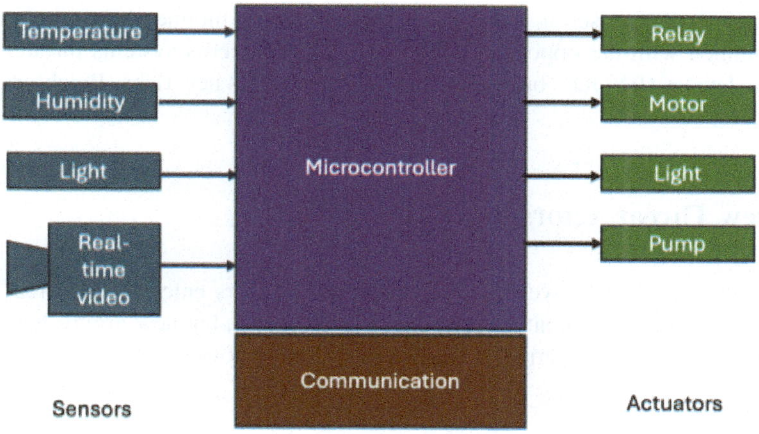

Fig. 2 IoT device

the number of connected devices to 19 billion at the end of 2024, and the number is increasing (Carrol 2024).

An IoT device typically consists of a set of sensors or actuators or a combination of both. Sensors provide information about the physical world around the IoT device, from simple measurements of temperature, humidity, and light to more complex data such as real-time video. Actuators make some impact on the physical environment around the IoT device. Typical actuators are lights, relays, motors, and pumps (Arshi and Mondal 2023). The IoT device also includes some kind of communication with the Internet, such as Wi-Fi, 5G, or LoRa. Since IoT devices are connected to the Internet, they are hackable. It means that threat actors may eavesdrop on the transmission of sensor data and may also take over actuators (Fig. 2).

4 New Cyberattack Threat Landscape

The threat landscape has changed—several developments have caused this.

- Smart cities are relying on IoT devices to collect information and control city functions (Syed et al. 2021).
- Systems are becoming more interconnected and complex. Combined with an increasing number of external partners and systems, the systems of smart cities have become more vulnerable to threats.
- Systems are no longer necessarily on-premises. The adoption of cloud services is increasing, and the systems are often run on computers in the cloud. Cloud services are vulnerable to cyberattacks (Mukhopadhyay 2022).

- Systems are accessed in new ways. The adoption of mobile phones and tablets, combined with the opportunity for employees to access systems through their own devices (BYOD—bring your own device), poses new risks (Ratchford et al. 2022).

5 New Threat Actors

During the last decade, we have seen new threat actors entering the threat landscape. The traditional threat actors were cybercriminals, but new groups have been appearing, such as cyberterrorists, hacktivists, foreign powers, and insiders.

5.1 Cybercriminals

Cybercriminals are mostly after financial gain. They aim to get the victim to pay. The primary forms of attacks are:

- Ransomware attacks where the victim's data is encrypted and thereby made unavailable to the city. The attacker asks for a ransom for a decryption key or some other way to restore the data.
- Attacks where privacy-related data is exfiltrated and the city is asked to pay for these data not to be released.
- Denial of service attacks make services unavailable to users, thereby causing problems for the citizens.
- IoT attacks where the infrastructure is attacked with the possibility to manipulate sensor data and control actuators, e.g., turning off power or changing traffic signals.

5.2 Cyberterrorists

According to Venkatachary et al. (2024), cyberterrorism involves the use of technology to inflict pain, devastation, or instability on society. This includes attacks on critical infrastructures. The difference between cyber criminals and cyber terrorists is their goals; cyber criminals want money, and cyber terrorists want to frighten citizens with their attacks.

5.3 *Hacktivists*

Hacktivists seek to raise attention to some cause. Hacktivists are driven by ideology or politics, often based on ethics or morals (Romagna and Leukfeldt 2024). One example is the current fight against tourism in several European cities. The protesters claim the city centers are taken over by tourists, or in fact, AirBnB, and that ordinary citizens cannot afford to live in the city centers anymore. Hacktivists are cyber protesters who use hacking to promote their cause. One common attack vector is to deface websites. Defacing means substituting an original landing page with a new one with a targeted message or hijacking the domain name server to send users to an alternative website.

5.4 *Foreign Powers*

Foreign powers want to demoralize or mislead the population by creating uncertainty or chaos. A recent report from the European Union Agency for Cybersecurity—ENISA (2024) reports on continuous campaigns against both member states and European Union institutions, bodies, and agencies. ENSIA points out that these campaigns remain a persistent and severe threat and observes that Russia-nexus and China-nexus threat groups remain prominent threats. The report mentions the European Parliamentary elections as a target for foreign powers trying to influence the public.

5.5 *Insiders*

Insiders are persons who already have access to systems that are to be attacked (Prabhu and Thompson 2021). Often, their rationale for the attack is a wish for revenge for some felt injustice at work. Some may be after financial gain. Insiders are that they know what is valuable and what is not. They also know the procedures.

6 Smart City Application Areas

A city is a complex organization with many responsibilities. A smart city uses technology to improve operations in many different sectors. Examples are technology used in healthcare and schools and administrative systems to handle day-to-day operations, such as resource planning or citizen relation management. However, in this chapter, we will focus on the definition by the European Commission, which mentions critical infrastructure such as traffic, energy, waste handling, and water supply.

6.1 Smart Traffic

Smart traffic is about optimizing roads and other infrastructure to avoid congestion and secure the flow of traffic. In most cities, traffic signals play an essential role. From a cybersecurity point of view, traffic control is a vulnerability. If all traffic lights are put on either red or green, chaos will happen. Cybersecurity will be even more critical when vehicles start communicating between themselves (Feng et al. 2022).

Public transport is also an important application area of smart cities. Public transport includes railways, metros, trams, and buses. Public transport is digitalized through monitoring and scheduling systems and payment systems. Disabling the public transport system functions will cause problems.

6.2 Smart Energy

Venkatachary et al. (2024) discusses the potential consequences of a successful cyberattack on the energy sector. Cyberattacks can be severe and long-lasting and may result in power outages, oil spills, or nuclear accidents, leading to economic loss, environmental damage, and even loss of life. He also observes that energy sector cyberattacks can also have cascading effects on other sectors, such as transportation, healthcare, and communication, leading to further disruptions and chaos.

Distribution grids have traditionally been protected by isolation from the Internet (separate networks). Energy distribution is a crucial infrastructure in an intelligent city. However, smart buildings make it possible to control end-user consumption through the Internet. Simultaneous attacks on several buildings at the same time may disrupt the energy distribution by triggering fault detection mechanisms that shut down power.

6.3 Smart Waste Handling

Also, waste handling has become digitalized. Smart waste bins inform when they are almost filled. Advanced scheduling systems optimize waste collection. Also, sorting waste into different fractions is part of smart waste handling. Naiho et al. (2024) made a critical review of cybersecurity considerations in waste management.

If waste management is targeted, the waste will soon pile up. We have seen from strikes that waste cause an outbreak of vermin and possible infections.

6.4 Smart Water

Water supply and energy distribution share some of the same characteristics. Water supply is a critical infrastructure and is essential to make society work. IoT devices control modern water supply systems and are vulnerable to cyberattacks. Tuptuk et al. (2021) made a systematic review of the state of cybersecurity in water systems.

Just as electricity, water is essential for the society to function. A loss of water supply may be fatal for a city. The water supply is increasingly digitalized. The water is tested, purified, and distributed. Most cities have some metering to measure the amount of water being used. Some possible attack vectors are:

- Manipulate sensor data to raise false alarms about water quality.
- Valves may be manipulated to close off the water supply.
- Manipulating purifying process, e.g., to introduce excess chlorine in the water.
- The meters may be manipulated to impose higher fees on citizens

7 Resilience and Mitigation

First, cities need to plan for cybersecurity attacks. Preparations can be done on different levels: Individual, organizational, and technological. On the individual level, both city employees and citizens should be aware of risks and possible threats. This requires basic digital competence enhanced with basic security training. On the organizational level, the city should have the necessary regulations, procedures, and guidelines in place. City employees, in particular, those who deal with information systems, should be trained for cybersecurity breaches. Finally, technology should be used to detect and prevent cyberattacks.

The first measure is to detect and stop cyberattacks through firewalls and intrusion detection and prevention systems. Secondly, it is necessary to have redundancy and manual procedures in case an attack should be successful (Fig. 3).

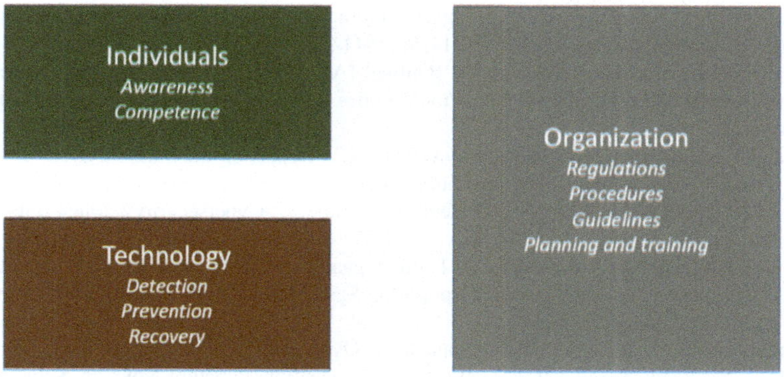

Fig. 3 Individual, organizational, and technological measures

7.1 Recommendations

- Establish a Security Operation Center (SOC) or collaborate with neighboring cities in establishing a SOC. Also, establish a high-level position as Chief Security Officer to plan and coordinate cybersecurity defenses. Incident response team with regular training
- Updated risk analysis and mitigation plans.
- The city should consider measures to protect critical infrastructure by isolating critical systems and creating necessary redundancy.
- Focus on targeted training at all organizational levels. Every city employee should know how to detect abnormal behavior and be aware of cybersecurity risks.
- The city should employ state-of-the-art intrusion detection and prevention.

8 Conclusion

Based on the European Commission's definition of smart cities, this chapter has focused on cybersecurity threats to essential city functions such as traffic control, energy, water supply, and waste handling. It also shows new threat actors and their motivations and methods. Finally, it provides a set of recommendations for smart cities.

References

Arshi O, Mondal S (2023) Advancements in sensors and actuators technologies for smart cities: a comprehensive review. Smart Constructions and Sustainable Cities, pp. 1–18. https://doi.org/10.1007/s44268-023-00022-2

Carrol M (2024) Connected iOT devices tipped to approach 19B in 2024. Mobile World Live. Sep. 3, 2024 [Online][Available from] https://www.mobileworldlive.com/industry/connected-iot-devices-tipped-to-approach-19b-in-2024/

Dameri RP (2013) Searching for Smart City definition: a comprehensive proposal. International Journal of Computers and Technology, 11(5): 2544-2551

European Commission (n.d.) Smart Cities. [Online] [Available from] https://commission.europa.eu/eu-regional-and-urban-development/topics/cities-and-urban-development/city-initiatives/smart-cities_en

European Union Agency for Cybersecurity (2024) 2024 Report on the State of Cybersecurity in the Union. [Online] [Available from] https://www.enisa.europa.eu/sites/default/files/2024-11/2024%20Report%20on%20the%20State%20of%20the%20Cybersecurity%20in%20the%20Union.pdf

Feng Y et al. (2022) On the Cybersecurity of Traffic Signal Control System With Connected Vehicles. IEEE Transactions on Intelligent Transportation Systems 23(9), pp. 16267-16279. https://doi.org/10.1109/TITS.2022.3149449

Mukhopadhyay I (2022). Cyber Threats Landscape Overview Under the New Normal. In: Fong, S., Dey, N., Joshi, A. (eds) ICT Analysis and Applications. Lecture Notes in Networks and

Systems, vol 314. Springer, Singapore. pp. 729–736. https://doi.org/10.1007/978-981-16-5655-2_70

Naiho HNN (2024). Cybersecurity considerations in the implementation of innovative waste management technologies: "A critical review". Computer Science & IT Research Journal, 5(6), 1408–1433. https://doi.org/10.51594/csitrj.v5i6.1225

Prabhu S, Thompson N (2021). A primer on insider threats in cybersecurity. Information Security Journal: A Global Perspective, 31(5) pp. 602–611. https://doi.org/10.1080/19393555.2021.1971802

Ratchford M et al. (2022) BYOD security issues: a systematic literature review. Information Security Journal 31(3), pp. 253-273. https://doi.org/10.1080/19393555.2021.1923873

Romagna M, Leukfeldt, RE (2024) Becoming a hacktivist. Examining the motivations and the processes that prompt an individual to engage in hacktivism. Journal of Crime and Justice 47(4), pp. 511–529. https://doi.org/10.1080/0735648X.2023.2216189

Syed A (2021) IoT in Smart Cities: A Survey of Technologies, Practices and Challenges. Smart Cities 4(2), pp 429-475. https://doi.org/10.3390/smartcities4020024

Tuptuk N (2021) A Systematic Review of the State of Cyber-Security in Water Systems. Water 13(1),81 https://doi.org/10.3390/w13010081

Venkatachary SK et al. (2024) Cybersecurity and cyber-terrorism challenges to energy-related infrastructures – Cybersecurity frameworks and economics – Comprehensive review, International Journal of Critical Infrastructure Protection 45:100677 https://doi.org/10.1016/j.ijcip.2024.100677.

Wu L et al. (2019) Misinformation in Social Media: Definition, Manipulation, and Detection. ACM SIGKDD Explorations Newsletter, 21(2), pp. 80-90.

South Tyrol—Smart Areas by SINFONIA Projects

Christian Schachtner

1 Introduction: Project Goals and Vision

Based on a guest visit to the Eurac Research Institute for Public Management in Bolzano, the following article was prepared.

At the heart of the SINFONIA project is the unique cooperation between the cities of Bolzano and Innsbruck. This cooperation is at the heart of the initiative and aims to achieve ambitious goals: (a) 40 to 50% reduction in primary energy consumption and a 20% increase in the share of renewable energy in two pilot districts. To achieve these goals, an integrated approach is being pursued that combines various measures. These include the energy-efficient renovation of over 100,000 m^2 of living space, the optimisation of the electricity grid and innovative solutions for district heating and cooling (Federal Ministry for Economic Affairs and Energy 2022).

A key aspect of the SINFONIA project is bridging the gap between demonstration and large-scale replication. To achieve this, much of the project will be dedicated to the transferability and scalability of the solutions deployed in the two pilot districts. A central goal is the definition of a limited number of district typologies and corresponding redevelopment models. These are intended to enable cities to easily assess their needs and develop efficient long-term renovation strategies (German Institute of Urban Affairs 2021).

To ensure the scalability and transferability of these models and typologies, they will be tested and validated not only in Innsbruck and Bolzano, but also in five "early adopter" cities. The inclusion of these additional cities makes it possible to test and adapt the solutions developed in different geographical, climatic and socio-economic conditions (Association of Municipal Enterprises 2023).

C. Schachtner (✉)
Vice President Education and Sustainability, RheinMain University of Applied Sciences, Wiesbaden, Germany
e-mail: Christian.schachtner@hs-rm.de

C. Schachtner (ed.), *The European Smart City Movement*,
https://doi.org/10.1007/978-3-031-96700-9_26

Another important goal of the SINFONIA project is the creation of a "replication cluster". This cluster is intended to serve as a platform for the exchange of knowledge and the dissemination of good practice. Cities interested in smart solutions for the redevelopment of urban districts can join this cluster and benefit from the experience gained from Bolzano and Innsbruck as well as other participating cities. By joining the replication cluster, cities become part of a community of like-minded actors and are given the opportunity to actively participate in knowledge exchange (Federal Environment Agency 2024).

The vision of the SINFONIA project goes beyond the immediate goals of saving energy and increasing the share of renewable energy. It strives to initiate a paradigm shift in urban development where sustainability and quality of life go hand in hand. By integrating different technological solutions and involving all relevant stakeholders—from citizens to energy regulators—the project aims to develop holistic and sustainable city models (Fraunhofer Institute for Solar Energy Systems ISE 2023).

The importance of the SINFONIA project is underlined by the context of increasing urbanisation in Europe. With 80% of European citizens living in urban areas, cities play a crucial role in the transition to a low-carbon economy. They are faced with the challenge of ensuring the quality of life of their citizens and at the same time becoming more energy-efficient. The SINFONIA project addresses exactly this challenge by encouraging cities to think at the system level and to develop integrated urban development strategies (Federal Institute for Research on Building, Urban Affairs and Spatial Development 2022).

2 Project Partners and Co-Creation

The main actors of the SINFONIA project are the cities of Bolzano and Innsbruck, which are pioneers and working closely together to implement and test innovative energy solutions (European Commission 2020). This collaboration is a prime example of co-creation in the field of sustainable urban development, where different stakeholders work together to develop and implement solutions.

The measures in the pioneer cities include the renovation of more than 100,000 m^2 of living space, the optimisation of the electricity grid and solutions for district heating and cooling. These integrated approaches demonstrate the importance of taking a systemic approach to the energy transition in cities, taking into account different sectors (Loorbach et al. 2016).

A key aspect of the SINFONIA project is the transferability and scalability of the solutions used in the two pioneer districts. To achieve this, SINFONIA defines a limited number of district typologies and corresponding renovation models. These are intended to enable cities to easily assess their needs and define efficient long-term renovation strategies (Mosannenzadeh et al. 2017).

The validation of these models and typologies is not only taking place in Innsbruck and Bolzano, but also in five "early adopter" cities that are actively participating in the SINFONIA project: Paphos (Cyprus), Rosenheim (Germany), Seville (Spain), La Rochelle (France) and Borås (Sweden). This broad geographical distribution makes it possible to test and adapt the solutions under different climatic, cultural and regulatory conditions (Puerari et al. 2018). The inclusion of these "early adopter" cities is a crucial step in bridging the gap between demonstration and large-scale replication. Through the active participation of these cities, potential obstacles and challenges in the transfer of the solutions can be identified and addressed at an early stage (Nevens et al. 2013).

Another important aspect of the co-creation process in the SINFONIA project is the involvement of all relevant stakeholders—from public and private actors to citizens and energy regulators. This participatory approach ensures that the solutions developed are not only technically feasible, but also socially acceptable and economically viable (Frantzeskaki and Kabisch 2016).

To further promote knowledge exchange and replication, SINFONIA has launched a "Replication Cluster". Cities interested in smart neighbourhood redevelopment solutions can join this cluster and benefit from the experience gained in Bolzano and Innsbruck (and beyond). This knowledge-sharing platform is an essential part of the project's co-creation strategy, as it allows insights and best practices to be disseminated beyond the immediate project partners (Voytenko et al. 2016).

The cooperation between the various partner cities and stakeholders within the framework of the SINFONIA project clearly shows how important co-creation is for the successful implementation of sustainable urban development projects. By co-developing and testing solutions, cities can learn from each other, use resources efficiently, and put innovative approaches into practice more quickly (Bulkeley et al. 2016).

3 The SINFONIA Project in Bolzano

The SINFONIA project in Bolzano represents a pioneering approach to sustainable and intelligent urban development. Bolzano, the capital of South Tyrol, has set itself the ambitious goal of becoming a pioneer in smart urban development with the SINFONIA project (European Commission 2020).

The project focuses on improving energy efficiency and reducing CO_2 emissions, taking a holistic approach that includes both technological and social innovations (City of Bolzano 2023). One focus of the project is the energy-efficient renovation of residential buildings. Innovative technologies and materials were used to significantly reduce energy consumption and at the same time increase living comfort (Housing Institute South Tyrol 2024).

Particularly noteworthy is the installation of photovoltaic systems on the roofs of the renovated buildings, which make a significant contribution to self-generation of electricity (Renewable Energy Agency 2022). Another central aspect is the optimisation of the urban energy network. This enables better adaptation to fluctuating energy requirements and contributes to the stability of the power grid. By integrating smart grid technologies, a more efficient distribution and use of energy could be achieved (Association of Electrical, Electronic and Information Technologies 2023). Progress has also been made in the field of renewable energies, in particular through the integration of solar thermal and geothermal energy into the urban energy concept (German Federal Association of Renewable Energy 2023).

Particular attention was paid to the involvement of citizens. Workshops and information events helped to raise awareness of energy efficiency and sustainable living among the population. Through extensive information and participation campaigns, a high level of acceptance for the measures was achieved (Bolzano City Planning Office 2023).

Numerous innovative technologies were tested and implemented as part of the project. These include intelligent building control systems that optimize energy consumption in real time, as well as advanced thermal insulation materials that drastically reduce heating energy requirements (Fraunhofer Institute for Building Physics 2024).

Of course, the implementation of such a comprehensive project also brought challenges. One of the biggest was the coordination of the various stakeholders, from apartment owners to energy suppliers and municipal authorities (German Institute of Urban Affairs 2022). However, efficient decision-making processes have been established through the establishment of a central project steering group and regular coordination rounds. Another challenge was the integration of the new technologies into the existing infrastructure. Innovative solutions were needed here, for example, to renovate historic buildings in terms of energy efficiency without impairing their architectural character (Federal Monuments Office Austria 2023).

4 Results and Perspectives

The results of the SINFONIA project so far are extremely promising. The energy-efficient renovation of residential buildings has led to an average energy saving of 40% (Energy Agency South Tyrol 2024b, a). CO_2 emissions have been reduced by around 30% in the project area, which represents a significant contribution to climate protection (Federal Environment Agency 2023).

In addition, the project has provided important insights for future urban development. The successful integration of renewable energies, energy efficiency measures and intelligent control systems shows how modern cities can be made more sustainable and livable (German Institute of Urban Affairs 2024).

5 Conclusion

To sum up, the goals and visions of the SINFONIA project go far beyond the technical aspects of energy saving and the promotion of renewable energies. It strives to bring about a comprehensive change in the way we plan, build and live our cities. The focus on partnership and co-creation can make an important contribution to the development of sustainable and liveable cities in Europe. The experience and lessons learned from this project will undoubtedly make a valuable contribution to addressing the challenges of urbanisation and climate change. By combining technological innovation, citizen participation and strategic planning, the project aims to create a model for sustainable urban development that can be replicated across Europe and beyond (German Energy Agency 2025).

The SINFONIA project in Bolzano impressively demonstrates how sustainable urban development can be achieved through the combination of technological innovations, citizen-oriented planning and holistic concepts. The experience gained and the solutions developed serve as a model and can serve as a blueprint for similar projects in other European cities. SINFONIA shows that the path to a climate-neutral and liveable city is challenging, but quite feasible.

References

Agency for Renewable Energies. (2022). *Potentials of photovoltaics in urban areas*. Berlin: AEE.

Association of Electrical, Electronic and Information Technology. (2023). *Smart Grids: Technologies for the Energy Transition*. Frankfurt am Main: VDE Verlag.

Association of Municipal Enterprises. (2023). Municipal Energy Supply in Transition: Challenges and Opportunities.

Bolzano City Planning Office. (2023). *Citizen participation in urban development: Experiences from the SINFONIA project*. Bolzano: City of Bolzano.

Bulkeley, H., Coenen, L., Frantzeskaki, N., Hartmann, C., Kronsell, A., Mai, L., & Voytenko Palgan, Y. (2016). Urban living labs: Governing urban sustainability transitions. Current Opinion in Environmental Sustainability, 22, 13-17

City of Bolzano. (2023). *SINFONIA: A milestone for Bolzano's sustainable future*. Bolzano: Stadtverlag.

Energy Agency South Tyrol. (2024). *Evaluation report: Energy savings through building renovation in Bolzano*. Bolzano: EAS.

European Commission. (2020). *Smart Cities and Communities: The European Way*. Brussels: EU publications.

German Federal Association of Renewable Energy. (2023). *Integration of renewable energies into urban energy concepts*. Berlin: BEE.

Federal Environment Agency. (2023). *CO2 balancing of urban renovation projects*. Dessau-Roßlau: UBA.

Federal Environment Agency. (2024). Urban Transformation: Pathways to Sustainable Cities.

Federal Institute for Research on Building, Urban Affairs and Spatial Development. (2022). Integrated urban development in the face of climate change.

Federal Ministry for Economic Affairs and Energy. (2022). Energy efficiency in buildings: key to the energy transition.

Federal Monuments Office Austria. (2023). *Guidelines for the energy-efficient renovation of historic buildings*. Vienna: BDA.
Frantzeskaki, N., & Kabisch, N. (2016). Designing a knowledge co-production operating space for urban environmental governance—Lessons from Rotterdam, Netherlands and Berlin, Germany. Environmental Science & Policy, 62, 90-98.
Fraunhofer Institute for Building Physics. (2024). *Innovative insulation materials and their application in residential construction*. Stuttgart: Fraunhofer IRB Verlag.
Fraunhofer Institute for Solar Energy Systems ISE. (2023). Integrated Energy Systems for the City of the Future.
German Energy Agency. (2025). Future of cities: Energy-efficient and liveable neighbourhoods.
German Institute of Urban Affairs. (2021). Sustainable Urban Development: Concepts and Strategies for the Future.
German Institute of Urban Affairs. (2022). *Challenges in the implementation of smart city projects*. Berlin: Difu.
German Institute of Urban Affairs. (2024). *Forward-looking urban development: lessons from the SINFONIA project*. Berlin: Difu.
Housing Institute South Tyrol. (2024). *Energy-efficient renovation in social housing: best practices from Bolzano*. Bolzano: WOBI.
Loorbach, D., Frantzeskaki, N., & Avelino, F. (2016). Sustainability transitions research: Transforming science and practice for societal change. Annual Review of Environment and Resources, 41, 441-466.
Mosannenzadeh, F., Di Nucci, M. R., & Vettorato, D. (2017). Identifying and prioritizing barriers to implementation of smart energy city projects in Europe: An empirical approach. Energy Policy, 105, 191-201.
Nevens, F., Frantzeskaki, N., Gorissen, L., & Loorbach, D. (2013). Urban Transition Labs: co-creating transformative action for sustainable cities. Journal of Cleaner Production, 50, 111-122.
Puerari, E., De Koning, J. I., Von Wirth, T., Karré, P. M., Mulder, I. J., & Loorbach, D. A. (2018). Co-creation dynamics in urban living labs. Sustainability, 10(6), 1893.
Voytenko, Y., McCormick, K., Evans, J., & Schliwa, G. (2016). Urban living labs for sustainability and low carbon cities in Europe: Towards a research agenda. Journal of Cleaner Production, 123, 45-54.

Greece—Smart District Development Projects

Christian Schachtner

1 Introduction: Project Objectives

This article was based on information presented by representatives of the procet at the SCCON Germany Conference in 2023.

The main goal of the Ellinikon project is to create a sustainable and smart city that balances the needs of people, the environment and the economy. A central concept in this is the "15-minute city", where all essential facilities and services are within 15 min on foot or by bike (Amoledo 2024). This not only improves the quality of life of the residents, but also reduces traffic and the associated environmental pollution.

Another important goal is the integration of state-of-the-art technologies to improve the daily lives of residents and optimize the efficiency of urban infrastructure. This includes smart energy management systems, advanced connectivity through 5G and fiber optic networks, and the use of IoT technologies to manage resources such as water and waste (Digital Chiefs 2024).

The first phase of the project, which is expected to be completed in late 2026 or early 2027, includes key infrastructure and utility projects. These include the laying of the main road underground and the construction of Greece's first green high-rise. The entire project is expected to be completed by 2036, which means a gradual transformation of the area over a period of 15 years. Job creation and the promotion of the local economy are also central concerns of the project. Ellinikon is expected to create up to 85,000 new jobs and attract over a million tourists annually, resulting in a significant increase in tax revenues (Handelsblatt Global Edition 2024).

C. Schachtner (✉)
Vice President Education and Sustainability, RheinMain University of Applied Sciences, Wiesbaden, Germany
e-mail: Christian.schachtner@hs-rm.de

C. Schachtner (ed.), *The European Smart City Movement*,
https://doi.org/10.1007/978-3-031-96700-9_27

2 Partners

The Ellinikon project is supported by a network of various partners and stakeholders. At the forefront is Lamda Development, which is responsible for the overall conception and implementation of the project as the main developer. The company works closely with international architecture and planning firms to realize the vision of a forward-looking smart city. The design of the green spaces, in particular Ellinikon Park, which is to become the largest coastal park in Europe, is the responsibility of the renowned landscape architecture firm Sasaki in collaboration with the local office doxiadis + (Designboom 2024). This partnership ensures that international expertise is combined with local knowledge to develop a sustainable and ecologically valuable green space concept.

In the field of technological infrastructure, the project cooperates with leading technology companies. Although specific partnerships are not mentioned in the available sources, it is safe to assume that companies such as Axians, which have experience in implementing smart city solutions, could play an important role. The Greek government is also an important partner in the project. Their support is crucial for the realization of a project of this magnitude, especially in terms of legal frameworks and integration into existing urban structures (Tomorrow City 2024).

In addition, various social institutions are involved in the project. An example of this is the recently completed Centre for the Care of Persons with Disabilities, which aims to serve as a model for similar facilities across Europe. These partnerships underline the goal of making Ellinikon an inclusive city for all population groups. Cooperation with educational institutions and research institutes is another important aspect of the project. Although no specific partnerships are mentioned in the available sources, it is likely that collaborations with universities and research centers exist to develop and implement innovative solutions to the challenges of urban development (Wiwo 2024).

3 Vision

Ellinikon's vision is based on the concept of a 15-minute city, where all essential facilities and services are within a quarter of an hour on foot or by bike. This approach not only promotes the physical activity of the residents, but also contributes to the reduction of CO_2 emissions. Street design prioritizes pedestrians and cyclists, which represents a paradigm shift in urban mobility. A central element of the project is the planned coastal park, which will be Europe's largest of its kind with an area of 2 million square meters. This park will not only be a green lung for the city, but will also serve as a model for smart park management. Equipped with adaptive lighting systems and environmental monitoring technologies, the park will be an example of the integration of nature and technology in urban space (Falstaff Travel 2022).

Ellinikon's technological infrastructure sets new standards for smart cities. A nationwide 5G network and public Wi-Fi, combined with a state-of-the-art fiber optic network, form the backbone for unparalleled connectivity. An innovative element of the project is the development of an AI-powered, user-friendly app that bundles all digital and physical offerings in Ellinikon. From bike rental to bill payment, all services are to be easily accessible via smartphone, which should significantly increase the quality of life and efficiency for the residents. Smart buildings equipped with IoT applications promise to improve the daily lives of residents through automated and demand-based control of energy consumption, lighting and other home technologies (Lamda Development 2024).

A special focus is on sustainability and resource efficiency. Smart grid infrastructures to monitor and control energy consumption aim to reduce consumption and carbon emissions. IoT technologies are being used to efficiently manage water and waste, maximize recycling, and support renewable energy integration. The goal is to create a city that is self-sufficient in terms of irrigation and electricity demand and has a net-zero carbon footprint. Mobility in Ellinikon will be characterized by a wide range of environmentally friendly options. In addition to smart parking aids, there will be a special focus on micromobility, with a comprehensive range of bicycles and electric scooters. The nationwide implementation of electric vehicles and the corresponding charging infrastructure underline the commitment to a low-emission future (Redaktionsnetzwerk Deutschland 2024).

Ellinikon promises to be more than just a city—it is intended to become a model for future urban developments worldwide. With its combination of advanced technology, sustainable planning and human-centered design, Ellinikon could indeed pave the way for a new era of urban development where quality of life, environmental protection and economic prosperity go hand in hand (Schäfer 2024).

4 Smart Sustainability: A Key Element

The integration of smart technologies and sustainability concepts is at the heart of this in order to create a city that is not only technologically advanced, but also environmentally responsible. A central element of smart sustainability in Ellinikon is intelligent energy management. Through the use of smart grids and IoT technologies, energy consumption is monitored and controlled to maximize efficiency and reduce carbon emissions. These approaches are in line with current trends in smart city development, which are increasingly relying on the integration of sustainability goals and digital solutions (Koch 2023).

Mobility in Ellinikon is another focus of the smart sustainability strategy. Widespread use of electric vehicles, smart parking aids, and micromobility options such as bicycles and electric scooters will promote environmentally friendly transportation. These measures aim to reduce CO_2 emissions while improving the quality of life of residents, which is in line with the goals of many smart city initiatives in Germany (German Federal Government 2024).

A special feature of Ellinikon is the "smart" park, the first of its kind in Greece. Equipped with adaptive lighting systems and environmental monitoring technologies, it embodies the symbiosis of technology and nature. Such innovative green spaces play an important role in sustainable urban development and can serve as a model for other European cities (Bowen et al. 2017).

Resource efficiency is promoted in Ellinikon through the use of IoT technologies to manage water and waste, as well as to maximize recycling and renewable energy integration Amoledo (2024). These approaches are in line with the recommendations of recent studies highlighting the importance of digital solutions for improving urban climate protection (BBSR 2024).

An often overlooked aspect of smart sustainability is the social dimension. Ellinikon addresses this by creating barrier-free access to all digital and physical offerings via an AI-powered app. This concept of digital inclusion is an important part of sustainable smart city strategies and contributes to the creation of a more equitable urban environment (German Energy Agency 2023).

Ellinikon's development as a sustainable smart city is in line with global efforts to achieve the United Nations' Sustainable Development Goals (SDGs). The project's integrative approach, which takes into account ecological, economic and social aspects, is in line with the holistic understanding of sustainability in the SDGs (Koch 2023).

Despite the promising concept, projects like Ellinikon face challenges. The shortage of skilled workers, which has been observed in similar projects in Germany, could also be an obstacle here (German Federal Government 2024). In addition, it must be ensured that technological innovations actually lead to measurable improvements in terms of sustainability and quality of life and do not just serve as "greenwashing". The success of the project will depend on how well it succeeds in reconciling technological innovations with the needs of residents and the requirements of sustainable development.

5 Conclusion

In conclusion, Ellinikon is an ambitious example of the integration of smart technology and sustainability in urban development. Overall, it is clear that the Ellinikon project is based on a broad network of partners and stakeholders who contribute their respective expertise to realize the vision of a forward-looking, sustainable and liveable smart city. The combination of private sector engagement, government support and the involvement of various social groups forms the basis for the ambitious attempt to reshape urban life in Greece and beyond. As Europe's largest smart city, Ellinikon has the potential to provide important insights for the future development of sustainable urban spaces. The project's comprehensive approach, incorporating advanced technologies across various sectors such as energy management, transportation, and waste management, sets a new benchmark for urban planning.

References

Amoledo. (2024). Greece is building Europe's largest smart city: Ellinikon. https://amoledo.com/de/news/5224-griechenland-baut-die-groesste-smart-city-europas-ellinikon

Digital Chiefs. (2024). Europe's largest smart city is being built in Greece. https://www.digital-chiefs.de/europas-groesste-smart-city-entsteht-in-griechenland/

Handelsblatt Global Edition (2024). Insight Innovation: Europe's largest smart city is being built in Athens. https://www.handelsblatt.com/technik/forschung-innovation/insight-innovation-europas-groesste-smart-city-entsteht-in-athen-01/100042244.html

Tomorrow City. (2024). Ellinikon, Greece's futuristic smart city project in an abandoned airport. https://www.tomorrow.city/ellinikon-greeces-futuristic-smart-city-project-in-an-abandoned-airport/

Designboom. (2024). The ellinikon takes shape: what we know so far about greece's smart city. https://www.designboom.com/architecture/the-ellinikon-takes-shape-greece-15-minute-smart-city-athens-05-18-2024/

Wiwo. (2024). A ghost town becomes Europe's largest smart city. https://www.wiwo.de/technologie/wirtschaft-von-oben/wirtschaft-von-oben-263-smart-city-in-griechenland-aus-einer-geisterstadt-wird-europas-groesste-smart-city/29812946.html

Falstaff Travel. (2022). Ellinikon: Greece plans "Smart City" on the Athens coast. https://www.falstaff-travel.com/news/ellinikon-griechenland-plant-smart-city-an-der-athener-kueste/

Lamda Development. (2024). Europas größte Smart City—Ellinikon. Smart Country Berlin. https://www.smartcountry.berlin/de/newsblog/europas-groesste-smart-city-%E2%80%93-ellinikon.html

Redaktionsnetzwerk Deutschland. (2024). Athens gets a green heart: The Ellinikon development project. https://www.rnd.de/wirtschaft/athen-bekommt-ein-gruenes-herz-das-entwicklungsprojekt-ellinikon-ECB7NDNY2RDORL6XI3LENUKPL4.html

Schäfer, S. (2024). Ellinikon: Groundbreaking ceremony for skyscraper Riviera Tower. bi-medien.de. https://bi-medien.de/fachzeitschriften/baumagazin/hochbau/bauprojekt-ellinikon-sieben-fakten-ueber-die-neue-smart-city-an-athens-kueste-b15547

Bowen, K. J., Cradock-Henry, N. A., Koch, F., Patterson, J., Häyhä, T., Vogt, J., & Barbi, F. (2017). Implementing the "Sustainable Development Goals": towards addressing three key governance challenges—collective action, trade-offs, and accountability. Current Opinion in Environmental Sustainability, 26, 90-96.

German Federal government. (2024). Answer of the Federal Government to the minor question of the CDU/CSU parliamentary group (20/13972). Printed matter 20/14210. Berlin: German Bundestag.

German Energy Agency. (2023). International Environment Analysis on Smart Cities. Berlin: dena.

Koch, F. (2023). Actors and their contributions to the great transformation in selected fields of action. In: Sustainable Spatial Development for the Great Transformation (pp. 153–176). Hannover: Academy for Spatial Development in the Leibniz Association.

German Federal Institute for Research on Building, Urban Affairs and Spatial Development (BBSR). (2024). Information on spatial development: Smart city development in Germany. Bonn: BBSR.

France—Smart City Cluster Based on Pioneer Cities

Christian Schachtner

1 Introduction

The Smart City movement in France was brought closer to this through contact with the Leo Portal in a podcast conversation in exchange, parts of which were incorporated into this.

The main objective of this paper is to develop a comprehensive understanding of smart city initiatives in France and to identify best practices that can serve as a model for other cities (Giffinger et al., 2007). This is done through five cities that stand out for their innovative approaches and successful implementations of smart city concepts: Montpellier, Bordeaux, Strasbourg, Roubaix and Nice (Ministère de la Transition écologique et de la Cohésion des territoires, 2024).

Another important goal is to analyze the specific technologies and strategies that these cities use to improve the quality of life of their citizens while increasing the efficiency of urban services (Caragliu et al., 2011). Areas such as intelligent mobility, sustainable energy supply, digital administration and citizen participation are to be addressed.

Research into the success factors and challenges in the implementation of smart city projects is another focus of this work. It examines which framework conditions and resources are necessary to successfully transform a city into a smart city (Nam & Pardo, 2011). Technological, social and economic aspects must be taken into account.

Special attention is paid to the sustainability and environmental friendliness of the implemented solutions. Smart cities should not only be technologically advanced, but also make a positive contribution to climate protection and resource efficiency (Bibri & Krogstie, 2017).

The promotion of knowledge exchange and cooperation between cities is another central concern of our project. Creating a platform where the participating cities

C. Schachtner (✉)
RheinMain University of Applied Sciences, Wiesbaden, Germany
e-mail: Christian.schachtner@hs-rm.de

C. Schachtner (ed.), *The European Smart City Movement*,
https://doi.org/10.1007/978-3-031-96700-9_28

can share their experiences and learn from each other seems expedient (Angelidou, 2014). This is intended to help accelerate and optimize the development of smart cities across France.

2 Project Partners and Co-creation

To achieve these ambitious goals, a large number of partners are working together. First and foremost, of course, are the city administrations of the five selected cities. Their support and commitment are crucial to the success of the project. They provide us with insight into their strategies, challenges and successes, enabling an in-depth analysis of their smart city initiatives.

In addition, they cooperate closely with leading technology companies involved in the development and implementation of smart city solutions. These partners bring valuable technical know-how and help us understand the latest trends and innovations in smart urban development (Hollands, 2008).

Research institutions and universities are also important partners in the project. They support us in the scientific analysis and evaluation of the collected data and help to place our results in a broader academic context (Albino et al., 2015).

Last but not least, civic organizations and local communities are also involved. Their perspectives and experiences are essential to understand how smart city initiatives affect the daily lives of city dwellers and what citizens' needs and expectations are for their city (Cardullo & Kitchin, 2019).

Through the cooperation of these diverse partners, a holistic picture of smart city development in France can be drawn. In this respect, the aim is not only to document the status quo, but also to provide impetus for the future development of smart cities in France and beyond.

3 Challenges

One of the biggest challenges in the implementation of smart city concepts is the integration of different technological systems into the existing urban infrastructure (Giffinger et al., 2023). This requires not only significant financial investment, but also close cooperation between public and private actors. In Montpellier, for example, an innovative traffic management system has been implemented that uses real-time data to optimize traffic flow (Ville de Montpellier, 2024). The challenge was to integrate different data sources while ensuring the privacy of citizens.

Another critical aspect is the digital divide that can arise between different populations (Caragliu & Del Bo, 2022). While smart city technologies have the potential to improve the quality of life, there is a risk that certain groups, especially the elderly or socio-economically disadvantaged citizens, will be excluded from these benefits.

In Bordeaux, this problem has been addressed with targeted training programmes and the creation of digital learning centres (Métropole de Bordeaux, 2024).

Cybersecurity is another significant challenge. As urban systems become more connected, so does vulnerability to cyberattacks (European Union Agency for Cybersecurity, 2023). In this context, Strasbourg has set up an advanced cybersecurity centre that not only protects urban systems, but also serves as a centre of excellence for other French cities (Eurométropole de Strasbourg, 2024).

Despite these challenges, smart cities offer enormous opportunities for sustainable and liveable urban development. A key benefit is the ability to use resources more efficiently and reduce the environmental footprint (United Nations, 2023). In Roubaix, for example, a smart energy management system was introduced that was able to reduce the energy consumption of public buildings by 30% (Ville de Roubaix, 2024). Such initiatives not only contribute to climate protection, but also reduce urban operating costs.

Improving citizen participation is another important aspect of smart cities. Through digital platforms and apps, citizens can participate more directly in decision-making processes and provide feedback on urban services (Meijer & Bolívar, 2022). Dijon has pioneered this area with the launch of a comprehensive citizen app, ranging from reporting infrastructure issues to online voting on local projects (Métropole de Dijon, 2024).

In the healthcare sector, smart city technologies are opening up new possibilities for improved medical care. Telemedicine and networked health systems can improve access to medical expertise, especially in rural areas (World Health Organization, 2024). In Montpellier, a pilot project has been launched for a networked healthcare system that digitally connects hospitals, doctors' surgeries and nursing services, enabling more efficient patient care (Agence Régionale de Santé Occitanie, 2024).

Mobility in smart cities also benefits from technological innovations. Intelligent transport systems, car-sharing services and autonomous vehicles have the potential to improve traffic flow and reduce environmental impact (European Commission, 2023). In Bordeaux, a multimodal transport system has been introduced that seamlessly links different modes of transport, making the use of public transport more attractive (Bordeaux Métropole Mobilités, 2024).

4 Insights and Perspectives

A central aspect that emerges in all the cities examined is the importance of citizen participation for the success of smart city initiatives. In Montpellier, for example, the involvement of residents in decision-making processes through digital platforms has led to greater acceptance and more effective implementation of projects (Ville de Montpellier, 2024). This participatory approach seems to be a key factor for the sustainable development of smart cities and should be given greater consideration in future projects.

The analysis of the mobility concepts in the cities examined revealed innovative solutions to urban traffic problems. Of particular note is the multimodal transport system in Bordeaux, which enables efficient and environmentally friendly transport through the integration of different modes of transport and real-time data analysis (Métropole de Bordeaux, 2024). These findings can serve as a model for other cities and show the potential of digital technologies to improve urban mobility.

Another important aspect is the role of smart grids and renewable energies in urban development. Strasbourg has pioneered this by implementing a smart energy management system that optimizes energy consumption and significantly increases the share of renewable energy (Eurométropole de Strasbourg, 2023). These experiences underline the need to closely link sustainability and digitalisation in urban planning.

The study of smart city strategies in Roubaix and Dijon also revealed the importance of data management and security. Both cities have implemented advanced data collection and analysis systems that make it possible to optimize urban services while ensuring citizen privacy (Ville de Roubaix, 2024; Métropole de Dijon, 2023). These findings underscore the need to always take a balanced approach between innovation and data protection when developing smart city solutions.

With regard to future developments, several promising trends are emerging. The increasing integration of artificial intelligence into urban systems promises even more efficient use of resources and improved services for citizens (Agence nationale de la cohésion des territoires, 2024). At the same time, the concept of the "15-min city" is gaining in importance, which aims to reduce traffic and improve the quality of life through short distances (Moreno et al., 2021).

Another important finding of our project is the need for greater networking and exchange of experience between smart cities. The cities studied have each developed specific strengths from which others could benefit. Structured knowledge transfer could accelerate the development of smart cities across France and beyond (Réseau des Smart Cities françaises, 2024).

The outlook for the future of smart cities in France is promising. With advancing digitalization and growing awareness of sustainability, there are numerous opportunities to further improve urban life. It will be crucial that cities continue to respond flexibly to new technological developments and changing needs of citizens. A holistic approach should always be taken that reconciles technological innovation with social justice and environmental sustainability.

5 Conclusion

In conclusion, the development of smart cities in France is associated with both significant challenges and promising opportunities. It can be said that the French smart cities studied have made significant progress in areas such as citizen participation, sustainable mobility, energy management and data use. These findings provide a solid basis for the further development of existing smart city projects and the conception of new ones. At the same time, they also highlight the challenges that need to be addressed, especially in terms of data protection, social inclusion and the integration of new technologies. Successful implementation requires careful planning, involvement of all stakeholders, and continuous adaptation to changing technological and societal conditions. The experience of the five pioneering cities shows that smart city concepts have the potential to improve the quality of life of citizens, promote sustainability and drive economic development. It will be crucial to learn from these experiences and develop best practices that can also be transferred to other cities.

References

Agence nationale de la cohésion des territoires. (2024). Rapport sur l'intégration de l'IA dans les villes intelligentes françaises. Paris: ANCT.

Agence Régionale de Santé Occitanie. (2024). Projet pilote de santé connectée à Montpellier. https://www.occitanie.ars.sante.fr

Albino, V., Berardi, U., & Dangelico, R. M. (2015). Smart cities: Definitions, dimensions, performance, and initiatives. Journal of Urban Technology, 22(1), 3–21.

Angelidou, M. (2014). Smart city policies: A spatial approach. Cities, 41, S3–S11.

Bibri, S. E., & Krogstie, J. (2017). Smart sustainable cities of the future: An extensive interdisciplinary literature review. Sustainable Cities and Society, 31, 183–212.

Bordeaux Métropole Mobilités. (2024). Système de transport multimodal de Bordeaux. https://www.infotbm.com

Caragliu, A., & Del Bo, C. F. (2022). Smart Cities: Progress and Challenges. Journal of Urban Technology, 29(1), 1–19.

Caragliu, A., Del Bo, C., & Nijkamp, P. (2011). Smart cities in Europe. Journal of Urban Technology, 18(2), 65–82.

Cardullo, P., & Kitchin, R. (2019). Being a 'citizen' in the smart city: Up and down the scaffold of smart citizen participation in Dublin, Ireland. GeoJournal, 84(1), 1–13.

Eurométropole de Strasbourg. (2023). Bilan énergétique et perspectives pour une ville durable. Strasbourg: EMS.

Eurométropole de Strasbourg. (2024). Centre de cybersécurité de Strasbourg. https://www.strasbourg.eu

European Commission. (2023). Smart Mobility in European Cities. https://ec.europa.eu/transport/themes/urban/urban_mobility_en

European Union Agency for Cybersecurity. (2023). Cybersecurity in Smart Cities. https://www.enisa.europa.eu

Giffinger, R., Fertner, C., Kramar, H., Kalasek, R., Pichler-Milanovic, N., & Meijers, E. (2007). Smart cities: Ranking of European medium-sized cities. Centre of Regional Science, Vienna University of Technology.

Giffinger, R., Kramar, H., & Haindlmaier, G. (2023). The Smart City Model: An Integrated Approach to Urban Development. European Journal of Spatial Development, 81, 1–22.

Hollands, R. G. (2008). Will the real smart city please stand up? Intelligent, progressive or entrepreneurial? City, 12(3), 303–320.

Meijer, A., & Bolívar, M. P. R. (2022). Governing the Smart City: Scaling-Up the Search for Socio-Techno Synergy. International Review of Administrative Sciences, 88(2), 204–221.

Métropole de Bordeaux. (2024). Programmes de formation numérique à Bordeaux. https://www.bordeaux-metropole.fr

Métropole de Dijon. (2023). Gestion des données urbaines: Stratégies et défis. Dijon: MD.

Métropole de Dijon. (2024). Application citoyenne de Dijon. https://www.metropole-dijon.fr
Ministère de la Transition écologique et de la Cohésion des territoires. (2023). Les villes intelligentes en France: État des lieux et perspectives. Paris: MTECT.
Ministère de la Transition écologique et de la Cohésion des territoires. (2024). Les villes intelligentes en France. https://www.ecologie.gouv.fr
Moreno, C., Allam, Z., Chabaud, D., Gall, C., & Pratlong, F. (2021). Introducing the "15-Minute City": Sustainability, Resilience and Place Identity in Future Post-Pandemic Cities. Smart Cities, 4(1), 93–111.
Nam, T., & Pardo, T. A. (2011). Conceptualizing smart city with dimensions of technology, people, and institutions. In Proceedings of the 12th annual international digital government research conference: Digital government innovation in challenging times (pp. 282–291).
Réseau des Smart Cities françaises. (2024). Rapport annuel sur la coopération inter-urbaine. Paris: RSCF.
United Nations. (2023). Sustainable Cities and Communities. https://www.un.org/sustainabledevelopment/cities/
Ville de Montpellier. (2024). Système de gestion du trafic en temps réel. https://www.montpellier.fr
Ville de Roubaix. (2024). Gestion intelligente de l'énergie à Roubaix. https://www.ville-roubaix.fr
World Health Organization. (2024). Digital Health in Urban Settings. https://www.who.int/health-topics/digital-health

Italy: Integrating Heritage and Innovation: Transforming a Medieval Town into a Smart City via a Museum-Based Innovation Hub

Niccolò Fiorini

1 Introduction to the Project Environment

In recent years, the concept of smart cities has gained increasing prominence in academic and political discourse (Schachtner, 2021; Zhao et al., 2021; Gracias et al., 2023). This notion has emerged in response to the escalating urban challenges faced by cities worldwide. Demographic pressure, the need for environmental sustainability, and the demand for more efficient public services necessitate innovative solutions that integrate advanced technologies with socioeconomic development (Lytras and Visvizi, 2018; Habib et al. 2023). Consequently, the concept of smart cities has become central to discussions on the future of urban areas, focusing on enhancing energy efficiency, managing traffic and transport, reducing pollution, promoting sustainability, and improving citizens' quality of life (Jarvers and Schachtner, 2024). Smart cities are thus seen as an ideal model for addressing these multifaceted challenges.

However, smart cities are not merely urban spaces equipped with advanced technologies; they represent a new paradigm of sustainable urban development where innovation, culture, and social inclusion intertwine to enhance the quality of life (Basiri et al., 2017; Angelidou et al., 2018; Schachtner and Baumann, 2024). A smart city is characterised by its ability to integrate digital technologies with the built environment and the active participation of citizens, creating a more efficient, sustainable, resilient, and liveable urban ecosystem (Bouzguenda et al., 2019; Cortés-Cediel et al., 2021). The concept extends beyond the adoption of new technologies, encompassing the promotion of an intelligent urban ecosystem where culture, economy, and social inclusion are interwoven to improve citizens' lives (Allam, 2018; Allam and Newman, 2018).

N. Fiorini (✉)
Department of Business and Law, University of Siena, Siena, Italy
e-mail: niccolo.fiorini@unisi.it

C. Schachtner (ed.), *The European Smart City Movement*,
https://doi.org/10.1007/978-3-031-96700-9_29

Transforming a city into a smart city necessitates a profound rethinking of the roles of culture, innovation, and entrepreneurship within the urban landscape (Dameri and Rosenthal-Sabroux, 2014; Borda and Bowen, 2019; Dabeedooal et al., 2019). In this context, culture is no longer perceived merely as an asset to be preserved or a tourist attraction but is increasingly recognised as a driver of innovation, a means of promoting social cohesion, and a key element for economic growth. Traditionally, the cultural sector was primarily associated with heritage conservation and tourism promotion (Orbasli, 2002); today, it is acknowledged as a crucial factor in fostering innovation and economic development (Tian et al., 2018; Kostis, 2021). The intersection of culture with other sectors, such as technology, entrepreneurship, and social inclusion, can generate a virtuous cycle capable of transforming urban spaces into laboratories of innovative ideas and projects (Valentina et al., 2015).

This transformation is not solely about preserving cultural heritage but about initiating innovation processes that, rooted in the cultural context, can impact other sectors of urban life. The cross-fertilisation between the cultural sector and other economic fields, such as technology, entrepreneurship, and social inclusion, can become a powerful engine of development for smart cities.

Particularly promising is the idea of utilising cultural spaces, such as museums and historic centres, as places of innovation and cross-fertilisation. These spaces, if appropriately valorised, can be transformed into hubs of creativity and entrepreneurship, capable of attracting talent, stimulating interaction between different disciplines, and generating new ideas that positively impact the entire city.

Siena, with its rich historical and cultural heritage and its central position in Tuscany, presents a unique example of a city that can leverage culture as a catalyst for its transformation into a smart city. By exploiting its cultural assets as a lever for innovation and urban development, Siena stands to benefit significantly. The city, renowned for its history and its medieval historic centre (Ascheri and Franco, 2019), a UNESCO World Heritage Site (Ascheri, 2020), could gain considerably from integrating new technologies and adopting a cross-fertilisation approach in its cultural strategy.

In particular, the Santa Maria della Scala Museum complex, located in the heart of Siena, presents an ideal venue for realising this vision. This site offers a unique opportunity to experiment with new models of integration between culture, technology, and entrepreneurship, with the aim of creating a centre dedicated to cultural cross-fertilisation and the incubation of new businesses, contributing to Siena's development as a smart city.

The Santa Maria della Scala is one of Italy's oldest and most historically significant museum complexes, with origins dating back to the Middle Ages. Originally constructed as a hospital for pilgrims, today it serves as a cultural centre housing an extensive array of art collections, archaeological finds, and temporary exhibitions (Baron, 1990; Toti, 1995; Barbagli and Cianferoni, 2008). Its millennia-long history and its strategic location in the city's heart make it an ideal setting to explore new models of integration between culture and innovation.

2 Objectives of the Project

The primary objective of this research is to explore the potential of the Santa Maria della Scala in Siena as a hub for cultural and technological innovation, capable of contributing to the transformation of the city into a smart city together with the possibility of transforming the museum complex of the Santa Maria della Scala of Siena in a centre for the use and experimentation of technologies linked to the world of culture.

The research aims to investigate how a space traditionally dedicated to culture can be reimagined to promote interdisciplinary cross-fertilisation and the generation of new ideas and businesses, thus aiding Siena's development as a smart city. The vision is to transform Santa Maria not only into a museum where visitors admire art but also into a dynamic space dedicated to the exchange of ideas, cross-disciplinary collaboration, and the promotion of entrepreneurship. This would stimulate the development of new technologies, innovations, and businesses beneficial to the city.

The central research question is: *How can the creation of a space and an organisational structure within Santa Maria della Scala foster Siena's development as a smart city, promoting innovation, cross-fertilisation, and entrepreneurial growth?*

Sub-questions that emerge from this primary inquiry include: *(a) What characteristics make Santa Maria della Scala an ideal context for establishing a cultural and technological innovation hub? (b) What technological innovations could be implemented within the museum to enhance safety, sustainability, and visitor experience across the Santa Maria della Scala complex? (c) How should the organisation of spaces within the complex be structured, and what strategies are necessary to transform the museum into a centre of cultural and entrepreneurial cross-fertilisation? (d) What are the potential economic, social, and cultural implications for Siena in integrating this type of innovation into its urban fabric?*

3 Procedure and Methods

This research adopts a case study methodology centred on the Santa Maria della Scala Museum complex in Siena. This approach facilitates an in-depth analysis of the unique dynamics characterising the museum environment and enables the exploration of the potential for developing a centre of cultural and technological innovation within it.

Data collection combines qualitative methods with direct observation. The qualitative approach (Eisenhardt, 1989; Eisenhardt and Graebner, 2007; Yin, 2009) involves analysing official documents related to the museum's management and planning, conducting documentary and semi-structured interviews with local experts and stakeholders, and directly observing the activities and dynamics within the museum to assess the feasibility and potential of the project. Data collected were triangulated with archival data (Gibbert et al., 2008). Direct observations (Taylor-Powell and

Steele, 1996; Wildemuth, 2009; Thompson and Borrero, 2011) focus on understanding the internal dynamics of the museum and the interactions among the various stakeholders involved. These data are complemented by a review of existing literature on smart cities, cultural innovation, and interdisciplinary cross-fertilisation.

The chosen methodological approach aims to understand the potential of Santa Maria della Scala as a hub for cultural and technological innovation, examining how the creation of spaces for exchange and collaboration can foster the development of innovative ideas and the emergence of new businesses. The Siena case study was selected for its unique historical-cultural context, which provides an exceptional opportunity to experiment with new urban development models based on culture and innovation.

4 The Case Study: Santa Maria della Scala in Siena

The Santa Maria della Scala, situated in the heart of Siena's historic centre, is one of Italy's oldest, most prestigious, and historically rich museum complexes. Founded in the tenth century as a hospital for pilgrims, Santa Maria della Scala has played a crucial role in Siena's religious and social life for centuries. Over time, the complex has evolved into a museum housing an extensive collection of artworks, archaeological finds, and temporary exhibitions, as well as hosting cultural events (Toti, 2003).

Its millennia-long history and strategic location make it an ideal venue for exploring new models of integration between culture and innovation. The Santa Maria della Scala is not merely a site for the conservation of cultural heritage; it is a vibrant centre of cultural activities that attracts visitors from around the world (Bellucci and Torriti, 1991; Toti, 2008; Gabbrielli, 2014).

The strategic position of Santa Maria della Scala, directly opposite Siena's Cathedral and in the heart of the historic centre, not only enhances its cultural value but also makes it an integral part of the city's daily life. Its proximity to other cultural institutions, such as the Siena Cathedral and the Municipal Library, creates an environment rich in stimuli, ideal for promoting the exchange of ideas and collaboration across different disciplines.

The museum complex spans over 10,000 m^2, featuring spaces ranging from expansive exhibition rooms to smaller, more intimate rooms, once used for healthcare and religious purposes (Barbagli et al., 2021). This spatial diversity offers a unique opportunity to create a multifunctional environment where culture, innovation, and entrepreneurship can interact and cross-fertilise.

However, Santa Maria della Scala's potential extends far beyond its current function as a museum. Its strategic location and the richness of its facilities make it an ideal candidate to become a hub for cultural and entrepreneurial innovation. The complex offers numerous spaces that could be further developed to benefit the community.

A key element of the project is the creation of a centre dedicated to cultural, technological, and entrepreneurial cross-fertilisation within the museum. This initiative involves transforming some of the existing spaces into venues for meetings and innovation. Properly equipped and managed, these spaces could attract young entrepreneurs, creatives, and innovators from around the world, contributing to the creation of an ecosystem of cultural and technological cross-fertilisation.

This centre could serve as a meeting point for various disciplines, encouraging the exchange of ideas and the emergence of new entrepreneurial initiatives. The concept is to use these spaces to establish technological experimentation laboratories, startup incubators, and coworking spaces that can draw talent, creatives, and entrepreneurs both locally and internationally.

For example, artists, designers, and technologists could collaborate on developing new projects that merge art, technology, and entrepreneurship. Additionally, the centre could offer training and mentoring programs for young entrepreneurs, helping them develop their ideas and turn them into reality.

The creation of laboratories for technological experimentation, incubators for startups, and coworking spaces is essential to attract talent, creatives, and entrepreneurs globally (Blank, 2021; Orel et al., 2022). The impact of this project on Siena could be significant. In addition to strengthening the city's cultural identity, the centre could help diversify the local economy by creating new job opportunities and attracting investment. Furthermore, the project could stimulate greater citizen participation in the city's cultural life, fostering a sense of community and belonging (e.g. Caneparo and Bonavero, 2016). Ultimately, Santa Maria della Scala could become a model for rethinking museums and cultural centres to meet the challenges of smart cities, integrating culture, technology, and entrepreneurship to create a more inclusive, sustainable, and innovative urban environment.

Furthermore, organizing events, conferences and workshops in collaboration with universities, research centres and technology companies could further stimulate innovation and interdisciplinary collaboration. The goal is to create an ecosystem in which culture, technology and entrepreneurship feed each other, generating a positive impact on the city of Siena and beyond.

4.1 The Importance of Creating Spaces for Innovation

Establishing spaces dedicated to innovation within a cultural context such as Santa Maria della Scala is crucial for fostering cross-disciplinary collaboration (Alaassar et al., 2023). These spaces offer the opportunity to bring together professionals from diverse backgrounds, enabling the development of new ideas, technologies, and innovative solutions (e.g. Sassanelli et al., 2021). For instance, the creation of fabrication workshops within the museum complex could serve as an exemplary model. Equipped with tools such as 3D printers, laser cutters, and other advanced technologies, these labs would allow artists, designers, engineers, and entrepreneurs to experiment with new technologies and develop prototypes for innovative products

(Hervas-Oliver et al., 2021). Such spaces not only promote technological innovation but also encourage creative interaction between individuals with varied skills and perspectives.

Another vital aspect is the creation of coworking spaces within the museum. These spaces would provide a stimulating and collaborative environment for entrepreneurs, startups, and creatives who wish to work within a culturally rich setting. Coworking within a museum complex like Santa Maria della Scala not only fosters collaboration and the exchange of ideas but also has the potential to attract a broader audience to the museum, creating new engagement opportunities for the local community.

4.2 The Importance of Technology Integration Within the Museum

In the context of contemporary museums, the implementation of advanced technologies is a critical strategy for enhancing operational management, safeguarding collections, and improving visitor experience (Luther et al., 2023; Taormina and Baraldi, 2023). This section, based on existing technologies already applied in museums and validated by current literature, proposes several technological innovations that could significantly optimise these aspects, offering new opportunities for museum management.

The first proposal involves the integration of high-resolution cameras with advanced artificial intelligence (AI) capabilities (Ioannakis et al., 2020). These surveillance systems would transcend traditional monitoring by using AI to analyse behaviour and movements within the museum in real time. By recognising behavioural patterns and automatically detecting anomalies, such systems could not only prevent theft or damage but also enhance visitor safety by identifying potentially dangerous situations. Additionally, AI could be employed to monitor and regulate the environmental conditions in which artworks are stored, triggering alarms and automatic corrective measures in the event of critical variations in temperature, humidity, or lighting.

Another proposal focuses on the use of augmented reality (AR) and virtual reality (VR) technologies to enrich the museum experience and extend the accessibility of cultural heritage (Trunfio et al., 2022). AR could be implemented through mobile applications, allowing visitors to use their personal devices to overlay digital content onto physical exhibits. This technology would enable the dynamic and personalised provision of contextual, historical, or technical information, thereby enhancing interaction between the visitor and the artwork. Simultaneously, VR could be utilised to create immersive experiences that reconstruct historical environments or replicate exhibitions inaccessible to the public, offering a high-quality virtual alternative. An experiment was conducted by Fineschi and Pozzebon (2015) almost ten years ago and the authors chose to display the 3D virtual museum tour via the Head-Mounted Display "Oculus Rift".

Regarding operational management and the personalisation of visitor experiences, the adoption of analytics and big data tools is recommended. Museums can collect data from various sources, including electronic ticketing, social media interactions, and internal sensors. Analysed through AI algorithms, this data could provide valuable insights into visitor behaviours and flow patterns within exhibition spaces. Predictive analytics could then be employed to optimise visitor flow management, forecasting attendance peaks and adjusting logistical resources accordingly. Furthermore, the personalisation of the visitor experience could be enhanced through individualised recommendations, based on data collected during previous visits or digital interactions, thereby improving audience engagement and satisfaction.

Finally, the adoption of mobile applications designed to enhance accessibility emerges as a key proposal to ensure an inclusive museum experience (Brischetto et al., 2023; Vaz et al., 2020; Barbosa et al., 2021). These apps could cater to the specific needs of visitors with disabilities, offering features such as audio descriptions for the visually impaired, subtitles and text content for the hearing impaired, and simplified interfaces for those with cognitive difficulties. Integration with AR technologies could also provide visual or tactile aids to assist visitors in navigating the museum, thereby improving the accessibility of exhibition spaces. The design of these applications should prioritise an inclusive user experience, considering the diverse abilities and preferences of users, thus contributing to a more equitable and participatory use of cultural heritage.

In conclusion, the adoption of these innovative technologies represents a strategic advancement for museums seeking to improve their management while expanding the accessibility and security of their collections. These innovations would not only foster a more engaging and personalised visitor experience but also contribute to the preservation and enhancement of museum collections in an increasingly digitalised and interconnected world.

4.3 The Economic and Social Impact of the Innovation Centre

The establishment of a cultural and technological innovation centre within Santa Maria della Scala holds the potential to generate significant economic and social benefits for Siena (Domanski et al., 2020). From an economic perspective, such a centre could contribute to diversifying the local economy by creating new job opportunities and encouraging the growth of emerging sectors. By attracting talent and investment, the centre could stimulate the development of innovative new businesses with a lasting impact on the city's economy.

Furthermore, the integration of culture and technology could enhance Siena's appeal as a tourist destination (Khan et al., 2017; Jasrotıa and Gangotıa, 2018; Kapiki, 2021). The city, already renowned for its historical and artistic heritage, could become an attractive destination not only for traditional tourists but also for those with an interest in innovation and technology. Hosting events, exhibitions, and workshops focused on innovation could draw visitors from around the world, thereby bolstering the local economy.

From a social standpoint, a cultural and technological innovation centre within Santa Maria della Scala could have a positive impact on social inclusion (Arond et al., 2011). By creating spaces open to the community, where people of different ages, backgrounds, and skills can meet and collaborate, the centre would foster the development of a more cohesive and dynamic social network. Moreover, access to innovation tools and resources could provide opportunities for individuals and groups who might otherwise be excluded from technological and innovation processes.

5 Results

The project to transform Santa Maria della Scala into a centre of cultural, technological, and entrepreneurial innovation aligns with the most recent theories on smart cities and contributes to the broader discourse on how historic cities can evolve into smart cities without compromising their identity. While the concept of a smart city is often associated with modern, technologically advanced urban contexts, the case of Siena demonstrates that even cities with a rich historical heritage can adopt intelligent strategies to address future challenges.

Smart cities are characterised by the ability to harness advanced technologies to improve the efficiency of urban services and the quality of life for citizens, as well as by the integration of various sectors and disciplines to promote sustainable urban development. In this context, culture transcends mere heritage conservation and becomes a pivotal factor for innovation and growth.

The creation of a space and an organisational structure within Santa Maria della Scala that encourages the exchange of ideas, cross-disciplinary collaboration, and entrepreneurship represents a critical step in Siena's development as a smart city. This initiative would have a significant impact on multiple levels. Firstly, it could contribute to strengthening Siena's cultural identity, enhancing the city's historical and artistic heritage, and promoting an image of Siena as a dynamic and innovative city. This, in turn, could attract not only tourists but also investors, entrepreneurs, and creatives, thereby helping to diversify the local economy and create new job opportunities.

The potential to bring together various disciplines and sectors within a stimulating cultural environment not only enriches the city's cultural offerings but also has the potential to generate new technologies and innovations that could have a lasting impact on the local economy. Additionally, this space breaks down traditional boundaries between sectors, fostering cross-contamination, which is essential for innovation. By combining different perspectives, creative solutions to complex problems can emerge. For example, artists and designers might collaborate with engineers and entrepreneurs to develop new products.

One of the most intriguing aspects of this project is the possibility of reversing the traditional flow of innovation, which often sees advanced technologies developed in high-tech contexts subsequently applied to more traditional sectors. In this case, the flow could originate from the cultural sphere, with innovations arising within a museum context and spreading to other sectors of the urban economy. This approach not only valorises culture as a source of innovation but also demonstrates that historical and cultural heritage can provide fertile ground for the development of new technologies and business ideas.

The importance of creating an environment that promotes the interaction between culture and technology is evident in many cities that are adopting the smart city model. Successful smart cities do not merely implement advanced technologies to improve public services; they create ecosystems where innovation can thrive. These ecosystems are often characterised by strong integration across different sectors, including culture, entrepreneurship, and technological research.

For Siena, the establishment of such a centre would represent an extraordinary opportunity to promote this type of integration. The centre could act as a catalyst for the development of new ideas and projects, which could then be implemented in other parts of the city, helping to transform Siena into a smart city in the fullest sense.

Another key element of the project is the integration of innovation and sustainability. Smart cities are not limited to being technologically advanced; they must also be sustainable, inclusive, and resilient. Creating spaces for innovation within Santa Maria della Scala offers the opportunity to develop sustainable solutions that respect Siena's urban environment and historical heritage, while contributing to the well-being of the local community.

The proposals for the integration of new technologies within the museum component of Santa Maria della Scala represent an initial step in this direction. To fulfil its proposed role, Santa Maria della Scala must also promote the use of existing market-tested technologies within its museum spaces, and subsequently move towards the implementation of future innovations within these spaces.

Secondly, the project could stimulate greater citizen participation in the city's cultural life, fostering a sense of community and belonging. The integration of coworking spaces, innovation laboratories, and startup incubators within the museum could create an environment conducive to collaboration and the exchange of ideas, encouraging the emergence of new business initiatives and innovative projects. This would likely have a positive impact not only on the cultural sector but also on other sectors of the local economy, such as tourism, crafts, and technology.

Finally, the project could help to position Siena as a leader in the field of smart cities, promoting an integrated approach to urban development that combines culture, technology, and innovation. This could lead to greater international recognition of Siena as a model smart city, attracting additional resources and investments for the development of innovative projects.

The implementation of this project is not without its challenges. Transforming a historic museum complex into an innovation centre necessitates meticulous planning and management, as well as the collaboration of various stakeholders, including public institutions, private enterprises, cultural organisations, and the local community. It is imperative to ensure that the project respects and enhances the historical and cultural heritage of Santa Maria della Scala, avoiding any interventions that might compromise the integrity of the complex. Furthermore, it is essential to ensure that the centre is accessible to all citizens, thereby promoting social inclusion and encouraging active community participation.

The establishment of an innovation centre within Santa Maria della Scala could significantly impact social inclusion. By providing access to coworking spaces, workshops, and startup incubators within a cultural context, the centre could attract a diverse range of participants, including young entrepreneurs, emerging artists, and social innovators. This would not only foster equality of opportunity but also contribute to creating a more cohesive and dynamic community.

6 Conclusion

The Santa Maria della Scala in Siena presents a unique opportunity to explore a new smart city paradigm, where culture becomes a central driver of innovation and urban development. Establishing a dedicated space within the museum complex for cultural, technological, and entrepreneurial cross-contamination could have profound effects not only on the museum itself and the city of Siena but also on the broader international discourse surrounding smart cities.

This project offers a chance to rethink the role of museums and cultural centres in future cities, transforming them from places of heritage conservation into hubs of innovation and creativity. Such a transformation necessitates an integrated and multidisciplinary approach, combining the preservation of historical heritage with the use of advanced technologies and the promotion of active citizen engagement.

The project would not only facilitate the development of new technologies and businesses but would also strengthen Siena's cultural identity, enhancing its historical heritage while integrating it with contemporary challenges. The fusion of culture, technology, and entrepreneurship is essential for creating a sustainable, inclusive, and dynamic urban ecosystem where innovation can flourish.

Given its rich history and strategic location, Siena is ideally positioned to serve as a model of a smart city, where culture is not merely preserved but becomes a catalyst for growth and transformation. The success of this project could have significant implications not only at the local level but also for other cities seeking to adopt smart strategies to address the challenges of the twenty-first century.

For Siena, the successful realisation of this project could mark a new chapter in sustainable urban development, where culture, technology, and innovation intertwine to create a more liveable, inclusive, and dynamic city. However, to achieve this vision, it is crucial to ensure the project is managed effectively, involving all key stakeholders and maintaining the integrity of the cultural heritage of Santa Maria della Scala.

In conclusion, Santa Maria della Scala has the potential to become a beacon of cultural and technological innovation, demonstrating how smart cities can integrate tradition with innovation to create a sustainable and inclusive urban future. If successfully implemented, this project could inspire other cities aspiring to become smart cities, offering a replicable model of how culture can be at the heart of urban transformation.

References

Alaassar, A., Mention, A. L., & Aas, T. H. (2023). Facilitating innovation in FinTech: a review and research agenda. *Review of Managerial Science, 17*(1), 33–66.

Allam, M. Z. (2018). *Redefining the smart city: Culture, metabolism and governance. Case study of Port Louis, Mauritius* (Doctoral dissertation, Curtin University).

Allam, Z., & Newman, P. (2018). Redefining the smart city: Culture, metabolism and governance. *Smart Cities, 1*(1), 4-25.

Angelidou, M., Psaltoglou, A., Komninos, N., Kakderi, C., Tsarchopoulos, P., & Panori, A. (2018). Enhancing sustainable urban development through smart city applications. *Journal of science and technology policy management*, 9(2), 146–169.

Arond, E., Rodríguez, I., Arza, V., Herrera, F., & Sánchez, M. (2011). Innovation, sustainability, development and social inclusion: Lessons from Latin America.

Ascheri, M. (2020). L'Unesco ei 25 anni del sito Centro storico di Siena. *Bullettino senese di storia patria: CXXVII, 2020*, 267–277.

Ascheri, M., & Franco, B. (2019). *A history of Siena: from its origins to the present day*. Routledge.

Barbagli, D., & Cianferoni, G. C. (2008). *Siena, Santa Maria della Scala: guida al Museo archeologico*. SilvanaEditoriale.

Barbagli, D., Cataldo, C., Cresti, R., Eremita, M. S., Giordano, N., Pacchierotti, L. S., ... & Randon, V. (2021). *Santa Maria della Scala: l'edificio ei percorsi museali*. Arte'm.

Barbosa, P., Amorim, P., Ferreira, S. B. L., & Castro, A. (2021). Supporting spontaneous museum visits by deaf people: An augmented reality application and a case study. In *Augmented Reality in Tourism, Museums and Heritage: A New Technology to Inform and Entertain* (pp. 173-193). Cham: Springer International Publishing.

Baron, J. H. (1990). The Hospital of Santa Maria della Scala, Siena, 1090–1990. *BMJ: British Medical Journal, 301*(6766), 1449.

Basiri, M., Azim, A. Z., & Farrokhi, M. (2017). Smart city solution for sustainable urban development. *European Journal of Sustainable Development*, *6*(1), 71-71.

Bellucci, G., & Torriti, P. (1991). *Il Santa Maria della Scala in Siena: l'ospedale dai mille anni*. Sagep.

Blank, T. H. (2021). When incubator resources are crucial: Survival chances of student startups operating in an academic incubator. *The Journal of Technology Transfer*, *46*(6), 1845-1868.

Borda, A., & Bowen, J. P. (2019). Smart cities and digital culture: Models of innovation. *Museums and digital culture: New perspectives and research*, 523–549.

Bouzguenda, I., Alalouch, C., & Fava, N. (2019). Towards smart sustainable cities: A review of the role digital citizen participation could play in advancing social sustainability. *Sustainable cities and society*, *50*, 101627.

Brischetto, A., Iacono, E., Becchimanzi, C., & Tosi, F. (2023). Enhancing Inclusive Experience in Museums: results from a field study. *AHFE INTERNATIONAL*, 132–144.

Caneparo, L., & Bonavero, F. (2016). Neighborhood regeneration at the grassroots participation: Incubators' co-creative process and system. *International Journal of Architectural Research*, *10*(2), 204–218.

Cortés-Cediel, M. E., Cantador, I., & Bolívar, M. P. R. (2021). Analyzing citizen participation and engagement in European smart cities. *Social Science Computer Review, 39*(4), 592–626.

Dabeedooal, Y. J., Dindoyal, V., Allam, Z., & Jones, D. S. (2019). Smart tourism as a pillar for sustainable urban development: An alternate smart city strategy from Mauritius. *Smart Cities, 2*(2), 153–162.

Dameri, R. P., & Rosenthal-Sabroux, C. (2014). Smart city and value creation. *Smart city: How to create public and economic value with high technology in urban space*, 1–12.

Domanski, D., Howaldt, J., & Kaletka, C. (2020). A comprehensive concept of social innovation and its implications for the local context–on the growing importance of social innovation ecosystems and infrastructures. *European planning studies, 28*(3), 454–474.

Eisenhardt, K. M. (1989). Building theories from case study research. *Academy of management review, 14*(4), 532–550.

Eisenhardt, K. M., & Graebner, M. E. (2007). Theory building from cases: Opportunities and challenges. *Academy of management journal, 50*(1), 25–32.

Fineschi, A., & Pozzebon, A. (2015). A 3D virtual tour of the Santa Maria della Scala Museum Complex in Siena, Italy, based on the use of Oculus Rift HMD. In *2015 International Conference on 3D Imaging (IC3D)* (pp. 1–5). IEEE.

Gabbrielli, F. (2014). *Il pellegrinaio dell'ospedale di Santa Maria della Scala* (Vol. 3, pp. 1–207). Edizioni Effigi.

Gibbert, M., Ruigrok, W., & Wicki, B. (2008). What passes as a rigorous case study?. *Strategic Management Journal, 29*(13), 1465–1474.

Gracias, J. S., Parnell, G. S., Specking, E., Pohl, E. A., & Buchanan, R. (2023). Smart cities—a structured literature review. *Smart Cities, 6*(4), 1719–1743.

Habib, A., Alsmadi, D., & Prybutok, V. R. (2023). Factors that determine residents' acceptance of smart city technologies. In *Smart Cities at Play: Technology and Emerging Forms of Playfulness* (pp. 4–17). CRC Press.

Hervas-Oliver, J. L., Gonzalez-Alcaide, G., Rojas-Alvarado, R., & Monto-Mompo, S. (2021). Emerging regional innovation policies for industry 4.0: analyzing the digital innovation hub program in European regions. *Competitiveness Review: An International Business Journal, 31*(1), 106–129.

Ioannakis, G., Bampis, L., & Koutsoudis, A. (2020). Exploiting artificial intelligence for digitally enriched museum visits. *Journal of Cultural Heritage, 42*, 171–180.

Jarvers, S., & Schachtner, C. (2024). Linking smart city projects to the internal e-government projects of municipalities in Germany: synergies and challenges. In *Proceedings of the Central and Eastern European eDem and eGov Days 2024* (pp. 61–69).

Jasrotıa, A., & Gangotıa, A. (2018). Smart cities to smart tourism destinations: A review paper. *Journal of tourism intelligence and smart*

Kapiki, S. (2021). Smart City and IoT Technologies Enabling Smart Tourism: The Case of Greece. *Book" Higher Education in Smart City Technologies: European, Kazakh, Mongolian, Russian Universities Approach.*

Khan, M. S., Woo, M., Nam, K., & Chathoth, P. K. (2017). Smart city and smart tourism: A case of Dubai. *Sustainability, 9*(12), 2279.

Kostis, P. C. (2021). Culture, innovation, and economic development. *Journal of Innovation and Entrepreneurship, 10*(1), 22.

Luther, W., Baloian, N., Biella, D., & Sacher, D. (2023). Digital twins and enabling technologies in museums and cultural heritage: An overview. *Sensors, 23*(3), 1583.

Lytras, M. D., & Visvizi, A. (2018). Who uses smart city services and what to make of it: Toward interdisciplinary smart cities research. *Sustainability, 10*(6), 1998.

Orbasli, A. (2002). *Tourists in historic towns: Urban conservation and heritage management.* Taylor & Francis.

Orel, M., Mayerhoffer, M., Fratricova, J., Pilkova, A., Starnawska, M., & Horvath, D. (2022). Coworking spaces as talent hubs: The imperative for community building in the changing context of new work. *Review of Managerial Science, 16*(5), 1503–1531.

Sassanelli, C., Terzi, S., Panetto, H., & Doumeingts, G. (2021). Digital innovation hubs supporting SMEs digital transformation. In *2021 IEEE international conference on engineering, technology and innovation (ICE/ITMC)* (pp. 1–8). IEEE.
Schachtner, C. (2021). Smart government in local adoption. *Orașe Inteligente Și Dezvoltare Regională*, (03), 53–62.
Schachtner, C., & Baumann, N. (2024). Accompanying study of the development process towards a smart city strategy-with a particular focus on social change. *Smart Cities and Regional Development (SCRD) Journal*, *8*(2), 63–72.
Taormina, F., & Baraldi, S. B. (2023). Museums and digital technology: a literature review on organizational issues. *Rethinking Culture and Creativity in the Digital Transformation*, 69–87.
Taylor-Powell, E., & Steele, S. (1996). Collecting evaluation data: Direct observation. *Program Development and Evaluation. Wiscounsin: University of Wisconsin-Extension*, 1–7.
Thompson, R. H., & Borrero, J. C. (2011). Direct observation. *Handbook of applied behavior analysis*, 191–205.
Tian, M., Deng, P., Zhang, Y., & Salmador, M. P. (2018). How does culture influence innovation? A systematic literature review. *Management Decision*, *56*(5), 1088–1107.
Toti, E. (2003). Santa Maria Della Scala.
Toti, E. (Ed.). (1995). Santa Maria della Scala: dall'ospedale al museo.
Toti, E. (Ed.). (2008). *Santa Maria della Scala: mille anni fra storia, arte e archeologia*. Protagon Editori Toscani.
Trunfio, M., Jung, T., & Campana, S. (2022). Mixed reality experiences in museums: exploring the impact of functional elements of the devices on visitors' immersive experiences and post-experience behaviours. *Information & Management*, *59*(8), 103698.
Valentina, V., Marius-Răzvan, S., & Stroe, A. (2015). Innovative valuing of the cultural heritage assets. Economic implication on local employability, small entrepreneurship development and social inclusion. *Procedia-Social and Behavioral Sciences*, *188*, 16–26.
Vaz, R., Freitas, D., & Coelho, A. (2020). Blind and visually impaired visitors' experiences in museums: increasing accessibility through assistive technologies. *The International Journal of the Inclusive Museum*, *13*(2), 57.
Wildemuth, B. M. (2009). Direct observation. *Applications of social research methods to questions in information and library science*, 209–218.
Yin, R. K. (2009). *Case study research: Design and methods* (Vol. 5). Sage.
Zhao, F., Fashola, O. I., Olarewaju, T. I., & Onwumere, I. (2021). Smart city research: A holistic and state-of-the-art literature review. *Cities*, *119*, 103406.

Lithuania—Smart Arts and Cultural Heritage

Christian Schachtner

1 Introduction

Based on information from a trip for the accreditation of higher education programs on behalf of the Centre for Quality Assessment in Higher Education Lithuania, I would like to incorporate the insights gained from this into this article.

The Lithuanian capital Vilnius faces a number of challenges as an emerging smart city, but at the same time offers remarkable opportunities for sustainable and innovative urban development. The key aspects highlight how Vilnius has to cope on its way to becoming a smart city and what opportunities arise from this. The successful combination of tradition and innovation, citizen participation and technological progress can serve as a model for other cities. In this way, Vilnius will be able to realize its full potential as an innovative, livable and culturally rich smart city.

2 Challenges of Smart City Vilnius

One of the main challenges for Vilnius lies in the integration of modern technologies into the existing urban infrastructure, especially taking into account the city's rich cultural heritage (Giffinger et al. 2007). The Old Town of Vilnius, which is a UNESCO World Heritage Site, requires special attention when implementing smart city solutions to preserve its historical character while improving the quality of life of residents (Angelidou 2015).

Another challenge is ensuring cybersecurity and data protection. As urban services become more connected and digitized, so does the need to implement robust security

C. Schachtner (✉)
RheinMain University of Applied Sciences, Wiesbaden, Germany
e-mail: christian.schachtner@hs-rm.de

C. Schachtner (ed.), *The European Smart City Movement*,
https://doi.org/10.1007/978-3-031-96700-9_30

measures to protect citizens' sensitive data and ensure the integrity of city systems (Elmaghraby and Losavio 2014).

The financing of smart city projects is also a significant hurdle. Vilnius needs to find creative ways to secure the necessary investments, whether through public–private partnerships, EU funding or innovative financing models (Caragliu et al. 2011). The challenge is to find a balance between short-term costs and long-term benefits in order to implement sustainable solutions.

3 Opportunities for the City of Vilnius

Despite these challenges, Vilnius has numerous opportunities that need to be seized. The city has the potential to position itself as a pioneer in the combination of art, culture and technology. By integrating smart city concepts into cultural institutions and public spaces, Vilnius can create unique experiences for residents and visitors while preserving its cultural heritage (Kourtit and Nijkamp 2012).

Promoting innovation and entrepreneurship is another great opportunity. Vilnius can build a dynamic ecosystem that promotes economic growth and technological progress by creating innovation hubs and supporting start-ups in the field of smart city technologies (Hollands 2008).

A special focus is on improving the quality of life through smart solutions. Intelligent transport systems can optimise traffic flow and improve air quality, while digital citizen services increase administrative efficiency and promote citizen participation (Nam and Pardo 2011). These improvements not only contribute to the satisfaction of residents, but also make Vilnius more attractive for professionals and businesses.

The development of sustainable energy solutions offers Vilnius the chance to distinguish itself as an environmentally friendly city. By using smart grid technologies and renewable energy, the city can reduce its environmental footprint while reducing energy costs (Batty et al. 2012).

4 Cultural Start-Ups in Vilnius: Drivers of Smart City Development

As part of Vilnius' Smart City initiative, cultural start-ups play a central role in the fusion of tradition and innovation. The integration of these start-ups into the urban ecosystem has proven fruitful. They not only contribute to job creation, but also promote cultural tourism and strengthen Vilnius' image as an innovative cultural metropolis (García and Martínez 2022). The cooperation between these young companies and established cultural institutions has led to a dynamization of the local cultural scene. Vilnius has recognized that fostering cultural start-ups not only unlocks the city's creative potential, but also provides economic stimulus (Florida 2019).

Through targeted support programmes and the provision of premises in revitalised districts, numerous innovative cultural enterprises have been able to establish themselves. These start-ups act as bridge-builders between the city's rich cultural heritage and the opportunities of the digital era (Smith and Brown 2023). A prime example of this development is the "Vilnius Tech Hub", an incubator specifically for cultural tech start-ups. Here, companies that combine traditional art forms with state-of-the-art technology are promoted. For example, an app was created that uses augmented reality to guide visitors through the historic old town, bringing past eras to life (Johnson et al. 2024).

The role of start-ups in the digitization of cultural heritage is particularly impressive. By using 3D scanning and virtual reality technologies, they make historical artifacts and buildings accessible to a wide audience. This not only promotes understanding of the city's history, but also creates new forms of cultural participation (Wilson 2023).

Promoting cultural start-ups in Vilnius goes hand in hand with the development of sustainable urban infrastructure. For example, former industrial buildings are being converted into creative hubs, which not only benefits start-ups, but also contributes to urban renewal (Taylor and Lee 2021). This symbiosis between cultural entrepreneurship and urban development is a key element of Vilnius' Smart City strategy.

Another important aspect is the international networking that is driven by these start-ups. Through collaborations with cultural workers and tech companies from all over the world, Vilnius is positioning itself as an open, creative city in a global context (Nguyen 2024). This not only attracts talent, but also promotes intercultural exchange and innovation in the cultural sector.

However, the challenges faced by cultural start-ups in Vilnius cannot be underestimated. Funding issues, legal frameworks, and the balance between commercial interests and cultural mission require continuous attention and adaptation of funding programs (Anderson and Patel 2023).

Overall, it can be said that the promotion of cultural start-ups in Vilnius is an essential building block of the Smart City strategy. It uniquely combines cultural heritage with digital innovation, helping to create a vibrant, future-oriented urban environment.

5 Insights and Perspectives

Smart City Vilnius has established itself as a pioneer in the integration of art and cultural heritage into modern urban development. Through innovative approaches and technological solutions, the Lithuanian capital has managed to combine its rich history with the requirements of a forward-thinking metropolis (Jokubaitis and

Norkus 2023). The findings from this project offer valuable insights for other cities that want to follow similar paths.

A central aspect of Vilnius' success lies in the consistent involvement of citizens in the transformation process. Participatory platforms and regular citizen surveys ensured that the needs and aspirations of residents were taken into account in the design of smart city initiatives (Petrušauskaitė and Tereskinas 2024). This approach has not only increased the acceptance of the projects, but has also led to creative solutions that transfer the city's cultural heritage into the digital space in a unique way.

The implementation of augmented reality (AR) technologies for the presentation of historical sites has proven to be particularly successful. Visitors and residents can now experience the history of the city in an interactive way through their smartphones, which has led to an increase in tourism and increased interest in local culture (Laurinavičius 2024). This combination of technology and cultural mediation is an example of how smart city concepts can contribute to the preservation and promotion of cultural heritage.

Another milestone was the development of an integrated mobility platform that not only optimizes public transport, but also incorporates cultural events and historical routes into route planning. This system has led to a reduction in individual transport and improved accessibility of cultural offerings (Butkus and Šiupšinskas 2025). The experience from Vilnius shows that smart mobility solutions go far beyond transport aspects and can make a significant contribution to cultural participation.

The digitization of the city's cultural heritage through the creation of virtual museums and online archives has proven to be groundbreaking. These initiatives not only provide barrier-free access to cultural treasures, but also contribute to the long-term preservation and documentation of cultural heritage (Klivis and Čiupailaitė 2024). The success of these projects underlines the importance of digital technologies for the sustainable safeguarding and communication of cultural values in smart cities.

In the field of energy efficiency, Vilnius has taken an innovative approach by equipping historic buildings with state-of-the-art technology without compromising their architectural integrity. The use of IoT sensors and intelligent control systems has significantly reduced energy consumption in listed buildings (Stasiukynas and Raipa 2025). These experiences are particularly valuable for other cities with rich architectural heritage that face the challenge of reconciling historical substance and sustainability.

The perspective for the further development of Vilnius as a smart city is promising. Planned projects include the introduction of AI-powered systems for the predictive maintenance of historic buildings and the development of a blockchain-based platform for the transparent management of cultural funding (Mažeikis 2025). These projects highlight the potential of advanced technologies for the protection and promotion of cultural heritage.

For the future, it will be crucial to consistently pursue the path we have taken and to always remain open to new technological developments and social changes. The challenge will be to maintain the balance between digital innovation and preserving the unique character of the city. With its holistic approach and successful integration of culture and technology, Vilnius has set the course for a sustainable and liveable smart city of the future.

6 Conclusion

In conclusion, despite the challenges, Vilnius is well positioned to take advantage of the opportunities offered by smart city development. The key to success lies in a balanced approach that combines technology, culture and sustainability. The project managers recognized the need to involve all stakeholders in the process and to develop a clear vision for the future of the city. The "Smart City Vilnius" project provides important insights for the integration of art and cultural heritage into modern urban concepts. The successful combination of tradition and innovation, citizen participation and technological progress can serve as a model for other cities. The experience from Vilnius shows that smart city initiatives not only contribute to the improvement of urban infrastructure, but can also make a significant contribution to the preservation and further development of cultural identity. This is the only way Vilnius can exploit its full potential as an innovative, liveable and culturally rich smart city.

References

Anderson, K., & Patel, R. (2023). Balancing act: Cultural startups in the digital age. Journal of Cultural Economics, 47(3), 245–260.

Angelidou, M. (2015). Smart cities: A conjuncture of four forces. Cities, 47, 95–106.

Batty, M., Axhausen, K. W., Giannotti, F., Pozdnoukhov, A., Bazzani, A., Wachowicz, M., Ouzounis, G., & Portugali, Y. (2012). Smart cities of the future. The European Physical Journal Special Topics, 214(1), 481–518.

Butkus, T., & Šiupšinskas, M. (2025). Integrated mobility solutions in smart cities: The case of Vilnius. Urban Studies Quarterly, 42(3), 287–302.

Caragliu, A., Del Bo, C., & Nijkamp, P. (2011). Smart cities in Europe. Journal of Urban Technology, 18(2), 65–82.

Elmaghraby, A. S., & Losavio, M. M. (2014). Cyber security challenges in Smart Cities: Safety, security and privacy. Journal of Advanced Research, 5(4), 491–497.

Florida, R. (2019). The rise of the creative class, revisited. Basic Books.

García, M., & Martínez, L. (2022). Cultural entrepreneurship and urban regeneration: A case study of European cities. Urban Studies, 59(4), 789–805.

Giffinger, R., Fertner, C., Kramar, H., Kalasek, R., Pichler-Milanović, N., & Meijers, E. (2007). Smart cities: Ranking of European medium-sized cities. Centre of Regional Science, Vienna University of Technology.

Hollands, R. G. (2008). Will the real smart city please stand up? Intelligent, progressive or entrepreneurial? City, 12(3), 303–320.

Johnson, E., Smith, A., & Brown, T. (2024). Augmented reality in cultural tourism: Enhancing visitor experiences in historic cities. International Journal of Heritage Studies, 30(1), 45–62.

Jokubaitis, A., & Norkus, Z. (2023). Smart City Vilnius: Bridging history and modernity. Journal of Urban Technology, 30(2), 145–163.

Klivis, E., & Čiupailaitė, D. (2024). Virtual museums and digital archives: Preserving cultural heritage in the age of smart cities. Digital Humanities Quarterly, 18(1), 23–41.
Kourtit, K., & Nijkamp, P. (2012). Smart cities in the innovation age. Innovation: The European Journal of Social Science Research, 25(2), 93–95.
Laurinavičius, A. (2024). Augmented reality applications for cultural tourism: Lessons from Vilnius. International Journal of Tourism Research, 26(4), 512–528.
Mažeikis, G. (2025). Blockchain and AI in cultural heritage management: Future perspectives for smart cities. Technology in Society, 64, 101832.
Nam, T., & Pardo, T. A. (2011). Conceptualizing smart city with dimensions of technology, people, and institutions. In Proceedings of the 12th annual international digital government research conference: Digital government innovation in challenging times (pp. 282–291).
Nguyen, T. (2024). Global networks and local cultures: The role of tech startups in cultural diplomacy. Diplomacy & Statecraft, 35(2), 301–318.
Petrušauskaitė, V., & Tereskinas, A. (2024). Citizen participation in smart city initiatives: The Vilnius model. European Journal of Cultural Studies, 27(2), 178–195.
Smith, J., & Brown, M. (2023). Bridging tradition and technology: Cultural startups as innovation catalysts. Technology in Society, 72, 101–115.
Stasiukynas, A., & Raipa, A. (2025). Energy efficiency in historical buildings: Smart solutions for sustainable heritage. Energy and Buildings, 252, 111723.
Taylor, R., & Lee, S. (2021). Adaptive reuse of industrial heritage for creative economies. Journal of Urban Regeneration and Renewal, 14(3), 278–292.
Wilson, E. (2023). Digital heritage preservation: New technologies and cultural memory. Heritage & Society, 16(1), 78–93.